Finding Oil and Gas
from Well Logs

Finding Oil and Gas from Well Logs

Lee M. Etnyre

 SPRINGER SCIENCE+BUSINESS MEDIA, LLC

Library of Congress Catalog Card Number 88-10795
ISBN 978-1-4757-5232-8

16 15 14 13 12 11 10 9 8 7 6 5 4 3 2 1

Library of Congress Cataloging-in-Publication Data
Etnyre, Lee M., 1937–
    Finding oil and gas from well logs.
    Includes bibliographies.
    1. Oil well logging. I. Title.
TN871.35.E86   1988   622'.1828   88-10795
ISBN 978-1-4757-5232-8   ISBN 978-1-4757-5230-4 (eBook)
DOI 10.1007/978-1-4757-5230-4

Contents

7. Radioactivity Logging   **233**

Answers to Problems   **273**

Preface

Several excellent books on well log interpretation have already been published. However, I feel that these books do not place enough emphasis on the inherent uncertainties in tool responses or on the related and very practical problem of selecting suitable data points for statistical or quantitative calculations. Thus, I have written this book not only to introduce the newcomer to this very complex art and science, but also to provide him or her with the necessary tools to produce better interpretations.

The problems at the end of each chapter are essential to a more complete understanding of the subject matter and include many practical notes based on problems I have encountered in actual applications. This book emphasizes that you develop your own concepts and understanding of the underlying principles, rather than acquiring a compendium of knowledge based on certain rules of thumb. If you are to successfully interpret well logs, you need to be able to apply your knowledge to new problems that may not follow the preconceived ideas and approaches you would follow if you approached well log analysis from a cookbook standpoint.

Although modern, sophisticated logging tools based on computer-operated logging systems provide much better quality data than was possible in the past, there are still inherent tool uncertainties that must be accounted for in any successful interpretation. This leads naturally to the need for more advanced cross-plot and pattern recognition techniques as advocated by the late George R. Pickett. Before you can learn these techniques and use them effectively, you need to understand the basic ideas. This book presents these basic concepts in a way that will lead naturally to the more advanced interpretation techniques.

Many well logs you will encounter were run before the advent of the modern logging systems. They will therefore contain more errors and have more data quality problems. Since there is an increasing emphasis on exploitation of existing fields and reviewing older logs to see if some hydrocarbon reservoirs were passed over, the hydrocarbon finder is going to have to deal with many of these older logs with their inferior data quality. Even in an exploration program, you may have only older well logs to look at when developing new drilling prospects.

Today, with the new computerized logging systems and analysis programs, data that is digitized in fixed increments is processed to obtain a foot-by-foot (or half-foot by half-foot, etc.) presentation of log calculations. The problem is that, even with modern tools, every point on a log is not suitable for quantitative or statistical calculations. This was pointed out many years ago by Louis Chombart and is still true today when logs are recorded opposite rock strata whose thickness is less than the logging tool is capable of resolving. Many tools will exhibit reversals of behavior and other characteristics that have no apparent relation to the actual rock properties in these situations. In a rather abrupt transition from one rock type to another, each logging tool has a unique way of averaging between the two strata on either side of the boundary that may be substantially different for each tool type. Trying to combine data from these different tools for quantitative calculations in a transition zone may lead to questionable results. These problems and their causes are illustrated in this book. This should put you in a better position to interpret and use computer-aided log calculations.

Many other commonly encountered pitfalls in well log interpretation are discussed. For example, the use of interpretation charts is illustrated as well as the pitfalls and possible misuse of these charts.

I have tried to simplify this introduction to well log interpretation by presenting the concepts of reservoir petrophysics, porosity, and saturation first before I deal with the complications of the individual tool response characteristics. Others may prefer a different order of presentation, however, I selected this approach feeling that it would be more appropriate to a systematic development. Many traditional approaches begin with the tool responses, but I feel you will be better motivated if you see where you are going first.

My own experience has been in large part with Schlumberger well logs, primarily because this company has dominated and led the logging service field in technology during the history of the petroleum industry. Thus, I have elected to use their charts in illustrating many of the concepts of log interpretation. However, you should become familiar with the charts and operational characteristics of the logging company whose logs you are working with. Although the basic principles of rock property logging are the same, each company may use a slightly different design and implementation of their tools in what is becoming a highly competitive field. Thus, you should not expect one company's logging tool to read identically to another company's version of the same tool opposite every depth in the same well, even though they may show identical readings for standard calibration points. Actual rocks are notorious for being ignorant of *standard* or *average* rock properties. In some cases, these departures are significant. For example, the beginner may be tempted to use typical quartz values for sandstones, but actual sandstones usually contain significant quantities of other minerals such as feldspars, calcite, clay minerals, and sometimes dolomite.

Readers familiar with the late G. R. Pickett's interpretation techniques, or who were privileged to have taken any of his petrophysics courses, will recognize the influence of his philosophy and approach to well log interpretation in my discussions of interpretation. Unfortunately, much of Pickett's material was never widely circulated or published. I have had to refer to my own class notes from the courses he taught at the Colorado School of Mines. I hope you will find this added knowledge both enlightening and useful to your own practical applications.

Finding Oil and Gas
from Well Logs

CHAPTER 1

Well Log Basics

The late George R. Pickett of the Colorado School of Mines once said that using well logs in oil and gas exploration was "like hunting on a game preserve." To some this may be a revelation. However, most people in the petroleum industry know that well logs play a key role in oil and gas exploration and reservoir evaluation. When a well drilling is finished, a decision must be made as to whether to complete the well or plug and abandon it. Well logs often provide the data that help make the correct decision. Well logs can sometimes be used to identify the presence of hydrocarbons where the quality of the reservoir rock is so good that nearly all traces of hydrocarbons have been flushed from the drilling cuttings circulated to the surface by the drilling fluid. It is a paradox that the best hydrocarbon "shows" occur in the poorer reservoir rocks, whereas the poorer shows occur in the best reservoir rocks. Well logs can be used to identify the best reservoirs.

Well logs do have their limitations, however, and some people tend to downgrade the important contribution of well logs to the oil and gas industry. On the other hand, people who know and understand logging tool responses have time and again used their knowledge and understanding to find new oil and gas fields. One of many classic examples is a major oil company's use of a specialty logging tool—the borehole gravimeter—to find large reef reservoirs in Michigan. This is typical of the history of using well logs in oil and gas exploration.

WHAT IS A WELL LOG?

Well logs are used to calculate the amount of oil and gas in the ground. A well log is a graph of depth in a well versus some characteristic or property of the rock. The rock property is derived from measurements made when instruments are lowered into the well on an electrical wireline or cable. Most measurements are actually recorded as the instruments are raised to the surface from the bottom depth in the well. This is the reason well logs are so important. Once a well is drilled, the only economical means of finding out what is down there is with a well log. The only other means is to drill out a core of the rock, pay for an expensive test, or pay for an even more expensive and unjustified completion, all of which are too costly to do on a routine basis and must be reserved for only a few wells.

Well logs are also known as electric logs, E-logs, wireline logs, and borehole measurements. In addition to measurements recorded in the borehole by sophisticated electronic instruments, the measurement can be something as simple as a geologist's description of the rock cuttings created by the action of the drill bit in the well. The drilling mud that is pumped down through the drill string and back up the borehole circulates these cuttings to the surface.

Measurements made downhole by instruments can be processed downhole and transmitted to the surface, or the unprocessed measurements can be transmitted to the surface for processing. Normally, data are recorded as the logging instrument is pulled up the borehole from the bottom. This process is referred to as *logging the well*.

Following is a list of many of the types of well logs that can be recorded:

- caliper (hole diameter measurement)
- acoustic (sound measurement)
- radioactivity (natural and induced)
- cutting samples and lithology
- dip of rock strata
- acoustic televiewer
- resistivity
- drilling time
- mud gas
- temperature
- dielectric constant
- gravimeter

In this book I primarily discuss resistivity logs, acoustic logs, radioactivity logs, and caliper logs. Some of the other logs and core measurements are also discussed to a limited extent. Logs such as the dip meter, televiewer, gravimeter, and dielectric constant are not commonly recorded in every well. However, these are important tools and should be studied after you are familiar with the more fundamental techniques and logs.

This book also concentrates on applications in the petroleum industry, although the principles apply equally to other uses of well logs. Well logs are important to the petroleum industry because they provide many rock property measurements that can be obtained at reservoir conditions of pressure, temperature, and fluid content. Other methods of measurement, such as coring, do not always reflect reservoir conditions. The coring operation itself can cause changes in the properties of the rock.

The quality of well logs can vary significantly from one well to the next. The type of log used in a well may also change from one well to another. Each type of log has its own volume of investigation within which it responds to the rock property it measures. This can vary dramatically from one logging tool to another. Some tools can resolve beds only a foot or two in thickness. Others respond to volumes of rock several feet thick. If the rock beds logged are thinner than what that tool can resolve, the recorded data will be a distorted representation of the actual rock properties. In some cases, well log data simply cannot be used to evaluate thin, laminated rock types.

Common uses for well logs in the petroleum industry include:

• estimate recoverable hydrocarbons

• calculate hydrocarbon saturation

• calculate porosity

• determine pore size distribution

• estimate net pay (thickness that will produce)

• determine water salinity

• locate oil–water contact in a reservoir

• monitor reservoir fluid movement

• type rock

• identify geologic environment

• detect fractures

• evaluate water flood feasibility

• evaluate cement bond isolation of zones

• determine mechanical rock properties

Most of these uses are directed toward estimating recoverable hydrocarbons. Some of the uses are related to well completion problems and exploration. Calculating both porosity and hydrocarbon saturation is most important because this information is necessary to estimate recoverable hydrocarbons.

WHAT IS WELL LOG INTERPRETATION?

The primary information supplied by well logs is porosity and hydrocarbon saturation. *Porosity* is the void space in the rock that can hold fluids or gas. It indicates what fraction of the rock volume could contain oil or gas. Porosity in rock is usually not visible to the naked eye because the pores are so small. Yet even the tiniest pores can make up over 40% of the rock volume. Once the volume that can hold hydrocarbons is known, the fraction of this pore volume that actually *does* contain oil or gas must be determined. This is called the *hydrocarbon saturation*.

However, porosity and hydrocarbon saturation cannot be measured directly with well logs. Equations based on empirical or mathematical models of rock behavior are used to convert measured characteristics into porosity and saturation. For instance, a logging tool can be used to measure the velocity of sound in the rock; this measurement is subsequently converted to porosity data using a physical model and some accompanying assumptions.

Interpretation, then, is the art and science of converting measured properties into the ones needed for reservoir evaluation. Fortunately, only simple mathematical equations are needed in most well log interpretations. The common logging tool response equations are simple linear relations that are easily solved or that can be quickly evaluated using charts and graphs. The saturation equations are only a little more complicated. The usual saturation relation presented in most publications can be broken down into two simple components that are easy to solve. In Chapter 3, I present a simple graphical technique for solving saturation equations. This technique not only makes things easy, it also provides a valuable crosscheck on calculator results that will help the novice with logging calculations. Once the simple arithmetic is dealt with, the more extensive task of relating well log responses to rock properties can be approached. This is the basis for well log interpretation.

Pickett defined the following four basic philosophies of well log interpretation: cookbook, log analyst, skeptic, and statistical.[1] The term *cookbook* approach refers to the process of plugging values into equations and following some fairly well-established procedure. In many practical applications, the cookbook interpretation is the only one accomplished. On the other hand, the *log analyst* approach makes use of supplemental data such as cutting samples, core data, and geological and petrophysical knowledge of the reservoir rock types in question to enhance the interpretation.

The *skeptic* approach is based on finding procedures that produce results that are independent of the many inherent log data errors and assumptions used in cookbook style interpretation methods. In this approach, the log analyst also makes use of data cross-plotting techniques and data quality checks. This is all done to reconcile the calculated results with all the supplemental information available. Rock typing may play an important role in this approach. Finally, the *statistical* approach is based on statistical methods that presume to accomplish the same error and parameter assumption independence achieved by the skeptic approach.

In this book, I dwell on the cookbook and log analyst approaches. It is necessary to master these approaches before proceeding to more advanced methods. Keep in mind that these four categories may have extensive overlap. This will depend on the approaches preferred by an individual as well as his or her knowledge of the various techniques.

WHO NEEDS TO KNOW HOW TO INTERPRET WELL LOGS?

Anyone who uses well log data in his or her work or who makes decisions based on well log data needs to know the basics of well log interpretation. The more that person knows, the more effective will be his or her use of such data. For example, most reservoir engineers and geologists make use of well log data in their work on a daily basis. Even in such an elemental task as constructing geologic cross sections from well logs, a knowledge of well log relations is very helpful in obtaining usable results. Well log data play a fundamental role in modern petroleum exploration and production. A manager will also find a knowledge of well log interpretation very helpful.

Well logging is a big business. Probably more than $2 billion are spent annually on well logging. There are more than a dozen measurements that can be made, and each one has a different cost and is affected by several different rock properties. Therefore the measurements that best serve in evaluation of the reservoir in question must be selected. This requires a knowledge of the various logging instruments and their response characteristics.

CHARACTERISTICS OF WELL LOG MEASUREMENTS

The characteristics of a well logging measurement system are shown in Figure 1–1. A measurement is made of some physical property of the rock, for example, the electron density of the rock or the velocity of sound in the rock. Note that the observed response of the logging tool will be a distorted

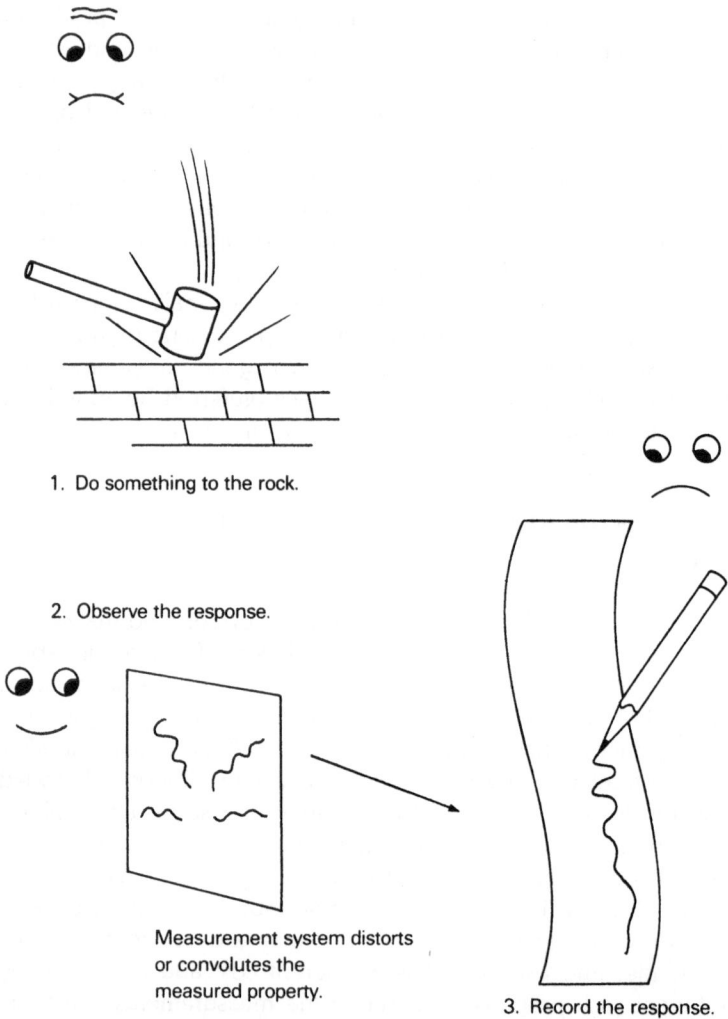

1. Do something to the rock.

2. Observe the response.

Measurement system distorts
or convolutes the
measured property.

3. Record the response.

Figure 1–1. Characteristics of a rock measurement system. The measurement system distorts or convolutes the measured property.

or convoluted version of the measured property. The distortion depends on the response characteristics of the particular logging tool. Finally, the recorded data can be converted from observed responses to the properties desired such as porosity and oil saturation. Once the measurement system distortion is accounted for and the mechanics of conversion are derived, *apparent* values of saturation and porosity are obtained.

Figure 1–2 illustrates a common assumption made about wireline measurements, that is, that a rock property measurement covering a radius from

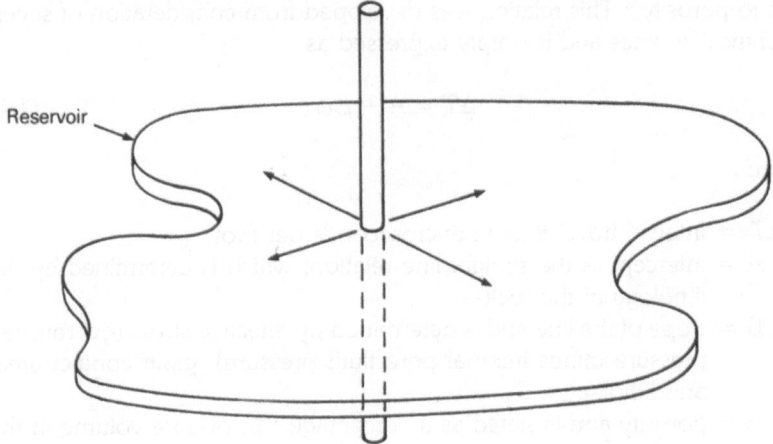

Figure 1–2. Borehole measurement assumption. Measurement valid for a few inches is assumed valid for a much wider area.

only a few inches to at most a few feet from the well bore is valid for a much larger radius around the well bore. This larger radius may be well beyond the radius of investigation of the logging tool. This assumption also applies to core measurements or other data and is necessary if reservoir calculations are to be made.

Consider an acoustic or sonic logging tool as an example of a typical well logging tool response relation. This tool can be used to measure the velocity of sound in a rock. Among other things, the velocity of sound in a rock depends on the porosity of the rock. The velocity data are usually recorded as the inverse or reciprocal of velocity. This is the travel time for a given distance and is referred to as the *interval travel time*. Convenient units for interval travel time are microseconds per foot (μsec/ft). These logs are recorded on a scale that typically ranges from 140 μsec/ft on the left side of the log to 40 μsec/ft on the right side of the log. That is, the slower velocities (long travel times) are recorded to the left side of the log and the faster velocities (short travel times) are recorded to the right. Other scales can also be used.

Examples of sonic logs are presented later. For now, it is only important to recognize that the interval travel time is determined by several factors in addition to porosity. Lithology, overburden stress, pore fluid pressure within the rock, grain contact areas, rock competency, and fluid content can all affect the travel time measurement. If porosity information is to be extracted from the acoustic log measurement of travel time, all these other factors have to be taken into account. Travel time is converted to porosity using a mathematical equation. The equation should account for the factors other than porosity. Pickett used one such relation that converts travel time to porosity.[2] This relation was developed from consideration of several theoretical sources and is simply expressed as

$$\Delta T = A + B\phi \qquad\qquad (1\text{–}1)$$

where

ΔT = interval travel time in microseconds per foot
A = intercept of the straight line relation, which is determined by the lithology of the rock
B = slope of the line and is determined by effective stress (overburden pressure minus internal pore fluid pressure), grain contact area, and lithology
ϕ = porosity and is stated as a percentage (%) of pore volume in the equation

Note that the constants A and B account for the factors other than porosity. The constant B must be multiplied by 100 if porosity is expressed as a fraction.

Figure 1–3 is a graph of Eq. 1–1. The line on the graph is defined by the intercept A and the slope B. For a rock with a given porosity ϕ, the formation travel time in the rock can readily be determined. Note that the same relation can be solved for porosity in terms of travel time ΔT. Each

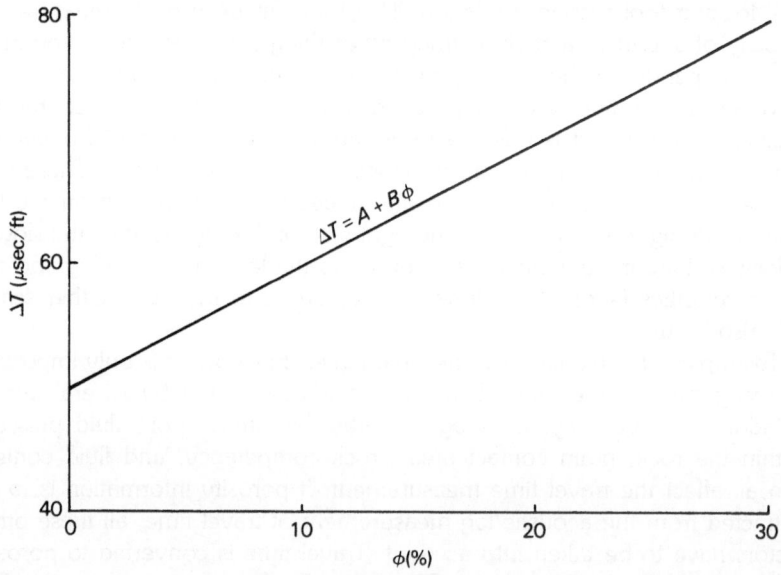

Figure 1–3.   Graph of sonic log response.

Table 1–1. Hypothetical Travel Time Versus Porosity

Travel Time (μsec/ft)	Porosity (fraction of rock volume)	
	Rock Type I	Rock Type II
50	0.00	0.03
55	0.05	0.10
60	0.10	0.18
Constants		
A (μsec/ft)	50	48
B (μsec/ft/fractional pore volume)	100	67

value of travel time corresponds to a unique value of porosity given a known pair of rock parameters A and B. Such a graph can be used to find porosity from travel time.

Table 1–1 shows hypothetical responses of a sonic log to two different rock types. Also shown are the values of constants A and B for the two rock types. Note that the same measured travel time may correspond to quite different porosity values. Thus, the accuracy of any porosity value determined from a sonic log will depend on the correct quantification of constants A and B, which underscores the importance of rock typing to well log interpretation.

It is also of interest to note that constant A has a particular physical meaning unique to the rock. ΔT is the matrix travel time of the rock, sometimes denoted by the symbol ΔTMA. This is consistent with Eq. 1–1 if porosity is set to zero. Then the rock matrix must be the only thing left to account for the observed travel time.

This example shows how to convert a measured rock property into a desired result by using a mathematical equation based on some physical or empirical description of rock property behavior. The result will be no better than the physical model used and your knowledge of any of the related constants such as A and B of Eq. 1–1.

CORE MEASUREMENTS

Core measurements can be used to advantage in formation evaluation if their limitations are understood. For example, they can be used to *calibrate* well logs.

Well log data usually provide more accurate estimates of fluid saturations in the reservoir than do core fluid analyses. In the coring operation, the rock and its relative fluid saturations are altered by the process of bringing the core to the surface. Figure 1–4 illustrates this process for a commonly used water-base mud system.

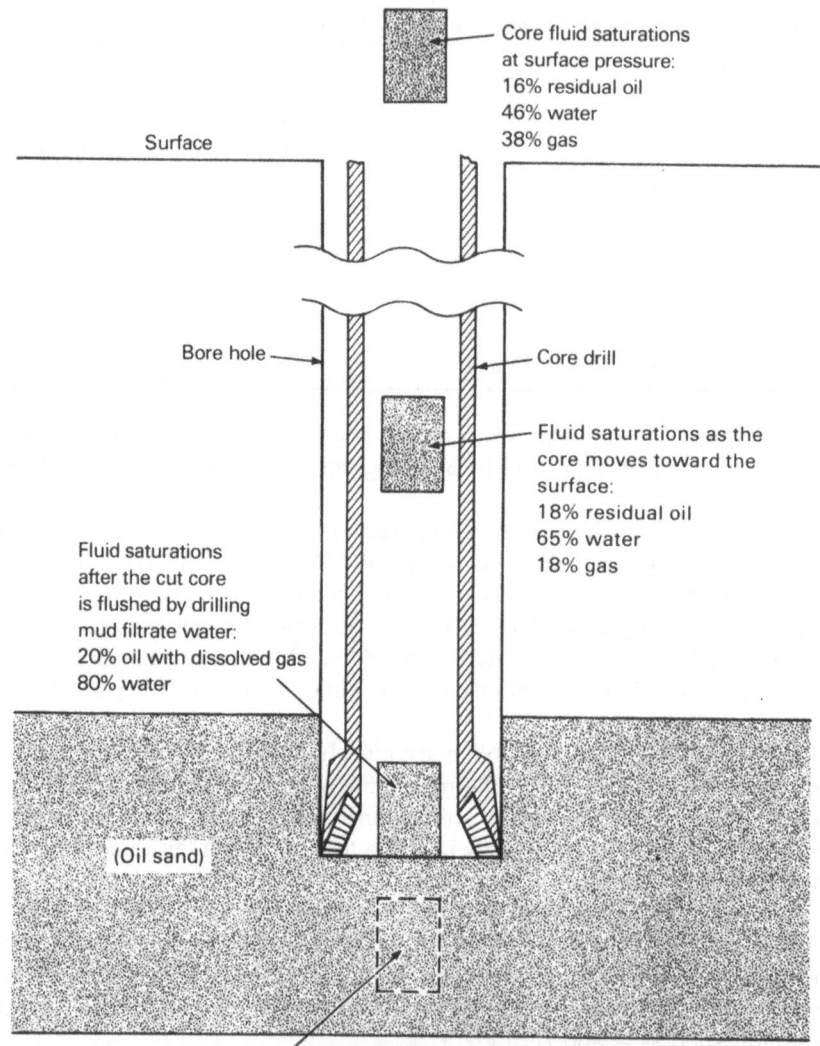

Core fluid saturations
at surface pressure:
16% residual oil
46% water
38% gas

Surface

Bore hole

Core drill

Fluid saturations as the
core moves toward the
surface:
18% residual oil
65% water
18% gas

Fluid saturations
after the cut core
is flushed by drilling
mud filtrate water:
20% oil with dissolved gas
80% water

(Oil sand)

Original conditions before penetration by the drill bit: 75% oil with dissolved
gas and 25% formation water

Figure 1–4.   Coring operation.

Immediately after cutting the core, the drilling mud filtrate water flushes most
of the formation water and quite a bit of the oil from the core. This is caused by
the excess pressure of the mud system compared to the normal formation fluid
pressures. The steadily decreasing pressure as the core is brought to the surface
causes expansion and release of the dissolved gas in the oil. This expanding
gas further removes more of the residual oil and a substantial amount of water.

Finally, at the surface the core contains mostly mud filtrate water instead of formation water. The residual oil saturation in the core is now a pessimistic representation of the residual oil saturation at depth immediately after flushing. Thus, the *core oil saturation at the surface does not represent the oil saturation in the reservoir* before penetration by the drill bit. In the more uncommonly used oil-base mud system, the result is a pessimistic representation of water saturation rather than oil saturation.

Well logs also have a sample size advantage over core data. The well log radius of investigation samples a much larger volume of rock than the core represents. As illustrated in Figure 1–5, a thumb-size plug sample is often drilled out of a core and used for measurements. One plug is taken for each foot, and that small thumb-size plug is assumed to represent the entire foot of core from which it is taken. If the rock is heterogeneous, the small plug may not represent the entire foot of core. Sometimes the plug will be selected from the most competent or best part of the core. Other parts of the core may be unsuitable for measurements. This only makes matters worse. It is sometimes referred to as the *sampling problem* with core data.

Another core-sampling problem occurs when the crew removing the pieces of core from the core barrel inadvertently places a piece of core bottom side up against the piece of core that came from the next shallower depth in the well.

Core sampling can be improved by *whole core analysis* where the whole core is used for measurements. This improves on the small-plug sample, but only a very tiny part of the total reservoir is being used. Consider a reservoir of constant thickness with a cross-sectional area of 80 a. A core

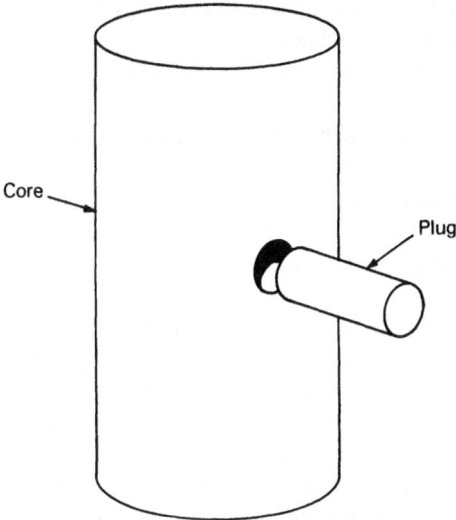

Core

Plug

Figure 1–5. Core sampling problem.

that is 7 7/8 in. in diameter represents a sample of just over .001% of the reservoir. That is only one part in a hundred thousand! Even if we have many wells of data over a field with tighter spacing of 40 a, we still have the small sample volume problem. We are not going to see very much of that reservoir represented by the cores we take!

Fortunately, some reservoirs are relatively homogeneous and exhibit nearly constant properties over their areal extent. The problem is more acute for heterogeneous carbonates and small reservoirs.

Well logs sample a somewhat larger volume of rock than do cores, but their sample volume still represents a very small part of the total reservoir. Heterogeneous rocks can present a real problem. The type of analysis selected should be determined by the quality and reliability of the data. Some statistical techniques permit evaluation of data quality as well as providing computed final results. Avoid using any statistical approach that does not provide for checking data quality and residual errors.

HYDROCARBON RESERVE ESTIMATION

The basic equation for hydrocarbon reserve estimation is

$$RH = c(DA)(RF) \sum (S_o \phi h) \qquad (1-2)$$

where

RH = recoverable hydrocarbons (barrels of oil)
DA = drainage area (acre-feet)
RF = recovery factor (fraction of hydrocarbons in the ground that can be recovered)
ϕ = porosity (what fraction of the rock volume can contain fluids: *void volume*)
S_o = oil saturation (fraction of porosity occupied by oil)
h = net pay in feet (interval thickness that can produce oil or gas)
c = 7,758 bbl per acre-foot (a volumetric conversion constant)

The summation of terms behind the summation sign Σ is made for depth intervals meeting the net pay criteria. Determining net pay criteria will be explained below. When using Eq. 1–2 for estimating gas volumes, constant c becomes 43,560 ft^3 of gas per acre-foot. Figure 1–6 illustrates the concepts of Eq. 1–2 for an idealized circular reservoir.

The symbol S_g is sometimes used for gas saturation. S_{hc}, which stands for hydrocarbon saturation, whether oil or gas, is another commonly used symbol instead of S_o or S_g. In some situations, both oil and gas reserves may have to be estimated, and it becomes necessary to separate Eq. 1–2

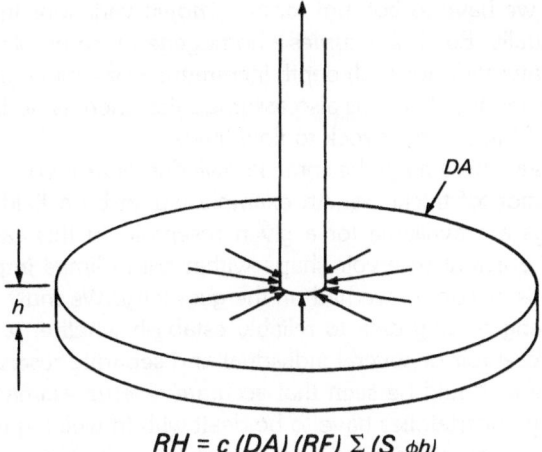

$$RH = c\,(DA)\,(RF)\,\Sigma\,(S_o \phi h)$$

Figure 1–6. Idealized circular reservoir.

into two component parts, each with its appropriate constant. An alternative would be to compute all recoverable hydrocarbon in terms of barrels of oil. In this event, the economic factors for the gas portion would still have to be sorted out.

Equation 1–2 defines the basic objective of formation evaluation: the determination of recoverable hydrocarbons. Well log evaluation plays a key role in solutions to this equation. The only known value is for the conversion constant c in Eq. 1–2. Thus, the data for the remaining five unknown variables has to be supplied. Drainage area (DA) and recovery factor (RF) are usually inferred from prior experience with other reservoirs of a similar type. The three remaining variables behind the summation sign $(S_o, \phi,$ and $h)$ are calculations made from well log data. Because of cost and data reliability problems, it is unusual to use any other source of information such as core data for these variables. Usually both hydrocarbon saturation and porosity are calculated from well logs. The net pay is then determined from these two calculations. The net pay will be those intervals meeting certain *cutoff* criteria for saturation and porosity. In Eq. 1–2, the summation of the product of $S_o, \phi,$ and h is formed for only those intervals meeting the cutoff criteria established for the reservoir. These cutoff values are usually determined from experience with other reservoirs of similar type or from the well log data. Sometimes petrophysical measurements made on cores can help in locating cutoff values. The relation of these cutoff values to reservoir performance will be described later.

It is fortunate if the reservoir thickness remains constant over the entire reservoir area. However, it is necessary to make this assumption in using Eq. 1–2 to calculate reserve estimates.

Otherwise, we have to account for any *known* variations in the reservoir geometry. Actually, Eq. 1–2 assumes a homogeneous reservoir with constant porosity and saturation for each depth increment as we move away from the borehole. In using Eq. 1–2, it is also assumed that there is no lateral change in the ability of the reservoir rock to flow fluids.

In some cases, there might be some knowledge of reservoir geometry and the areal variation of thickness. An example would be a field study where many well logs are available for a given reservoir. In this case, we could account for the actual reservoir shape within some limits imposed by the number of wells or control we had on the geometry. We must have enough geological or engineering data to reliably establish whether we are dealing with a single reservoir or several individual and separate reservoirs.

By this time it should be seen that accurate reserve estimation can be a problem. Many uncertainties have to be dealt with in well log interpretation. An estimate of the possible variation or uncertainty in a reserve estimate may turn out to be just as important as the reserve estimate itself. Statistics and computer methods are sometimes used to calculate estimates of the uncertainties.

I did not include a term to account for the volume shrinkage of oil with depth in a well in Eq. 1–2 because this is usually small. However, in estimating gas reserves, volume changes with depth are very significant due to the large pressure changes. A gas volume correction example is in one of the problems at the end of this chapter.

Brief mention was made above of the estimation of recovery factor and drainage area for Eq. 1–2. The extent of the reservoir drainage area can be difficult to determine. A common practice is to use the well spacing in existing, development fields. For example, if the wells are spaced one for every 80 a, you use 80 a for the drainage area.

For exploration wells, the geologist often estimates the reservoir size based on his or her experience with reservoirs of similar geometry and depositional environment. An engineer can sometimes calculate the reservoir area from productivity tests or drill stem tests. The recovery factor depends on the recovery mechanism or *drive* in the reservoir. Typical ranges for the recovery factor for oil reservoirs are: fluid expansion drive (in stratigraphic traps), 1–5%; solution gas drive, 15–25%; water drive (typical in structural reservoirs associated with extensive aquifers), 20–40%.[3]

In gas reservoirs, a significantly higher recovery of 50% to 70% of the gas in place can be expected. If the recovery factor is equal to 1 in Eq. 1–2, hydrocarbon in place (*HIP*) can be computed.

I mentioned earlier that the summation of $S_o \phi h$ is only taken over those intervals meeting minimum cutoff values for porosity and saturation. The ability of the reservoir rock to flow hydrocarbons establishes these cutoff values. This ability of the rock to flow fluids is called *permeability*. This important topic will be discussed in more detail in a later section.

However, I want to point out here the relation of permeability in a reservoir rock to the oil saturation in that rock. Figure 1–7 shows that below a certain critical value of hydrocarbon saturation S_c, the relative permeability to oil has declined to nearly nothing. Relative permeability to hydrocarbons K_{hc} changes according to the relative saturation of the fluids in the pore system. Below the critical hydrocarbon saturation, the pore system of the rock will tend to flow water instead of hydrocarbons. Figure 1–8 shows a typical porosity-versus-permeability relation for two different rock types. This figure shows absolute permeability, irrespective of fluid type. Note that the minimum cutoff porosity (ϕ_{ca} or ϕ_{cb}) for the two different rock types a and b may be different for the same minimum critical permeability K_c. Thus, the cutoff values of porosity and saturation that should be selected should ensure sufficient permeability to flow oil or gas from the reservoir rock. Figure 1–8 also illustrates the importance of knowing the characteristics of the rock type we are dealing with to establish cutoff values. Sometimes this information

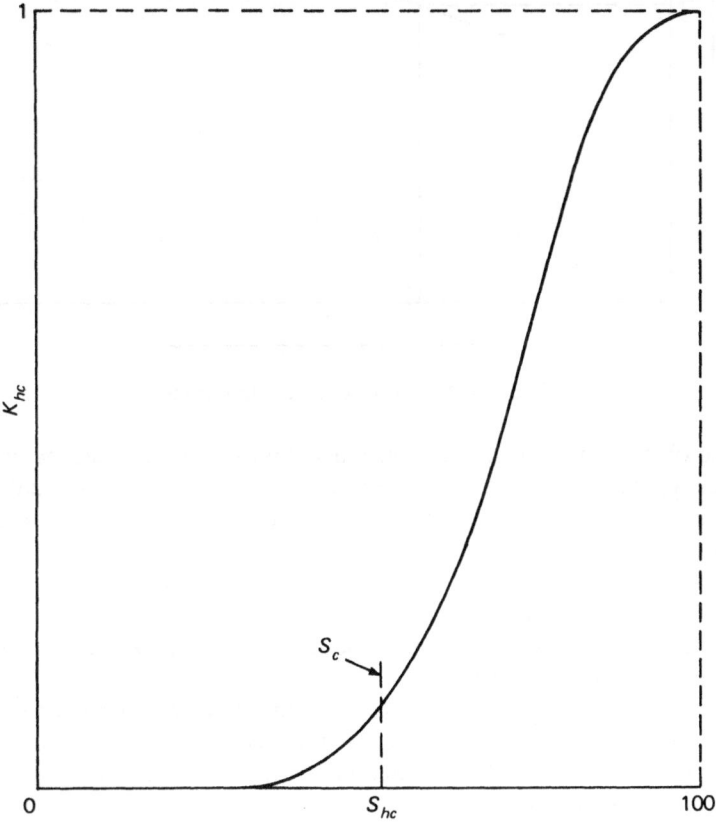

Figure 1–7. Oil saturation and permeability.

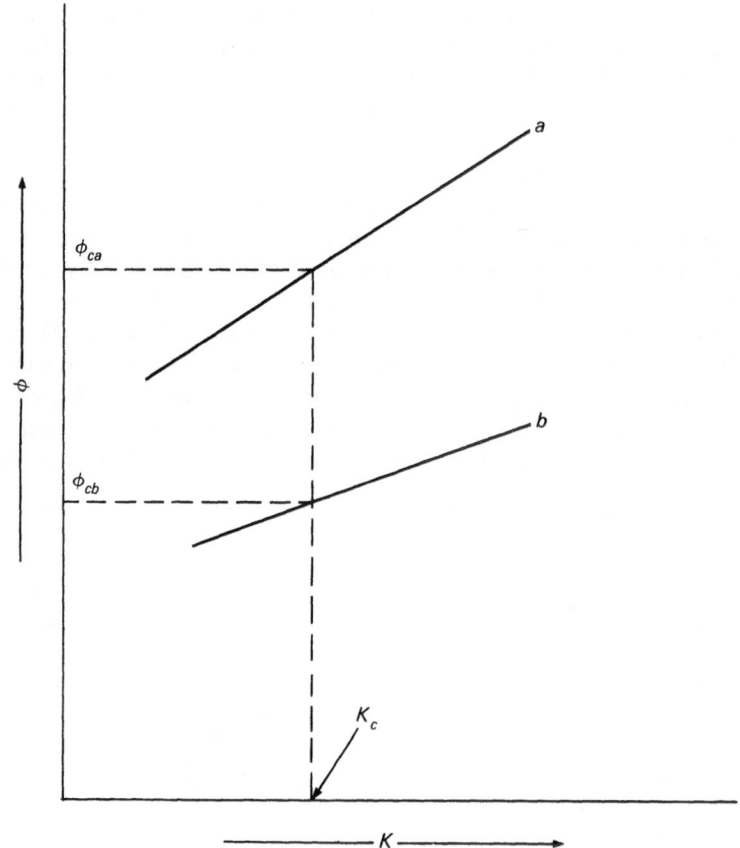

Figure 1–8.   Porosity and permeability.

can be obtained from well logs. Once again it can be seen that the concept of rock typing is very important to successful interpretation of log data.

To solve Eq. 1–2, consideration of the cutoff values leads to a sequential process:

1. Calculate porosity for the zones (depth intervals) of interest. If the porosity is less than the critical porosity, the zone has no potential for hydrocarbon production.
2. Find hydrocarbon saturation S_{hc} for zones having porosity greater than the critical value for porosity. If the hydrocarbon saturation is less than the critical value for hydrocarbon saturation, the zone has no productive potential.
3. For the intervals where hydrocarbon saturation and porosity both exceed the cutoff values, accumulate the sum of $S_o \phi h$. These are the

net pay intervals. On computer printouts of reservoir data this accumulation will be labeled in units of hydrocarbon feet.

4. If the drainage area *DA* can be estimated, HIP can be calculated using Eq. 1–2 with $RF = 1$.
5. If the recovery factor can be estimated, Eq. 1–2 can be further used to calculate recoverable hydrocarbons.

The basic approach to formation evaluation using well log data in conjunction with other frequently available data sources is:

1. Well log data provide S_{hc}, ϕ, and h.
2. Core data can sometimes be used to calibrate the well logs.
3. Drilling time logs, cuttings samples, and mud gas logs can be used to supplement well log data.
4. Productivity tests or drill stem tests can be used for important wells and as sources of formation water resistivity data.

Of course, items 2 and 4 add substantially to the cost of formation evaluation. Commonly run well logs cost from fifty cents to a dollar per foot. Cores cost hundreds of dollars per foot, and productivity tests cost from thousands to tens of thousands of dollars per foot. The decision to use these items depends on the particular situation. Cores are used more or less regularly for *wildcat* or exploration wells. They also may be used in *key* wells in a new development well program to establish calibration data for well logs.

Drilling time logs are a record of the time it takes to drill a given interval in the well. When drilling through more porous and permeable formations, the drilling time per foot of depth usually decreases markedly. This is referred to as a *drilling break*. Drilling breaks locate the porous and permeable zones in the well. During the drilling process small pieces of rock known as *cuttings* are broken off by the action of the drill bit. The circulating drilling mud system then moves these small pieces of rock to the surface. Determination of the lithology and sometimes identification of the presence of hydrocarbons and porosity can result from examination of these cuttings.

Mud gas logs are recorded at the surface using a hot wire type detector or other suitable instrument that can identify the presence of gas in the circulating mud system. In many cases, the mud gas log also gives information as to the types of gas present.

In productivity tests and drill stem tests, fluids are actually produced from the formation and pressure data are recorded. The pressure information can be used to establish the size of the reservoir drainage area. If the net pay can be established from either the drilling time log or the wireline logs, it is also possible to calculate the formation permeability from the pressure data. Drill stem tests that produce formation water also provide information on the resistivity of the formation water, which indicates the resistance the water offers to the flow of electrical current. In the following chapters, we see how this valuable piece of information can be used to estimate oil saturation from well logs.

There are pitfalls in all these supplemental sources of information. The drilling mud filtrate may completely flush the hydrocarbons from the most porous and permeable zones. When the mud weight is a little too high or the formation of interest is abnormally low pressured compared to the hydrostatic pressure for its depth, the detector at the surface may fail to identify the gas present in the formation. Normal formation pressure is about .433 pounds per square inch (psi) per foot of depth in the well. Another pitfall is errors in resistivity measurements of water produced during drill stem tests. Ions from the drilling mud system can contaminate the produced water, causing the measured water resistivity to be in error.

SUMMARY

Well logs play a central role in oil and gas exploration and reservoir evaluation. They are comparatively economical to run and provide measurements of rock properties made at reservoir conditions. A well log is a graph of some rock property recorded versus depth in a well.

The recorded data may be a distorted version of the actual rock properties due to the response characteristics of the logging tool. In addition, porosity and saturation cannot be directly measured with well logs. Therefore, equations must be used to convert the recorded data to the information that we would like to have in exploration or reservoir evaluation. This process is called *well log interpretation*.

Rock typing plays a very important part in correctly using the proper response relations or equations as part of the interpretation process.

Most of the data needed to calculate the amount of oil and gas that can be produced from a well is derived from well logs. The remaining data must be inferred from prior experience or expensive coring and testing operations.

Reserve estimation involves many uncertainties, both in the data quality and in the reliability of any unknowns that are estimated based on prior experience. This means there is an associated uncertainty with any reserve estimates made. The estimate of the amount of this uncertainty or the possible range in our answers can be as important as the actual reserve estimate itself.

PROBLEMS

1–1. Calculate the volume of HIP for an oil reservoir with the following parameters:

Drainage area = 80 a
S_o = 0.50
ϕ = 0.10
h = 10 ft

Hint: Remember to use Eq. 1–2 but set $RF = 1$ to calculate HIP.

1–2. Calculate the recoverable hydrocarbon volume for the following two zones, both having an 80 a drainage area and a recovery factor of 20%.

Zone	S_o	ϕ	h
1	0.60	0.20	20 ft
2	0.70	0.10	5 ft

1–3. What volume of gas can you expect to recover from the following reservoir? The zone of interest is at a depth of 10,000 ft in the well where the reservoir temperature (recorded from a maximum temperature device on a well log) is 170°F. Compute the gas volume at a surface temperature of 70°F.

DA = 640 a
S_g = 0.70
ϕ = 0.30
h = 20 ft

Hints: Since you are to calculate expected recovery, you will have to convert the gas volume at reservoir conditions of temperature and pressure to surface conditions of temperature and pressure. Normal atmospheric pressure at the surface will be somewhere near 14.7 psi. You can calculate the hydrostatic pressure in the well by multiplying the depth in feet by 0.433 psi/ft. Then you need to add 14.7 psi to this number to get the absolute pressure at depth for the following equation.

The gas volume at surface conditions is found by multiplying the volume at reservoir conditions by

$$((\text{pressure at depth}) \times (\text{surface temperature})) /$$
$$((\text{surface pressure}) \times (\text{reservoir temperature}))$$
$$= ((0.433 \text{ psi/ft} \times \text{depth in ft} + 14.7\text{psi})$$
$$\times (\text{surface temperature °F} + 460°)) /$$
$$((14.7 \text{ psi}) \times (\text{reservoir temperature °F} + 460°)).$$

The quantity of 460° is added to the Fahrenheit temperatures to scale the temperatures on the absolute basis required for the gas volume relations.

You also need a value to use for the recovery factor in this problem. Since it is not given, you have to use the typical range of recovery factor for gas reservoirs that was given in the text and express your answer as a range of recoverable gas volume. Express your answer in terms of the usual petroleum industry units of MCF for thousands of cubic feet or MMCF for millions of cubic feet. If large enough, you can express the answer in billions of cubic feet (BCF).

REFERENCES

1. Author's class notes from course on well log interpretation presented by George R. Pickett at the Colorado School of Mines, Golden, Colorado, 1975.
2. Ibid.
3. Author's class notes from course on advanced well log interpretation presented by George R. Pickett at the Colorado School of Mines, Golden, Colorado, 1976.

CHAPTER 2

Porosity and Saturation

Porosity is one of the most important rock properties that can be calculated from well log data. It gives the rock's capacity to store oil or gas. Thus far, I have referred to porosity as the *void* space in the rock. A rock such as chalk may be nearly half void space, but the pores are usually so small that they cannot be seen without suitable magnification.

Only total porosity can usually be calculated from well log data. However, only the effective porosity actually contributes to the flow of hydrocarbons. The effective porosity is that part of the pore system that is interconnected and permits fluid communication or flow. Fortunately, for many reservoir rocks, effective porosity is nearly the same as total porosity. However, in carbonates and some highly-cemented sands the total porosity can be much larger than the effective porosity.[1]

Porosity cannot be measured directly from well logs. Calculations are made from measurements of a rock property that in turn is related to porosity. The conversion of the log measurement to porosity is based on some response relation, much the same as for the acoustic or sonic log example in Chapter 1. Some assumptions usually must be made and the accuracy of the porosity calculations will be limited by these assumptions (how well do they describe reality?).

Although a number can be calculated for porosity, pore structure and pore size distribution also play important roles in reservoir performance. Generally, coarse-grained rocks have larger pores whereas fine-grained rocks have smaller pores. Pore structure is also affected by grain shape, which can range from almost perfect spheres to very flat, plateshaped grains.

Porosity is the fractional part of the rock that is not composed of solid rock matrix. I have referred to this as void space, but it is somewhat of a misnomer. The pore spaces are not void in the sense of being a vacuum. They are always occupied by a liquid or gas, even if that gas is air.

How large can porosity be in a reservoir rock? For spherical grains packed together in the least compact way, the porosity is nearly 48% of the total volume. For the most compact packing of spherical grains the porosity is almost 26%. Of course, in reservoir rocks the grains are not usually perfect spheres, so porosity will be something less than what it would be with perfect spheres. Note that these two theoretical porosities for spheres are not dependent on the particular size of the spheres. As long as the spheres are all the same size, these two porosity percentages apply. If rock grains are of different sizes, the porosity will be considerably less than these ideals. When the rock grains are nearly the same size they are said to be *well sorted*.

Porosity can be conveniently divided into two classes: primary and secondary. This division is based on the time of porosity formation in the rock. *Primary porosity* results after the sediments are initially laid down and is a function of the environment of deposition, which also affects grain sorting and, hence, the amount of porosity in the rock.

Secondary porosity occurs later in the development of the sediment. It results from processes such as recrystallization, fracturing of the rock, or solution of minerals in cavities and voids in carbonate rocks. These voids may be poorly connected. In such a case, total porosity may be large, whereas effective porosity is small. Fractures often contribute no more than 1% to total porosity, yet they can substantially improve reservoir performance because of the greatly increased ability of fluid to flow. Fractures can greatly enhance the connection of the pores by reducing the length of the often tortuous paths through the pore system. Petroleum people refer to rocks with good fluid communication as being *permeable*.

Expressed as an equation, porosity is given by

$$\phi = (\text{void volume})/(\text{total volume}) \qquad (2\text{--}1)$$

where the two volumes are measured in the same units, and porosity ϕ is a dimensionless fraction of the total rock volume. Two alternative ways of looking at this equation are helpful in grasping the concept of porosity as well as being useful in calculating porosity from practical laboratory measurements on cores. The first alternative is in terms of void volume and grain volume rather than total volume.

$$\phi = (\text{void volume})/(\text{void volume} + \text{grain volume}) \qquad (2\text{--}2)$$

The second alternative is in terms of grain volume and total volume.

$$\phi = (\text{total volume} - \text{grain volume})/(\text{total volume}) \qquad \textbf{(2–3)}$$

If these concepts are new to you, it may be helpful to spend some time verifying their equivalence. I sometimes refer to *grain* volume as *matrix* volume in this text.

Equation 2–1 could be rewritten to give the pore (void) volume in terms of porosity and total volume. This is what we would do to estimate the total volume in a reservoir that could contain fluids. For example, suppose we consider a rock volume of 1 mi^3 with a porosity of 40%. The pore volume must be .4 mi^3. This is the rock's storage capacity for fluids. If this were a homogeneous and uniform rock having the same porosity throughout, we could take any size piece out of it and still find that the ratio of void volume to total volume was 40%. It is irrelevant whether we say the porosity is .4 or 40%. Porosity is a dimensionless quantity always measured as a fraction or percentage of total volume. In some equations used in log analysis, however, you must carefully check to see whether porosity is specified as a fraction or as a percentage.

POROSITY IN SANDSTONES

Most porous reservoir rocks are either sedimentary sandstones or sedimentary carbonate rocks. Generally, sandstones have grains between 1/16 mm and 2 mm in size and can have porosities of 20%; sandstone porosity can be significantly larger if the grains are well sorted. However, sandstones usually have low permeability if the grains are small.

Sandstones are classified as orthoquartzite, arkose, or graywacke, according to their relative percentages of quartz, feldspar, and rock fragments.[2] Orthoquartzite consists of 80% or more quartz and may approach the almost universally assumed grain density of 2.65 grams per cubic centimeter (gm/cc). Arkosic sandstones have between 40% and 80% quartz, with the remaining percentage being predominantly feldspar and some rock fragments. Graywacke contains less than 20% quartz, with the balance being about equally divided between feldspar and rock fragments. Often sandstone grains are cemented by calcite and are referred to as calcareous sandstones. Most sandstones other than quartzite have grain densities greater than 2.65 gm/cc. Other commonly assumed sandstone grain densities are 2.68 gm/cc and 2.70 gm/cc. It is usually best to use grain density data from cores when available. Otherwise we must rely on our best guess. In some cases, we can use some of the more advanced log interpretation cross-plot methods to obtain usable grain density data.

Water drive is the usual mechanism for oil production in the quartzose or clean sandstones. The permeability to fluids in these rocks is usually the same either perpendicular to or parallel to the bedding plane of these sediments. Cutoff hydrocarbon saturation for hydrocarbon production may be greater than 50% in these rocks. Examples are the Wilcox and Woodbine sands of Oklahoma.[3]

Arkosic sands are usually poorly sorted and poorly rounded, which means they are likely to have much lower permeability than clean sands. Solution gas drive and water encroachment from aquifers of limited extent account for production in these rocks.[4] Permeability across bedding planes is greatly reduced. Cutoff hydrocarbon saturation for hydrocarbon production in these rocks will be lower than that for clean sands. Examples of arkosic sandstones, many of which are known by the term *granite wash*, are the Stevens sandstone of California and the granite wash of the Anadarko Basin in Oklahoma.[5]

Graywacke sediments are sometimes known as dirty sands or salt and pepper sands. When they contain clay minerals they are called shaly sands. Shaly sands can also form from what might have been relatively clean sands after their original deposition. The clay minerals may form by alteration of feldspar minerals in the rock. These *authigenic* clays may coat the pores, reduce porosity, and drastically reduce permeability. Shaly sands are becoming increasingly important as petroleum reservoirs. They sometimes present a difficult challenge to successful well log interpretation. This difficulty occurs because it is easy to erroneously calculate that these sands are very highly water saturated when they are actually hydrocarbon productive. This is due to the effect they have on the resistivity as measured by well logs. Well log measured resistivities read abnormally low under some conditions. Shaly sands normally consist of smaller-sized grains and have low permeability. They also have a very high ratio of surface area to volume compared to clean sands. This contributes significantly to their low permeability. In shaly sands, the effective porosity is the total porosity less the volume fraction of chemically bound water of the clay minerals of the shaly sand. This chemically bound water will be seen by the logging porosity tools as part of the total porosity. Because of the typically low permeability of the sands, they are often referred to as *tight* sands. Cutoff hydrocarbon saturation for hydrocarbon production is usually below 50%.

POROSITY IN CARBONATES

Porosity in carbonates can sometimes complicate calculations from well logs. Unlike sands that are usually more or less radially homogeneous from a borehole (at least within the radius of investigation of most well logging tools), carbonates tend to exhibit unpredictable heterogeneity of pore structure.[6] This heterogeneity of carbonate porosity has far-reaching consequences.

Not only are well log responses affected, but it becomes very difficult to obtain representative samples from core data. In many cases, the volume sampled by a core is substantially less than that sampled by a well logging tool. Core data may be measured on a foot-by-foot basis, whereas well log data are usually some type of average over several feet even though they can be recorded for every foot or half foot. It is well to remember these considerations when attempting to resolve apparent discrepancies between core data and well log data.

Carbonates exhibit both primary and secondary porosity, with secondary porosity being of far more importance. Secondary porosity in carbonates can be classified as intercrystalline, moldic, chalky, vug or channel, and fracture or breccia.

Primary porosity in carbonates is divided into several classes. We need not, however, concern ourselves with these divisions in detail except to mention that one classification—*interparticle porosity*—could be treated as a carbonate sand. These carbonate sands can have porosities and well-sorted pore size distributions similar to their clastic counterparts under the appropriate depositional conditions.

In North America, the most important type of porosity in carbonates for oil and gas production is the intercrystalline porosity found in dolomite. Dolomitization occurs when magnesium-rich waters circulate through the pore spaces of limestone and part of the original calcium in the limestone is replaced by magnesium.[7] The limestone is then changed to dolomite $(CaMg(CO_3)_2)$. Since dolomite molecules occupy only 88% of the volume of calcite molecules, a net increase in porosity results. Depending on the concentration of magnesium in the circulating waters, the result is a variable ratio of calcium to magnesium in the dolomite. The ratio of calcium to magnesium varies from 58:42 to 47.5:52.5.[8] This variation in turn results in a variation in dolomite's petrophysical properties, making it difficult to define a standard dolomite for laboratory purposes and to produce standardized tables of logging tool responses for dolomite. For example, calcite or limestone has a fixed grain density of 2.71 gm/cc, whereas the density of dolomite will vary. In *Dana's Manual of Mineralogy* the density of dolomite is listed as 2.85 gm/cc,[9] whereas most well logging references and chart books published by logging service companies give 2.87 gm/cc.

In practice, porosity may decrease in the early stages of dolomitization. This is followed by a rapid increase in porosity as the percentage of dolomite goes above 50. Figure 2–1 shows that pore size and porosity follow essentially the same changes with increasing dolomitization. Dolomite may occur in rare cases as an evaporite with primary porosity.[10]

The general mechanism for the formation of moldic porosity is the removal of a constituent such as a shell or oolite from the rock matrix. This removal is by dissolution in ground water solutions charged with carbon dioxide (CO_2) and organic acids. Molds can form in dolomite from selective

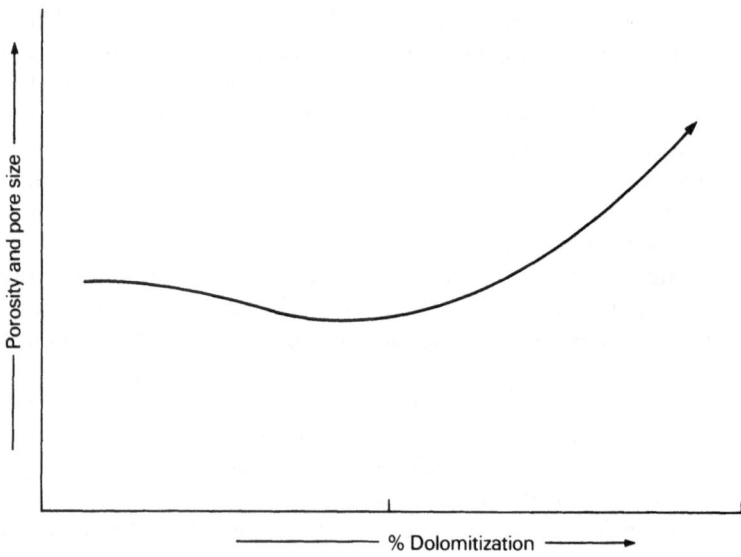

Figure 2–1. Porosity and pore size versus dolomitization.

solution of aragonite or calcite and less commonly by solution of anhydrite or halite. Sometimes these moldic pore spaces are poorly interconnected and have low permeability. They are easily damaged by formation treatments otherwise designed to enhance permeability in carbonates.[11]

Secondary pore space can be reduced or eliminated in some cases by deposition of crystals from some mineral, usually calcite. Sometimes pore space becomes plugged by salt (halite) or anhydrite.

The term *vug* is usually reserved for pores that are somewhat equant. Otherwise, they are termed *channel pores*. Vugs usually have a diameter greater than 1/16 mm and do not conform in position, shape, or boundary to particular fabric elements of the host rock. Dissolution is the dominant process in the formation of vugs.

Carbonates are inherently rigid and thus prone to fracturing. Fracturing can facilitate solution action in carbonate rocks. Total fracture porosity rarely exceeds 1%, yet it can be a very significant factor in hydrocarbon production because the fractures can greatly increase reservoir permeability.

Chalky porosity results in one of the largest porosities there is in rock. This is important in some reservoirs where porosities higher than 40% are possible. Despite this high porosity, the pore spaces are usually so small that the permeability of these rocks is very low. These chalky reservoir rocks produce small, yet commercial, quantities of gas from shallow, easy to drill wells. An example is the Smoky Hill member of the Cretaceous Niobrara formation in Colorado, Kansas, and Nebraska.

COMPUTATIONAL PROBLEMS WITH POROSITY AVERAGES

One would think that porosity calculations should be rather straightforward. In Chapter 1 an example with an acoustic, or sonic, log was presented where a simple, straight-line equation was used to convert the logging tool response into porosity data. However, it will not be long before you will be interested in calculating some type of average for porosity over some interval of interest in a well. At this point porosity calculations can become complicated.

A standard part of any complex analysis is to try to simplify the results with a good summary. Well log data are no exception. You can quickly calculate several pieces of data for several tens or even hundreds of feet of depth in a well. An important part of a good summary for this type of data is to use a single number to represent many individual and varied numbers. Simple examples of representative numbers that can summarize many data points are the median and average. These are used where there appears to be some predominant value and the individual measurements scatter about this one central value. It is tempting to form simple averages of porosity data. However, a simple average of porosity can be very misleading because of a frequent lack of a strong central tendency in these data. There may be two (bimodal) or even more (polymodal) predominant values rather than one central value.

However, we would like to use a good summary number to represent many porosity data points in a reservoir. Which number should be used? The correct number is a *weighted* average of the porosity data, where the weighting is the thickness of the depth interval(s) associated with the porosity values. Perhaps we could think of this one porosity number as a single number we would use in Eq. 1–2 for reserve estimation. We could use this single porosity summary number along with a corresponding summary number for saturation with the thickness *h* to calculate the reserves. Over the thickness *h*, we could have several tens of feet or even several hundred feet of data represented by the single summary numbers for porosity and saturation. Let us see how we can form this single average number for porosity.

Suppose that the cylindrical volume of Figure 2–2 represents a core, or maybe the cylindrical volume sampled by some porosity logging tool. Which one number would be most appropriate to represent the porosity for the entire 45 ft interval in the figure? If we assume that the rock thickness remains constant over the lateral extent of the reservoir, we can come up with a reasonable number.

We could simply average the three porosity values for the three sections shown, but something may prove to be amiss with this simple average. For example, most of the interval has a porosity of only 5% (35 ft out of the total of 45 ft). The arithmetic average of 15% porosity is not very representative.

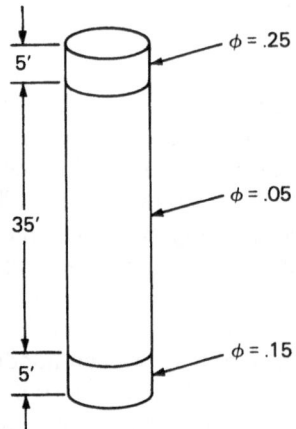

Average = (.25 + .05 + .15)/3 = .15.
Weighted by thickness (vol),
average porosity = [(5′ × .25) + (35′ × .05) + (5′ × .15)]/(5′ + 35′ + 5′)
            = .083

Figure 2–2.   Average porosity.

What we need is a single, representative number for porosity for the total volume of rock from which the cylindrical sample is taken. It does not matter whether we think of the cylinder as a core removed from some larger volume of rock, or whether we think of the cylinder as the volume of investigation of some logging tool. The cylinder is assumed to represent some larger volume of unknown lateral extent. That is, we will have to assume that the thicknesses of rock strata represented in the core remain constant over the areal extent of the reservoir if we are going to come up with any meaningful average. Likewise, we must assume that the rock properties represented by the cylindrical volume remain the same over the areal extent of the reservoir.

If we knew the total volume of rock and the total pore volume within this same volume of rock, we could simply divide the pore volume by the total volume and obtain one realistic, representative number. Recall from the earlier discussion of porosity that the pore volume is equal to the product of the total volume and the porosity. Here, the problem is that we do not know the radial extent of the total rock volume involved. We have only the core, which is a small sample.

If we assume that reservoir properties are constant as we move radially away from the well bore, we can calculate the total volume and pore volume in terms of the unknown reservoir area. Assume that the core or the cylindrical volume representing the radius of investigation of some logging tool is taken from a reservoir whose properties are constant over the interval analyzed. We can calculate the total pore volume by summing the separate pore volumes for each of the subintervals or segments of Figure 2–2, then divide by the total volume to get the porosity. The volume for each segment is the product of the thickness of the segment with the area of the segment. Note that the area is assumed to be the same for all three segments even though this area is unknown. For the three segments, starting at the top:

$$volume_1 = area \times 5 \text{ ft}$$
$$volume_2 = area \times 35 \text{ ft}$$
$$volume_3 = area \times 5 \text{ ft}$$

The total volume is the sum of the three separate volumes.

$$
\begin{aligned}
\text{Total volume} &= volume_1 + volume_2 + volume_3 \\
&= (area \times 5 \text{ ft}) + (area \times 35 \text{ ft}) + (area \times 5 \text{ ft}) \\
&= area \times (5 \text{ ft} + 35 \text{ ft} + 5 \text{ ft}) \qquad \textbf{(2–4)}
\end{aligned}
$$

Thus, the total volume is the sum of the product of the total interval thickness with the unknown area of the reservoir. This common unknown area can be factored out of the equation for total volume as was done in the bottom line of the expression above for total volume. This all may seem self-evident, but there is an important point to be made shortly that depends on this deceptive simplicity.

The pore volume for each element then is the product of the porosity of the element times the volume of the element. These individual pore volumes can be summed to get the total pore volume.

$$
\begin{aligned}
\text{pore volume} &= \phi_1 \times volume_1 + \phi_2 \times volume_2 + \phi_3 \times volume_3 \\
&= \phi_1 \times (area \times 5 \text{ ft}) + \phi_2 \times (area \times 35 \text{ ft}) + \phi_3 \times (area \times 5 \text{ ft}) \\
&= (\phi_1 \times 5 \text{ ft} + \phi_2 \times 35 \text{ ft} + \phi_3 \times 5 \text{ ft}) \times area \qquad \textbf{(2–5)}
\end{aligned}
$$

Again, the common area of the elements was factored out in the last line of this expression for pore volume.

Equation 2–5 contains a symbol for the porosity of each zone rather than the actual porosity values. The subscripts identify the three separate zones. This leads naturally to an expression that can be used in any problem. If we now divide the expression for pore volume by the expression for total volume, we obtain the porosity.

$$
\begin{aligned}
\text{porosity} &= \text{pore volume/total volume} \\
&= ((\phi_1 \times 5 \text{ ft} + \phi_2 \times 35 \text{ ft} + \phi_3 \times 5 \text{ ft}) \\
&\quad \times area)/((5 \text{ ft} + 35 \text{ ft} + 5 \text{ ft}) \times area) \qquad \textbf{(2–6)}
\end{aligned}
$$

The *area* term is common to both the numerator and denominator of the expression for porosity and thus cancels out in the division. This leaves the sum of the products of the porosities with the subinterval thicknesses divided by the total thickness of the interval considered. This is called a *weighted average*, where the weights are the thicknesses h of the subintervals involved. This weighted average porosity can be represented by the more general equation:

$$\phi' = \sum \phi_i h_i \Big/ \sum h_i \qquad\qquad \textbf{(2–7)}$$

where the primed symbol is used to distinguish the weighted average, and the subscript i refers to the ith subinterval. The summation is assumed to be over all subintervals of interest.

The porosities of the subintervals are weighted by their respective volumes. Since their areas are assumed to be common with the total reservoir area, this process reduces to a weighted-by-thickness process. If it turned out that the subintervals were in fact part of volumes having different areas, it would be impossible to come up with a representative average porosity without specific knowledge of the reservoir geometry. This probably hap-

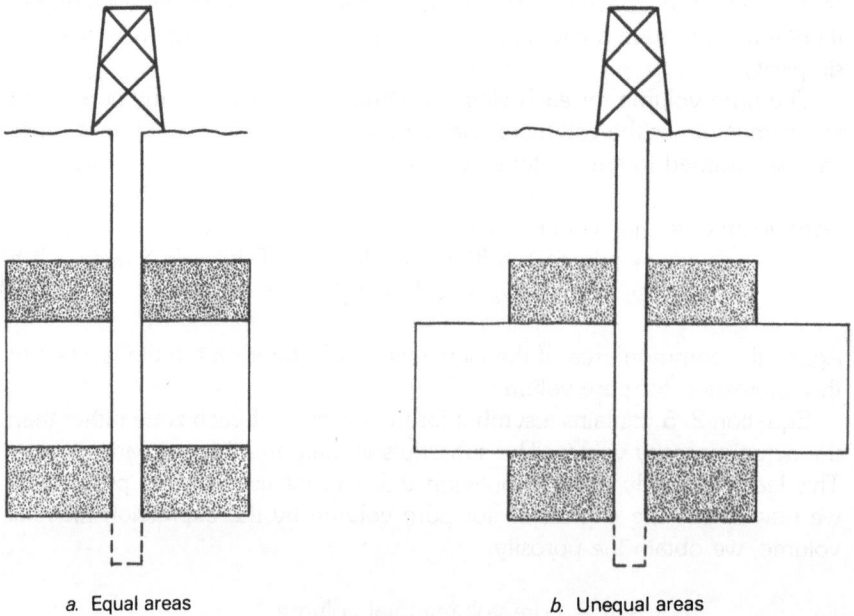

a. Equal areas   b. Unequal areas

Figure 2–3.   Errors in porosity averages.

pens to some extent in practice and could introduce significant error into the reserve estimates for the reservoir in question.

In Figure 2–3a, a borehole is shown penetrating a thick dolomite that is sandwiched between two sandstones. Since each of the three horizontal strata have the same area (assume that each has the same length in the dimension perpendicular to the page also), the weighted-by-thickness average porosity should represent the fractional fluid storage capacity of the entire block composed of the three layers. At least, it should if the three layers have a homogeneous distribution of porosity.

However, in Figure 2–3b, the dolomite zone has a significantly larger area. If we weight the three zones in Figure 2–3b by thickness, it is apparent that the volume of the dolomite zone will not receive enough weight or representation in the average. In this case, too much weight will be assigned to the two sand bodies. Their respective porosities will unduly influence the average, unless, fortuitously, all three strata have the same porosity! Otherwise, we would have to have some knowledge of the actual reservoir geometry to find a true average. For example, there may be many other boreholes drilled in the same area that would allow a reasonable estimation of the lateral extent of each of the three zones.

Now let us look at a final example of weighted average porosity in Figure 2–4, where I have elected to illustrate the volumes using gallons as a

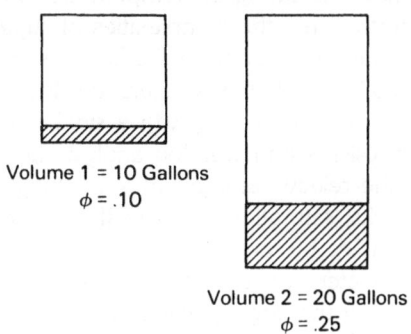

Volume 1 = 10 Gallons
$\phi = .10$

Volume 2 = 20 Gallons
$\phi = .25$

Pore volume 1 = 10 × .10 = 1 gallon
Pore volume 2 = 20 × .25 = 5 gallons
$$\phi = (.10 + .25)/2 = .175$$
Total pore volume = 1 gallon + 5 gallons
Total volume = 10 gallons + 20 gallons
Pore volume / total volume = 6/30 = .20

Figure 2–4.   Weighted average porosity.

measure of volume. Although gallon is a liquid measure, it serves well to illustrate that it does not really matter what units we use, since porosity is a dimensionless quantity. The volume units will cancel out in the arithmetic we use to calculate average porosity. Compare the straight average porosity of 0.175 with the weighted average porosity represented in the figure by the ratio of pore volume to total volume. This is 20% (0.2) porosity and represents a significant difference from the straight average. Thus, a straight average porosity is erroneous. To correctly average porosity, we must weight the porosity data by their corresponding interval thicknesses. Try it for this simple example but substitute the volumes for the thicknesses in Eq. 2–7.

SATURATION

Before we discuss hydrocarbon saturation or water saturation, let us examine the concept of saturation itself. What does it mean? If we look to elementary chemistry for an answer, we might consider a salt solution in water. A given volume of water can only hold so much salt in solution, with the amount being a function of the temperature of the solution. When that amount of salt is dissolved in solution, we say that the solution is *saturated*. What we mean is that no more salt can be added to the solution. Any additional grains of salt will simply go to the bottom of the solution. It would not be possible to dissolve them without increasing the temperature of the solution.

We need go no further into the complexities of supersaturated solutions and what happens if the temperature of the liquid is increased. The important point is that saturation in chemistry refers to salt solutions in electrolytes or liquids. With rocks, we are dealing with a similar, yet somewhat different meaning. Petrophysicists and well log analysts use the term *saturation* in rocks to refer to the relative amount of something that is contained in something else. In rocks, saturation refers to the fraction or percent of the pore space that is occupied by some fluid or gas. If the pore spaces in a rock are completely full of water, we say they are 100% saturated with water or simply *water saturated*. If the pore spaces have 40% of their volume occupied by gas and the rest by water, there is obviously 60% water saturation in the pore system. The total fraction of both gas and water is 100% or simply 1 if we are dealing in fractional parts instead of percentages.

Thus, saturation in rocks refers to the fraction of the pore volume that is filled with something. From the previous section on porosity, recall that porosity represents the fraction of the total rock volume that is void space. Now we are dealing with something that is, in turn, a fractional part of another fractional part. Water saturation is the fraction of the pore volume that is occupied with water, and the pore volume is, in turn, a fraction of the rock volume. If the pores are half filled with water (50% water saturation) and the pore space occupies 10% of the total rock volume (10% porosity), then the volume of water must make up only 5% of the volume of the rock. If we represent water saturation by the symbol S_w and porosity by ϕ, we can relate this fractional part of the pore space filled with water to the fractional part of the total rock volume filled with water. This fractional part of the rock volume that contains water is defined as the *bulk volume of water* (BVW) and we calculate it from the equation

$$\text{BVW} = \phi \times S_w \qquad \textbf{(2-8)}$$

Using the numbers just given for porosity (10%) and water saturation (50%), the substitution is straightforward.

$$\begin{aligned}\text{BVW} &= (0.10) \times (0.50) \\ &= 0.05\end{aligned}$$

Note that whereas the water saturation is 50%, the rock volume occupied by water is only 5%. Bulk volume oil or hydrocarbon can be defined analogously to BVW. Just use oil saturation (or hydrocarbon saturation) in place of water saturation in Eq. 2–8. Water saturation (or hydrocarbon saturation) always refers to a *fraction of the pore volume* and not the total rock volume. If we want to know the saturation of the liquid (or gas) in terms of total rock volume, we must calculate the BVW or liquid (or gas) that is the product of porosity and water saturation (or hydrocarbon saturation). Remember that porosity gives the fraction of the rock that could contain fluids, whereas the BVW (or hydrocarbon) gives the fraction of the rock volume that *actually* contains water (or hydrocarbons). The two numbers are identical only when water saturation (or hydrocarbon saturation) equals 100%. Then, the BVW (or hydrocarbon) equals the porosity. I prefer the term *unit* volume as more descriptive, but *bulk* volume prevails in well log literature.

Figure 2–5 shows an example of a BVW calculation. In this example, BVW is calculated by calculating the fraction of the total rock volume occupied by the volume of water. Note that we get the same answer if we use Eq. 2–8.

Figure 2–6 is another illustration of the BVW concept. In this figure, it is easy to see that, although 25% of the cross-sectional area of the rock represents void space that is in turn 40% saturated with water, only 10% of the total volume of the rock is filled with water.

I have concentrated on differentiating between water saturation as a percent of pore volume and the BVW as a percent of total rock volume so you will be able to use these concepts in estimating average water saturation or average hydrocarbon saturation. In Figure 2–7 you might wish to find one representative number for water saturation for the two separate volumes of

$\phi = .35$
$S_w = .40$

Volume = 10 gallons
Pore volume = 3.5 gallons
Grain volume = 6.5 gallons
Volume H_2O = S_w × pore volume
              = .4 × 3.5 = 1.4 gallons
Bulk volume H_2O = volume H_2O/bulk volume
              = 1.4 gallons / 10 gallons = .14
BVW = ϕ × S_w

Figure 2-5. Bulk volume water calculation. Water saturation S_w is a fraction of pore volume.

rock shown in cross section. The first one with a volume of 1 ft^3 and porosity of 25% has a water saturation of 40%. You immediately see that the water saturation fraction for the given pore volume is only 10% of the total rock volume. This is 1/10 ft^3. Similarly, for the second volume of rock (2 ft^3 volume, 10% porosity, and 75% water saturation), the water volume is 7.5% of that rock volume.

It is easy to see what might happen if we tried to take a straight average of

Bulk volume water = ϕ × S_w

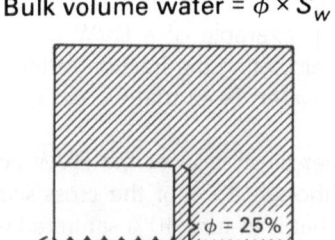

$\phi = 25\%$

40% of porosity filled with water, but this volume of water
is only 10% (.40 × 25) of the total bulk volume

Figure 2-6. Bulk volume water illustration.

$$\text{Average } S_w = \frac{S_w \times \phi \times h}{\phi \times h}$$

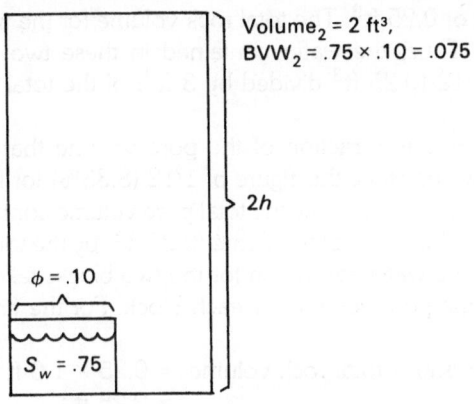

Volume$_1$ = 1 ft^3,
BVW$_1$ = .4 × .25 = .10

ϕ = .25

S_w = .4

Volume$_2$ = 2 ft^3,
BVW$_2$ = .75 × .10 = .075

ϕ = .10

S_w = .75

$$S_w = \frac{\text{BVW}_1 \times \text{volume}_1 + \text{BVW}_2 \times \text{volume}_2}{\text{volume}_1 \times \phi_1 + \text{volume}_2 \times \phi_2}$$

$$= \frac{1 \text{ ft}^3 \times .10 + 2 \text{ ft}^3 \times .075}{1 \text{ ft}^3 \times .25 + 2 \text{ ft}^3 \times .10}$$

$$= .56$$

Figure 2–7. Average water saturation. (Note: BVW $= \phi \times S_w$.)

the two water saturations in Figure 2–7. This straight average is 57.5% (40% plus 75% divided by 2). However, it has no relation to reality and in some cases such averages can be very misleading. Calculate the actual volumes of water for each block and see what fraction of the total rock volume is actually occupied by water. Since I have stated that the BVW is the fraction of the total rock volume occupied by water, you can find the volume of water for each block in the figure from

water volume = BVW × total rock volume

For the first block

$$\text{water volume} = 0.10 \times 1 \text{ ft}^3$$
$$= 0.10 \text{ ft}^3$$

For the second block

$$\text{water volume} = 0.075 \times 2 \text{ ft}^3$$
$$= 0.15 \text{ ft}^3$$

The total volume of water in the two blocks of rock is the sum of the separate volumes or 0.25 ft^3. The total rock volume for the two blocks taken together is 3 ft^3. Thus, the water contained in these two blocks together represents only 1/12 (0.25 ft^3 divided by 3 ft^3) of the total rock volume of the two blocks.

Water saturation is the fraction of the pore volume that is occupied by water. How can we translate this figure of 1/12 (8.33%) for BVW into water saturation? First, we need to know the total pore volume contained in the two blocks. If we divide the total water volume (0.25 ft^3) by the total pore volume, we should obtain the water saturation for the two blocks taken together.

First compute the pore volume for each block. For the first rock volume

$$\text{porosity} \times \text{total rock volume} = 0.25 \times 1.0 \text{ ft}^3$$
$$= 0.25 \text{ ft}^3$$

For the second rock volume

$$\text{porosity} \times \text{total rock volume} = 0.10 \times 2.0 \text{ ft}^3$$
$$= 0.20 \text{ ft}^3$$

The sum of these is the total void space for the two rock volumes: 0.45 ft^3. This 0.45 ft^3 of pore volume is occupied by 0.25 ft^3 of water. This volume of water represents, then, 56% of the pore volume (fortuitously, nearly the same as the straight average). This is the actual water saturation of the two blocks of rock taken together. It is instructive to rewrite this process using symbols instead of actual values. To calculate the total water volume, take the sum of the individual water volumes for each block. We did this above by summing the products of the BVW with the total rock volume for each block. Symbolically, we calculated

$$\text{water volume} = \sum (\text{BVW})_i \times (\text{volume})_i$$
$$= \sum \phi_i \times S_{w_i} \times (\text{volume})_i$$

making use of Eq. 2–8 to rewrite the BVW in terms of porosity and water saturation. The subscript *i* denotes the *i*th block. Next we calculate the total pore volume from the products of porosity and rock volume for each block.

$$\text{pore volume} = \sum \phi_i \times (\text{volume})_i$$

Before we divide the water volume above by the pore volume to obtain water saturation, it is necessary to decide how to obtain the volumes for each block. In practice, these volumes will not be known. The situation is similar to that of trying to average the porosities of Figure 2–2. Again, we must use the same assumption that we used then. That is, the individual volumes are made up of the product of a common area with the thickness of each separate volume element.

$$\text{volume} = \text{area} \times h$$

where *h* is the thickness of the volume element. If the area is the same for each volume element, it can be factored out of each of the summations above and will cancel when we divide water volume by pore volume to get

$$\text{average } S_w = \sum \phi_i S_{w_i} h_i / \sum \phi_i h_i \qquad \textbf{(2–9)}$$

The summation is taken for all elements of interest.

It is now apparent that we obtain a true average water saturation only when we weight each saturation number for each separate element by the product of porosity and thickness for that same element. Again, it is necessary to use the assumption that the lateral extent of the reservoir remains the same for each depth increment in the well. Without this assumption it is impossible to compute a *true* average saturation or porosity, unless we know the actual area of the reservoir as a function of depth in the well.

A key distinction to understanding water saturation is to remember that S_w is a fraction of *pore volume*. If we want to express the fraction of total rock volume that is occupied by water, we must use the BVW. From the example of Figure 2–7, we saw that the straight arithmetic average of S_w is not necessarily correct. The problem in using a straight average for S_w arises when the products of porosity and thickness for the intervals in the well are changing. We can expect to be correct in using a straight average S_w only if the porosity remains relatively constant as depth changes in the well. If the porosity changes significantly with depth, we can expect any volumetric calculations of hydrocarbon reserves to be in error if we use a straight, arithmetic average of water saturation.

As a final illustration, consider an example where hydrocarbon volumes are expressed in units of acre-feet. In Eq. 1–2 of Chapter 1 for hydrocarbon reserve estimation, the product of the summation of $S_o \phi h$ with the drainage area will give the hydrocarbon volume in place in the reservoir in units of acre-feet. In practice, we cannot compute hydrocarbon saturation S_h from well logs. What we *can* do is compute the water saturation S_w and use the assumption that what is left in the pores after water is accounted for must be hydrocarbons.

$$S_h = 1 - S_w \qquad (2\text{--}10)$$

Here I have used the more general symbol S_h for hydrocarbon saturation rather than S_o for oil saturation or S_g for gas saturation. Equation 2–10 is based on the assumption that the fractional parts of all the constituents of the pore volume must add up to one if each is expressed as a fraction of that pore volume. Of course it is always possible that what remains in the pore system after water is accounted for is some nonhydrocarbon gas such as carbon dioxide or even air. Rocks that outcrop not too far from a well sometimes contain air in the pore system, yet they seem to be large hydrocarbon reservoirs as far as can be ascertained from some well log responses.

Figure 2–8 shows a cylindrical-shaped reservoir with a cross-sectional area of 1 a. The reservoir is split into an upper block and a lower block. Call

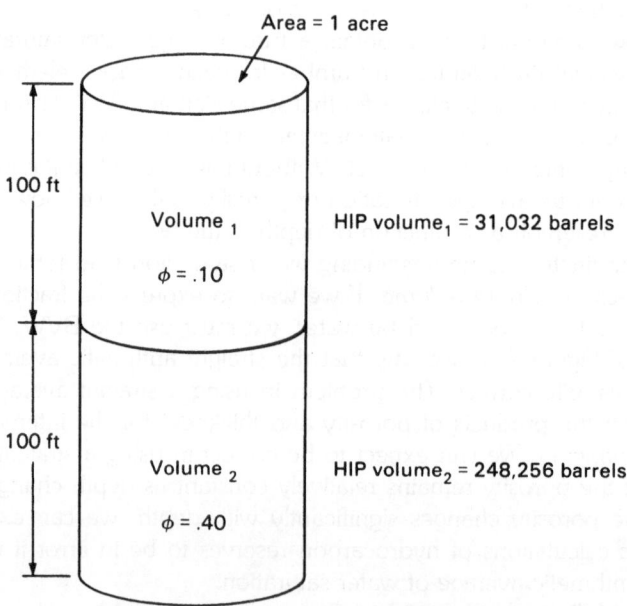

Figure 2–8. Cylindrical-shaped reservoir.

the upper block volume$_1$ and the lower block volume$_2$. Let each of the two blocks be 100 ft thick so that each block of rock has the same volume of 100 acre-feet. Of course, these volumes include both rock matrix as well as pore space. Figure 2–8 gives the porosity of volume$_1$ as 10% and that of volume$_2$ as 40%. You can easily verify by simple calculation that the pore volume in block 1 is 10 acre-feet and that the pore volume in block 2 is 40 acre-feet, a total of 50 acre-feet of pore volume out of a total volume of 200 acre-feet for the two blocks. This gives an overall (average) porosity for both blocks together of 25% (50 acre-feet of pore space divided by 200 acre-feet of total volume). Note that if we had used a straight average of the two porosities, we would luckily have arrived at the same 25% porosity figure. This luck comes only from the fact that the two blocks of rock have identical volumes.

Now suppose that the pore spaces in block 1 contain 31,032 bbl of oil and that the pore spaces in block 2 contain 248,256 bbl of oil. We can convert this total volume of 279,288 bbl of oil to units of acre-feet by using the conversion factor of 7,758 bbl/acre-foot. Dividing 279,288 bbl by 7,758 bbl per acre-foot, we get 36 acre-feet of oil.

What fraction of the pore space is represented by this 36 acre-feet of oil? Divide by the total pore volume of 50 acre-feet to get the answer.

$$36/50 = 0.72$$

What is the hydrocarbon saturation for each block taken separately? Divide the hydrocarbon in place (HIP) for each block by the respective pore volumes already determined. To simplify things, first convert the respective HIP figures to acre-feet by dividing each by the 7,758 bbl/acre-foot conversion factor.

For block one

$$31,032/7,758 = 4.0 \text{ acre-feet}$$

For block two

$$248,256/7,758 = 32.0 \text{ acre-feet}$$

These two volumes respectively occupy the 10 acre-feet pore volume in block 1 and the 40 acre-feet pore volume in block 2. Thus, the hydrocarbon saturation for the two pore volumes must be $4/10 = 0.40$ for volume$_1$ and $32/40 = 0.80$ for volume$_2$.

If we use a straight average hydrocarbon saturation of 0.60 (60%) for the two blocks, we find that this is significantly different from the actual hydrocarbon saturation of 72% determined above. Even though it was possible to use a straight average for the porosity in this example, we get trapped by the *differences in pore volumes* when we use a straight average for saturation.

Suppose we went ahead and used this straight average hydrocarbon saturation of 60% along with the 25% average porosity number and tried to calculate HIP in the reservoir? Instead of correctly summing the separate oil saturations and porosities for the two volumes, we substitute these averages for S_o and ϕ and use the overall reservoir thickness of 200 ft in Eq. 1–2 of Chapter 1.

$$\text{HIP} = (7,758 \text{ bbl/acre-foot}) \times (1) \times (1 \text{ a}) \times (0.60) \times (0.25) \times (200 \text{ ft})$$
$$= 232,740 \text{ bbl}$$

Note the 1 in the second term of the expression for recovery factor. Thus, the calculation gives HIP.

Note how using this estimate based on the straight average water saturation falls short of the given total of 279,288 bbl of oil for the example. Instead of using a straight average oil saturation, we can use Eq. 2–9 for average S_w, substituting hydrocarbon saturations S_h in place of S_w. If we do this, we will arrive at a correct average oil saturation of 0.72. If we use this correct average in Eq. 1–2 and set recovery factor equal to 1, we should obtain exact agreement with the 279,288 bbl of oil in place that was given for the example. This technique is mathematically correct and equivalent to summing the $S_o \phi h$ products for the two separate 100 ft intervals.

Another interesting question that you can examine in the problems at the end of this chapter is "what if one calculates the average water saturation first, then converts to average hydrocarbon saturation from Eq. 2–10?" That is, substitute the correct average S_w into Eq. 2–10 and find average hydrocarbon saturation. Would you get the same answer for hydrocarbon in place that you start with in the example?

DARCY'S LAW

If it were practical I would devote the rest of this book to the important subjects of permeability and capillary forces. However, in this introduction to well logging I can only highlight a few important concepts of permeability that relate to well log calculations.

Darcy's law explains the laminar flow of fluids in the pore system of a rock. It is expressed by the equation

$$Q = (K_a A \Delta P)/(\mu L) \qquad \textbf{(2–11)}$$

where

Q = fluid flow rate (cm^3/sec)

K_a = calibration constant of the equation, known as *permeability* and expressed in units of *darcies*

μ = viscosity of the flowing fluid in centipoise units

ΔP = pressure differential in units of atmospheres across the length L, expressed in centimeters in the direction of flow

A = cross-sectional area perpendicular to the direction of fluid flow, expressed in cm^2

Equation 2–11 could be used in the laboratory to calculate the permeability of a rock sample. This equation is for *single-phase fluid flow*, where only one fluid is in the pore space. Sometimes the permeability in Eq. 2–11 is called *absolute* permeability. Hence, I have used the subscript *a* with the symbol K. Absolute permeability may also be called *single-phase permeability*. The subscript *a* may also denote permeability to air. Air is a commonly used fluid for laboratory measurement of permeability.

Figure 2–9 represents these concepts. In practice, you will probably find the millidarcy unit (1/1000 of a darcy) more convenient than the darcy unit. The permeability of most reservoir rocks is usually only a small fraction of a darcy. This calibration constant called *permeability* provides a quantitative measure of the ability of the rock's pore system to flow fluids. If K_a is high, we say that the rock has *good* permeability. Figure 1–8 in Chapter 1 illustrated that K_a can be quite different for different rock types. Rocks with small pore throats, high grain surface area to volume ratios, and tortuous pore connections will have the lowest permeabilities for a given porosity.

Unfortunately, laboratory measurements of permeability should be regarded as one of the less useful quantitative measurements that can be made on a core. The action of the drill bit when drilling a core often alters a rock's permeability. In addition, petroleum reservoir rocks frequently contain more than one fluid or fluids and gas in their pore systems. When more than one fluid is present, we need to know the *relative* permeability, which is defined by the equation

$$k_i = K_i/K_a \qquad \qquad \textbf{(2–12)}$$

where

k_i = relative permeability to the *i*th fluid phase

K_i = effective permeability to the *i*th fluid phase (expressed in millidarcies or darcies) when more than one fluid phase occupies the pore system

K_a = absolute permeability (expressed in the same units as K_i)

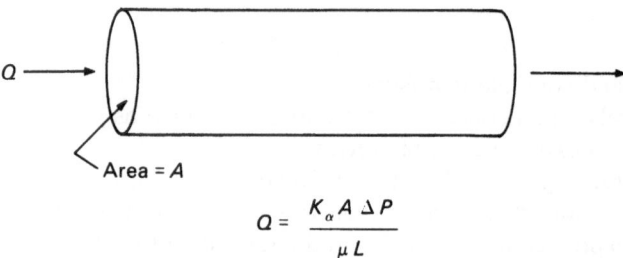

$$Q = \frac{K_\alpha A \, \Delta P}{\mu L}$$

Figure 2–9.   Darcy's law for fluid flow in porous media (single-phase flow).

In a multiple-phase fluid system, the effective permeability to a given fluid phase and the relative permeability will be functions of the relative proportions of the fluid phases that are present in the pore system. Relative permeability will always have a value between zero and one since the effective permeability to any single fluid phase cannot exceed the absolute permeability.

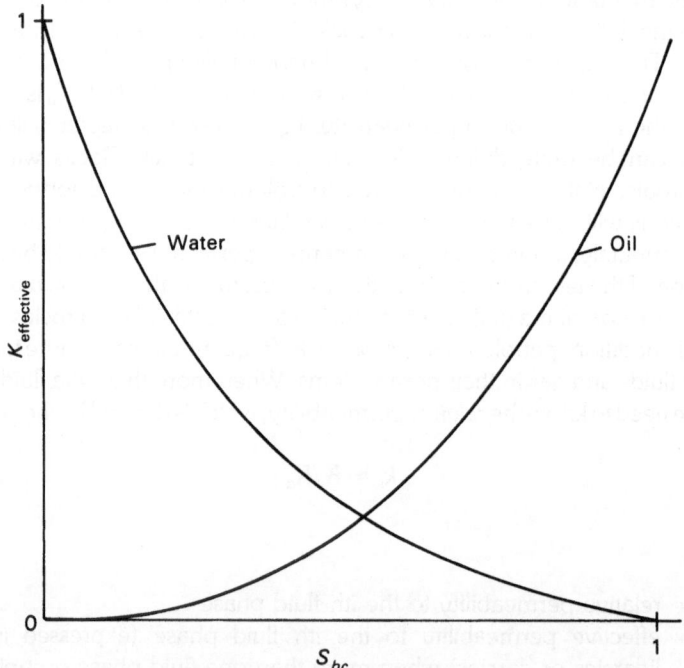

Figure 2–10.   Effective and relative permeability.

Relative permeability can be measured in the laboratory if the relative saturations of the fluid phases present in the sample are varied. Figure 2–10 illustrates typical curves for a two-phase system (see also Figure 1–7, Chapter 1). Note that in the two-phase oil and water system the relative permeability to water k_w steadily decreases with increasing oil saturation S_o and that at a certain value of oil saturation the relative permeability to oil becomes greater than the relative permeability to water. This is called the critical oil saturation. At this level of oil saturation, the rock would tend to flow oil instead of any water that is also present in its pores. This can also be expressed as a critical water saturation by using Eq. 2–10 as long as oil and water are the only two phases present. Note that the permeability to oil steadily increases with increasing oil saturation at the same time that the relative permeability to water decreases. Below the cutoff oil saturation, only water can flow in the pore system. Near the cutoff value, both fluid phases may flow simultaneously. This results in a *water cut* of the produced oil.

CAPILLARY PRESSURE

Strangers to the petroleum industry are often surprised to learn that only about one-tenth of the oil in the ground can be recovered. The remainder of the oil is trapped by capillary forces and cannot be recovered using current technology. You may recall that the oil recovery factor for various reservoir drive mechanisms was much less than one. This is the motivation for extensive research into secondary and tertiary recovery methods.

Capillary pressure is another topic on which I would like to spend a considerable amount of time but will have to be content with highlighting a few significant points. Capillary forces are a result of surface tension phenomena that occur at the interface between two fluid phases or a fluid and a gas phase. These forces control the movement of fluids in the pore systems of rocks. Many secondary recovery methods are based on using some chemical treatment to change the surface tension of the liquids in the pore system and hence lower these capillary forces.

Schowalter published an excellent paper on the role of capillary pressure in secondary hydrocarbon migration and entrapment based on research at Shell Development Research in Houston.[12] Berg presented an enlightening discussion of the role of capillary pressure in stratigraphic traps.[13] The latter paper indicates the sometimes little known fact that one can produce water-free hydrocarbons *down dip* from some water bearing zones! Both of these papers are recommended reading for supplemental study of capillary pressure.

Core measurements of capillary pressures recorded as a function of different relative fluid saturations can provide a wealth of information about pore size distribution and pore sizes in the reservoir rock. Pore structure and pore size distribution are fundamental characteristics of a rock.[14] The pore structure is the organization and geometry of the pore space. It is determined by the composition or texture of a rock. *Texture* is the size and shape of the grains or crystals making up the rock. Whenever two fluids are present in a rock and cannot be mixed together (immiscible), one is called the *wetting* fluid and the other is the *nonwetting* fluid. In most reservoirs, the wetting fluid is formation water and the nonwetting fluid is oil. Making capillary pressure measurements for different relative saturations of the fluids is a valuable tool for classifying reservoir rocks or *rock typing*.

Another rock-typing tool related to capillary pressure measurements is measurement of initial oil saturation versus residual oil saturation. The filtrate water from the drilling mud system acts under the pressure of the mud system weight to flush the formation water and some of the oil and gas out of the pore system in the rock adjacent to the well bore. This leads to a residual hydrocarbon saturation in the pores, and graphical plots of initial versus residual saturation are characteristic of particular reservoir rocks. I continually refer to rock typing throughout this book because of its importance to successful well log interpretation. Other rock-typing tools include lithology and texture differentiation. For example, there are chalky dolomites and granular sucrosic dolomites. Both are dolomites, but only the latter will have good permeability. Many well log responses are also characteristic of particular rock types and are useful in rock typing.

Earlier, I mentioned critical water saturation and critical oil saturation in relation to relative permeability. Critical oil saturation for rocks is a nominal 50%. Actually, there is a practical range of critical oil saturation, ranging from a low of maybe 30% to a high of nearly 70%. Some rocks may produce water with oil saturations greater than 60%. Others may produce water-free oil when the oil saturation is as low as 30%. Obviously, calculating water saturation or oil saturation is not enough. We must also know something about the reservoir rock's producibility if we plan to predict its performance.

Both permeability and critical oil saturation are controlled by pore throat size. Rocks with smaller pore throats have lower permeability, yet they will produce water-free oil even when oil saturations are much less than 50%. On the other hand, most rocks with larger pore throats have excellent permeability but may produce mostly water even when oil saturations are significantly greater than 50%. Figure 2–11 illustrates these effects.

This means that we must have higher oil saturations in the better reservoir rock types if we plan on producing oil. The lower quality reservoirs with low permeability may produce at significantly lower oil saturations, although production can be meager unless a successful stimulation method is found to enhance permeability. *Acidizing* and *fracturing* are common stimulation methods used, but when misapplied they can make things worse. For example, the permeability of a shaly sand that contains chlorite in the pore spaces can be reduced to nothing when the chlorite in the pores contacts the acidizing fluids. This means that we cannot always correlate actual production figures with well log calculations. Completion techniques can enhance or hinder production, depending on proper application.

Soon we'll see how to calculate reservoir properties and saturations from well logs, and that, at best, our calculations will sometimes be inexact and approximate. Therefore we must use care when establishing cut-off values

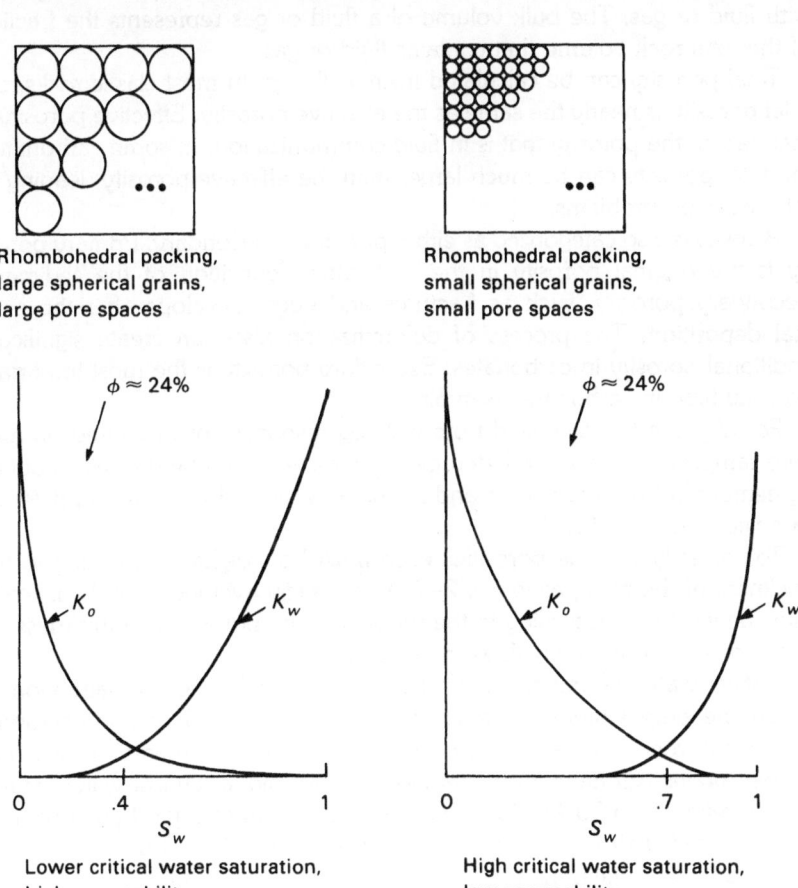

Figure 2–11.   Pore-size effects on producibility.

such as critical oil saturation. We cannot get too strict in using some cutoff value for porosity or saturation as though it were an exact dividing line. If we do, we may find ourselves passing over many potential reservoirs as well as occasionally placing our hopes too high on one that appears to be within the cutoff values.

SUMMARY

Porosity is the fraction of the total rock volume that is void space. Porosity in a rock represents the fraction of the rock available for storage of oil or gas. Saturation tells us what percentage or fraction of the porosity is actually filled with fluid or gas. The bulk volume of a fluid or gas represents the fraction of the total rock volume that contains fluid or gas.

Total porosity can be calculated from well logs. In most clastic rocks, the total porosity is nearly the same as the effective porosity. Effective porosity is that part of the porosity that is in fluid communication. In some carbonates, the total porosity can be much larger than the effective porosity, leading to interpretation problems.

Porosity is also categorized as either primary or secondary. Primary porosity is the original porosity in the rock after deposition of the sediment. Secondary porosity, such as fractures and vugs, develops after the original deposition. The process of dolomitization also can create significant additional porosity in carbonates. Secondary porosity is the most important porosity type in carbonate reservoirs.

Porosity can be calculated from well log responses or measured on rock core samples. However, in heterogeneous rocks it may be difficult to obtain agreement between core data and log data because they represent different sampled volumes of rock.

To correctly average porosities each must be weighted according to the thickness of the zone, as in Eq. 2–7. To correctly average saturations, each must be weighted according to the product of the porosity and thickness for each zone or depth interval, as in Eq. 2–9.

Critical water saturation or critical saturation refers to the saturation at which the pore system will tend to flow water instead of hydrocarbons. In general, rocks with small pore throats, poor grain size sorting, or plate-shaped grains will have lower permeability but higher critical water saturations (greater than 50%). Rocks with large pore throats, good grain sorting, and rounded grains will tend to have good permeability but lower critical water saturations (less than 50%).

Capillary pressure due to surface tension at the interfaces between fluids or fluid and gas in the pore system controls the movement of fluids in the pore system. Capillary pressure measurements on core samples can be used to gain valuable information about the pore size distribution, hence, quality of a reservoir rock.

PROBLEMS

2–1. Is the assumption that hydrocarbon saturation is equal to 1 minus the water saturation (expressed by Eq. 2–10) always valid? Explain your answer.

2–2. From the example given in Figure 2–8, the water saturation for volume$_1$ must be 0.60, and that for volume$_2$ must be 0.20 (using Eq. 2–10). If you now calculate a *correct* average water saturation for these two volumes and substitute this into Eq. 2–10 to find average hydrocarbon saturation, would you finally end up with the correct HIP estimate of 279,288 bbl? Use a recovery factor (*RF*) of 1 with the average porosity of 0.25 with the total interval thickness of 200 ft in your calculations.

2–3. Calculate the average porosity and the average water saturation for the two pieces of core.

Core 1	Core 2
1 ft. thick	9 ft thick
porosity = 0.50	porosity = 0.10
water saturation = 1.00	water saturation = 0.56

What is the average hydrocarbon saturation? Is 0.50 a realistic value for porosity of a typical sedimentary rock?

2–4. Assume you have made some calculations from well logs of a 10 ft section of Minnelusa sand (eolian sand of upper Pennsylvanian–Permian age in the Powder River Basin of the Rocky Mountain area). You have found porosity to be 30% and water saturation to be 25% over the entire 10 ft zone. What is the bulk volume oil in this sand? Since $S_o = 1 - S_w$, could you in a like manner calculate bulk volume oil as 1 less the BVW (BVO = 1 − BVW)?

2–5. Consider the well log calculations for the 20 ft interval listed (on the next page), where data have been recorded in a simplified form to ease calculations for this study exercise. How much error would you make in a reserve estimate if you used a straight average porosity with a straight average water saturation in Eq. 1–2 for the 20 ft interval? Assume drainage area is 80 a, recovery factor is 0.2, and that $S_o = 0.40$ is an acceptable minimum oil saturation for oil production (i.e., the entire 20 ft interval is net pay). Express your answer in barrels of oil and any fraction thereof.

Depth (ft)	Porosity ϕ	Oil Saturation S_o	ϕS_o
9,001	0.08	0.40	0.032
9,002	0.08	0.40	0.032
9,003	0.08	0.40	0.032
9,004	0.08	0.40	0.032
9,005	0.08	0.40	0.032
9,006	0.08	0.40	0.032
9,007	0.08	0.40	0.032
9,008	0.08	0.40	0.032
9,009	0.08	0.40	0.032
9,010	0.08	0.40	0.032
9,011	0.18	0.80	0.144
9,012	0.18	0.80	0.144
9,013	0.18	0.80	0.144
9,014	0.18	0.80	0.144
9,015	0.18	0.80	0.144
9,016	0.18	0.80	0.144
9,017	0.18	0.80	0.144
9,018	0.18	0.80	0.144
9,019	0.18	0.80	0.144
9,020	0.18	0.80	0.144

REFERENCES

1. Walter H. Fertl, "Knowing Basic Reservoir Parameters: First Step in Log Analysis," *The Oil and Gas Journal* (May 22, 1978):98.
2. W. G. Ernst, *Earth Materials* (Prentice-Hall, Englewood Cliffs, N.J., 1969), Chapter 6.
3. Sylvain J. Pirson, *Handbook of Well Log Analysis for Oil and Gas Formation Evaluation* (Prentice-Hall, Englewood Cliffs, N.J., 1963), p. 2.
4. Ibid., p. 3.
5. Ibid.
6. L. G. Chombart, "Well Logs in Carbonate Reservoirs," *Geophysics* **25**, no. 4 (1960):779–853.
7. Cornelius S. Hurlbut, Jr., *Dana's Manual of Mineralogy* (Wiley, New York, 1971), p. 497.
8. Ibid., p. 326.
9. Ibid.
10. Chombart, "Well Logs," p. 783.
11. Author's class notes from course on advanced well log interpretation presented by George R. Pickett at Colorado School of Mines, Golden, Colorado, 1976.
12. Tim T. Schowalter, "The Mechanics of Secondary Hydrocarbon Migration and Entrapment," *The Wyoming Geological Association Earth Science Bulletin* **9**, no. 4 (1976):1–43.
13. Robert R. Berg, "Capillary Pressures in Stratigraphic Traps," *The American Association of Petroleum Geologists Bulletin* **59**, no. 6 (1975): 939–956.
14. Chombart, "Well Logs," pp. 780–781.

CHAPTER **3**

Formation Factor and
Water Saturation

Almost all water saturation calculations using well log data are based on equations that make use of ratios of resistivities. The resistivity of a rock determines its resistance to the flow of electrical current. The flow of electrical current is essentially through the water in the rock's pore system. Oil and gas as well as solid rock do not conduct electrical current and have nearly infinite resistance. They could be termed *insulators*. If oil or gas replaces some of the water in the pore system, the rock's resistance to the flow of current should increase in comparison with a rock of the same porosity having only water in the pore system. The ratio of these resistivities between water-saturated and hydrocarbon-saturated rocks follows a well-established relationship. Thus, by making resistivity measurements, the water saturation of a rock can be established.

In some cases the relationships break down or must be modified. Shaly formations often require more complex relations to adequately relate water saturations to the observed resistivity measurements. In addition, there are some logging tool measurements that can be related to water saturation without regard to resistivity measurements. These exceptions are mentioned when the specific logging tools are discussed. The discussions that follow assume that you are somewhat conversant with the basic principles of electricity and Ohm's law.

RESISTIVITY AS A ROCK PROPERTY

Resistivity is a fundamental property of matter. That is, it is much like the density of a material. For example, all quartz has a density of 2.65 grams per cubic centimeter (gm/cc). Given an equal volume of quartz and limestone, the limestone will weigh more because its density is 2.71 gm/cc. Some minerals, such as dolomite, may have a variable density because of variations in the relative proportions of their chemical constituents.

Just as the density of a material reveals what a given volume of that material will weigh, so resistivity reveals what resistance a given volume of material will offer to the flow of current. For example, aluminum has a higher resistivity than copper. In addition to being directly proportional to resistivity, resistance to electric current flow is proportional to the length of a conductor and inversely proportional to the cross-sectional area of the conductor. Thus, the more resistive aluminum wire has to have a larger diameter than does a corresponding length of copper wire if it is intended to carry the same electric current.

In the following discussions, I spell out the word *resistance* rather than use a symbol. Some papers and publications use the symbol ρ, but well log analysts have adopted this symbol to represent density. Log analysts use the symbol R_t for resistivity. One could also use a symbol such as r for resistance to avoid confusion.

You must be careful to distinguish resistance from resistivity, although they are certainly related. Resistivity is a fundamental property that is constant for a given substance, whereas resistance varies according to the geometry and quantity of the substance.

Formation water found in the pore systems of rocks is an electrolyte solution. The ions that form when an electrolyte or salt is dissolved in a solution carry the current in a liquid. In electrolyte solutions, the resistance to current flow decreases as the temperature increases because of the increased mobility of the ions in solution. Thus, it is necessary to specify the temperature when quoting the resistivity of an electrolyte solution such as the formation water in the pore system of a rock. Since formation water is the source of current flow in rocks, it is likewise necessary to specify the temperature when giving the resistivity of a rock.

Resistivity is a constant for a given substance at a given temperature regardless of the volume of the substance. On the other hand, resistance may vary according to the amount of a substance and its geometry. Resistivity in an electrolyte is a constant *for given concentrations of the ions in solution.* Resistivity has also been called *specific resistance.*

Consider the solid rod in Figure 3–1. Assume that it is made up of a homogeneous and isotropic (having the same properties throughout regardless of the direction of measurement) substance. Let the resistivity be some

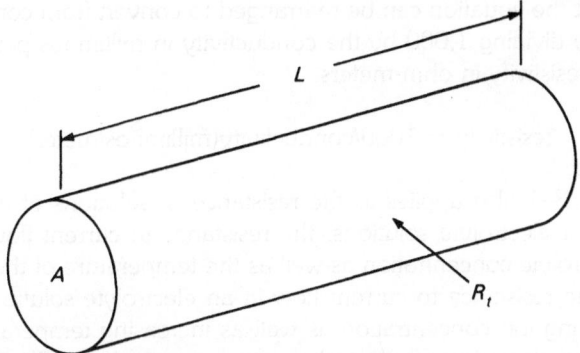

Figure 3–1.   Resistivity of a solid rod. (Note: Resistance $= (R_t)(L/A)$.)

value represented by R_t and let the cross-sectional area be A and the length be L, as shown in the figure. Its resistance to current flow along its major axis is given by the product

$$\text{resistance} = (R_t)(L/A) \qquad \qquad \textbf{(3–1)}$$

This equation says that the larger the cross-sectional area (A) becomes, the smaller will be the resistance to the flow of current. Also, the shorter the rod becomes (L), the smaller will be the resistance to current. This explains why wires that carry large amounts of current have large diameters, and electricians recommend that the length of extension cords used with heavy duty electrical appliances be kept as short as possible.

Resistance is usually expressed in units of *ohms*. Its reciprocal, the conductance, is expressed in units called *mhos*. Conductance expresses the reciprocal of resistance or the tendency of a substance to *allow* the flow of electrical current rather than to oppose it. If the length L in Eq. 3–1 is given in meters and the area A is expressed in square meters, the resistivity R_t must be expressed in units of ohm-meters if the equation is to be consistent regarding the units. It is possible to express the reciprocal of resistance or conductance as a function of the reciprocal of resistivity or conductivity of the formation. Conductivity would then be expressed in units of mhos per meter. In well logging practice, these units for conductivity are too large. A smaller unit, the *millimho* per meter, is used. It is 1/1000 of a mho per meter.

To convert from resistivity in ohm-meters to conductivity in millimhos per meter, simply divide 1,000 by the resistivity in ohm-meters.

$$\text{conductivity} = 1000/\text{resistivity(ohm-meters)} \qquad \qquad \textbf{(3–2)}$$

Note that the equation can be rearranged to convert from conductivity to resistivity by dividing 1,000 by the conductivity in millimhos per meter and obtain the resistivity in ohm-meters.

$$resistivity = 1000/conductivity(millimhos/meter) \qquad \textbf{(3–3)}$$

Equation 3–1 also applies to the resistance of solutions of electrolytes.[1] However, for electrolyte solutions, the resistance to current flow is a function of electrolyte concentration as well as the temperature of the electrolyte solution. The resistance to current flow in an electrolyte solution decreases with increasing ion concentration as well as increasing temperature. In *formation water*, the principal electrolyte is common salt (NaCl). The sodium (Na^+) and chloride (Cl^-) ions are formed when salt is dissolved in water. Lesser amounts of other electrolytes in formation water are mentioned later. The concentration of salt in the formation water is often referred to as *salinity*. Thus, the more saline the water, the less resistance it offers to the flow of electric current. Saline waters are conductive.

SOLVING PROBLEMS OF RESISTIVITY VERSUS TEMPERATURE OR SALINITY

Figure 3–2 is a resistivity graph used to solve many NaCl resistivity problems. Other types of nomographs are designed to solve the same types of problems, but the chart of Figure 3–2 is easy to use and provides a visual picture of the relationships of resistivities of solutions to ion concentrations and temperature. The chart can be used to solve two types of problems. First, it can be used to convert the resistivity of an NaCl solution at one temperature to the resistivity it has at another temperature. Second, given the ion concentration, the resistivity at any desired temperature can be found.

The chart is useful for finding resistivities and making temperature conversions of resistivity for formation water in a rock's pore system. However, it can also be used for solving resistivity problems for drilling mud system fluids. On the left side of the long, vertical axis of Figure 3–2, the resistivity of the solution is given in units of ohm-meters. The scale on the bottom gives the temperature of the solution in degrees Fahrenheit (°F) as well as degrees Celsius (°C). In this book, I assume that the reader has no difficulty in using both Celsius and Fahrenheit temperatures. However, most examples are illustrated in terms of the units that prevail in the petroleum industry today: degrees Fahrenheit.

The slanted lines across the graph represent different salinities or concentrations of NaCl ions. They are labeled on the right side of the graph in units of parts per million (ppm) and grains per gallon (g/gal) at 75°F.

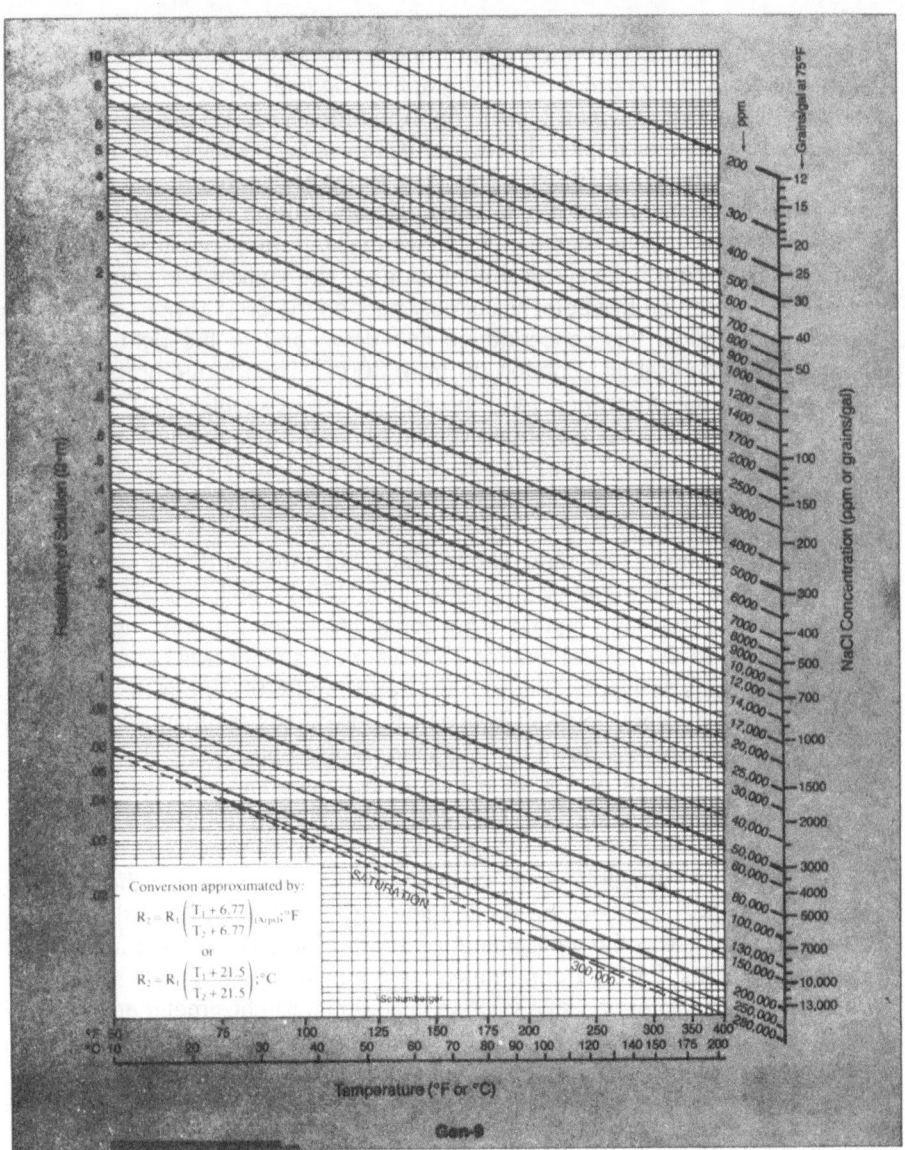

Figure 3–2. Resistivity of NaCl solutions. *(Courtesy of Schlumberger Ltd.)*

Note that the graph is to be used only for NaCl solutions. In fact, the relations for other ions may be somewhat different. If the graph is used with a solution that contains significant quantities of ions other than sodium or chloride, it may introduce some error into the results. I discuss this in more detail below. When examining the graph, the relation of resistivity in a solution to ion concentration is readily apparent. As ion concentration increases (moving down on the graph), the resistivity decreases. As temperature increases (moving right on the graph *along lines of constant concentration* or slantwise down and to the right) the resistivity also decreases.

Consider some examples. If the formation water resistivity (R_w) is 0.3 at 75°F, what is R_w at 300°F? This is a simple temperature conversion problem. It may help to use a straightedge when using the chart to solve these problems. Move to the right horizontally from a resistivity of 0.3 ohm-meters (about halfway up the left side) to the vertical line coming up from 75°F. Note that this point lies almost right on the 20,000 ppm concentration line. Follow this concentration line down and to the right until it intersects the vertical line coming up from 300°F. Then move horizontally to the left on the graph and read the answer: approximately 0.077 ohm-meters at 300°F. Figure 3–3 shows the graph with a heavy line illustrating how you move from 0.3 ohm-meters to the concentration line at 75°F and then along the slanted concentration line to the appropriate horizontal resistivity line at 300°F.

If the resistivity of the mud filtrate (R_{mf}) from the mud system is 1.0 ohm-meters at 75°F, what is the resistivity at 150°F? Follow the example procedure as before and extend a horizontal line from 1.0 ohm-meters on the left across to the vertical line for 75°F where this point falls between the two slanted lines corresponding to 5,000 ppm and 6,000 ppm. It appears that the point is nearly midway between the two lines, so maybe you will want to move slantwise down an imaginary 5,500 ppm line to the vertical line for a temperature of 150°F. If everything has gone well, you should find yourself on a horizontal line corresponding to about 0.51 ohm-meters.

A sample of formation water contains 20,000 ppm NaCl. What is its resistivity at 165°F? Follow the slanted line for 20,000 ppm NaCl down to a point that intersects the vertical line for 165°F. There you should intersect the horizontal line for 0.14 ohm-meters, which is the answer.

The conductivity of an NaCl solution is 10,000 millimhos/meter at 200°F. What is its conductivity at 75°F? Be sure to convert to resistivity so the chart can be used. The resistivity from Eq. 3–3 is 1,000 divided by the conductivity in millimhos per meter. That is the units you already have for conductivity, so

$$\text{resistivity} = 1,000/10,000 \text{ millimhos/meter}$$
$$= 0.1 \text{ ohm-meters}$$

Now, you should readily follow the procedures illustrated thus far to follow upward and to the left a slant line parallel to the 25,000 ppm line but

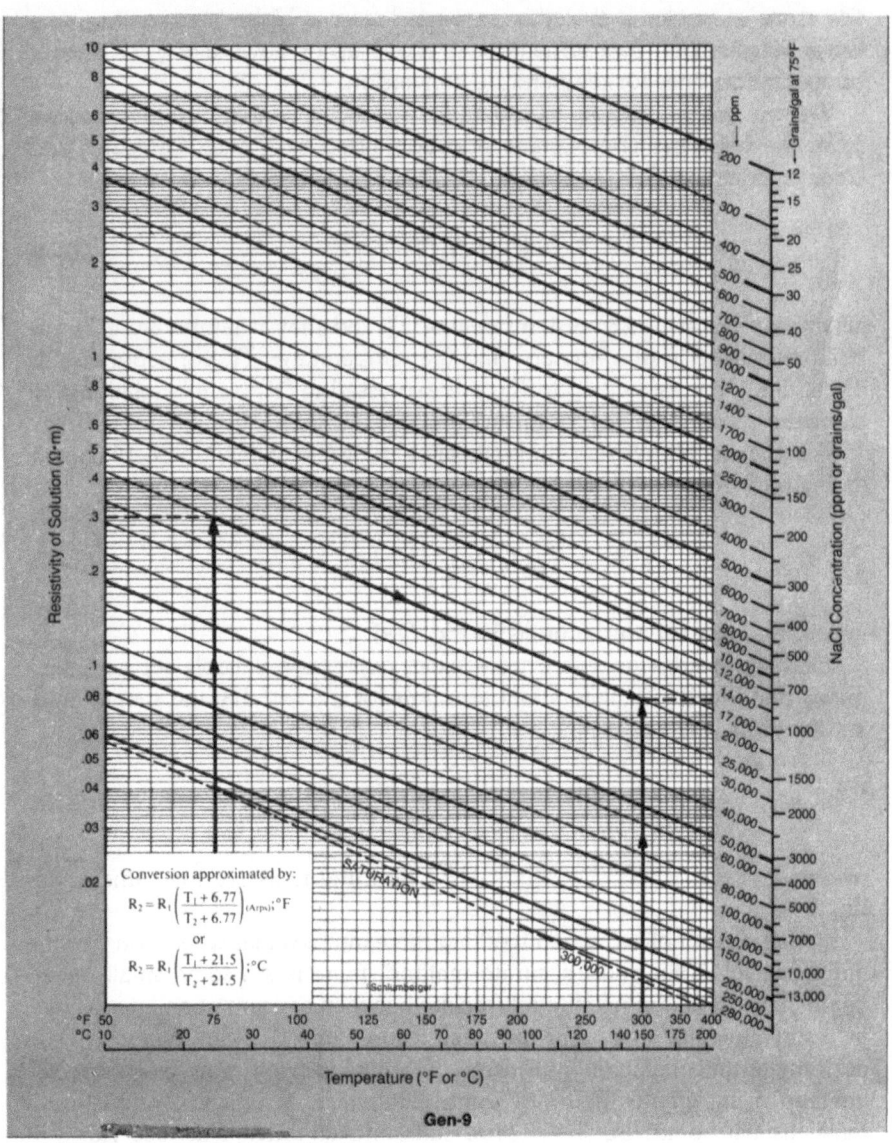

Figure 3–3. Example for the graph of NaCl solutions. *(Courtesy of Schlumberger Ltd.)*

somewhat above it. This should bring you to about 0.26 ohm-meters at the 75°F line. If you want the answer in conductivity units, divide 1,000 by this resistivity (Eq. 3–2) to obtain 3,846 millimhos/meter. As far as the use of the graph is concerned, the only difference in procedure here from the first three examples is that you converted from a higher temperature to a lower temperature. Just remember to move up or down along the slanted concentration lines for temperature conversion.

We can get satisfactory numbers for resistivity temperature conversions from the following approximate equation, used by Pickett,[2] Pirson,[3] and Dresser Industries.[4]

$$R_2 = R_1(T_1/T_2) \qquad (3\text{--}4)$$

where the resistivity (R) at temperature T_1 and that at temperature T_2 are both expressed in ohm-meters, and the temperatures T_1, corresponding to R_1, and T_2, corresponding to R_2, are expressed in °F. Try this relation and compare it with your results for the above examples.

A somewhat more exact approximation quoted by Schlumberger[5] (shown in Figure 3–2) is

$$R_2 = R_1(T_1 + 6.77)/(T_2 + 6.77) \qquad (3\text{--}5)$$

where the units are the same as for Eq. 3–4.

Hilchie provides an even more exact expression suitable for programmable hand-held calculators that accounts for the fact that the slanted lines on the graph are curves:[6]

$$R_2 = R_1(T_1 + X)/(T_2 + X) \qquad (3\text{--}6)$$

where $X = 10^{(-0.340396 \log_{10}(R_1) + 0.641427)}$, and the units are the same as for Eq. 3–4.

Perhaps you noticed when working the examples that when you use the graph, which is the most accurate method, there is still some small uncertainty because of the need to interpolate between concentration lines as well as resistivity value lines. This probably causes an 0.01 ohm-meter or 0.02 ohm-meter uncertainty in your results. Even in Hilchie's more exact approximation of the graph, there are some differences. In practical problems, I have usually found Eq. 3–4 to be sufficient, although I do use Eq. 3–5 in calculator and computer programs and may advance to Hilchie's newer expression for computer programs.

There are also approximate equations for converting concentrations to resistivities. One that seems fairly good over a wide temperature range is given by Dresser Industries.[7]

$$\text{ppm @ 75°F} = 10^x \qquad (3\text{--}7)$$

where

$$x = (3.562 - \log_{10}(R_{w75} - 0.0123))/0.955$$
R_{w75} = the solution resistivity in ohm-meters at 75°F
ppm = parts per million NaCl

Equation 3–7 can be solved for the solution resistivity

$$R_{w75} = 0.0123 + (3647.5/ppm^{0.955}) \qquad\qquad \textbf{(3–8)}$$

where the symbols and units are already described. If you are solving for resistivity from concentration, remember to convert the resistivity to the desired temperature. Equation 3–8 expresses the resistivity at 75°F.

There are two considerations to be kept in mind when using the graph in Figure 3–2. First, remember that it is applicable if only sodium and chlorine ions are present. It will not be valid if there are other ions present since they may have a different resistivity-versus-temperature behavior. If other ions are present in small amounts, it is probably safe to assume that sodium and chloride are the only ions present. The second consideration occurs when the party who measured the ion concentration reports only the *chlorides*, or the chlorine ion concentration. Then you must convert this to parts per million sodium *and* chlorine.

Consider the second possibility: only chloride ion concentration is reported. Many sources over the years, including Martin,[8] Pickett,[9] and Dresser Industries,[10] have stated that chloride ion concentration should be multiplied by 1.65 to convert to sodium chloride concentration. The factor 1.65 is the ratio of the gram molecular weight of sodium chloride to the gram atomic weight of chlorine.

If ions other than sodium and chloride are present, you can use a service company graph such as that shown in Figure 3–4. The graph is entered with the total solids concentration of all ions at the bottom on the horizontal axis and the multiplier is read from the vertical axes on the left or right. For example, if your sample contained 2,000 ppm of *all* ions, the multiplier for HCO_3 (bicarbonate) would be (entering the graph at the bottom at a concentration of 2k, where k is used to represent the multiplier 1,000), about 0.33. Note that the bicarbonate multiplier does not change much with changing concentration. So, 1,000 ppm of bicarbonate ion is only counted as (0.33 times 1,000) 330 ppm equivalent NaCl concentration. Repeat this process for all ions other than Na or Cl and add these equivalent NaCl concentrations to the concentration of sodium and chloride ions in the sample. Then take this adjusted total concentration and use the graph in Figure 3–2 to find the resistivity of the sample solution. Following is an example.

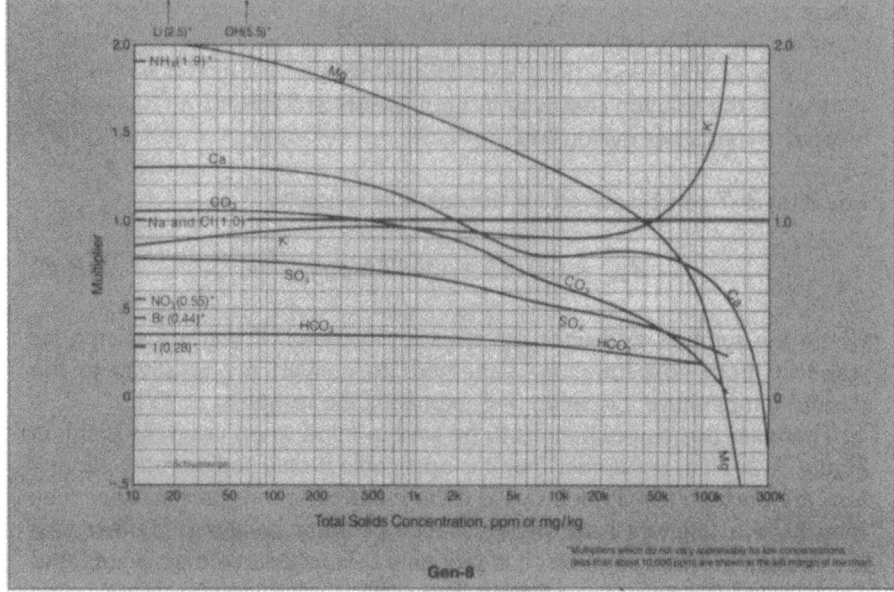

Figure 3–4. Multipliers for ion concentration. *(Courtesy of Schlumberger Ltd.)*

Calculate the resistivity at 75°F for this formation water sample analysis: 500 ppm Ca, 1,500 ppm SO_4, 1,200 ppm HCO_3, and 24,000 ppm NaCl (Na + Cl). Total solids: 27,200 ppm. From the graph in Figure 3–4, find the multipliers and adjust the concentrations:

$$
\begin{array}{ll}
\text{Ca: } 500 \times .81 & = 405 \text{ ppm (eq NaCl)} \\
SO_4\text{: } 1500 \times .43 & = 645 \text{ ppm (eq NaCl)} \\
HCO_3\text{: } 1200 \times .25 & = 300 \text{ ppm (eq NaCl)} \\
\text{Na + Cl: } 24,000 \times 1.00 & = 24,000 \text{ ppm NaCl} \\
\text{Total} & = 25,350 \text{ ppm NaCl (eq)}
\end{array}
$$

The total includes both NaCl and equivalent (eq) NaCl ion concentrations. Entering the graph in Figure 3–2, you should obtain 0.235 ohm-meters resistivity at 75°F.

As a further exercise, try using the total 25,350 ppm in Eq. 3–8 and see if you get somewhere near the same result as that from the graph in Figure 3–2 (using the equation, I calculated the resistivity to be 0.239 ohm-meters). Remember to convert any results from Eq. 3–8 to the desired temperature. In the above example, no conversion is necessary because the answer is desired at the temperature 75°F, which the equation provides.

Recall that in the graph of Figure 3–2 another system of units is used to express concentration. It is grains per gallon (g/gal). Although it is used for other solutions in the petroleum industry (e.g., grains of *lead* per gallon of gasoline), I have not often seen it used to report water salinity. Milligrams per liter (mg/l) is sometimes used, however, and we can convert mg/l to g/gal from the relation given as[11]

$$1.0 \text{ g/gal @ } 75°F = 17.118 \text{mg/l} \tag{3–9}$$

This conversion allows us to use the graph to solve water resistivity problems for concentrations reported in milligrams per liter. Note that these units (mg/l or g/gal) are expressed relative to a reference temperature since salt solutions change density with temperature.

Formation water salinities are reported as parts of NaCl by weight to million parts of formation water.[12] We can convert from milligrams per liter at 75°F to parts per million if we account for the fact that the more saline solutions will have a density greater than the density of water alone. The correct procedure is to divide the concentration in milligrams per liter by the density of the saline solution.

If we assume that the one liter of solution has a density of one and consider that the dissolved salts have not altered the *volume* of the solution, we can also convert to parts per million as follows: parts per million equals 1 million times the ratio of the weight of dissolved salt in milligrams to the weight of solvent plus dissolved salts in milligrams. To convert to parts per million use the relation

$$\text{ppm} = ((\text{mg/l})/(\text{mg/l} + 10^6)) \times 10^6 \tag{3–10}$$

Note also that the API procedures use the term *bulk volume* in a slightly different context from that defined earlier as simply the product of ϕS_w. In the API procedures, *bulk volume* refers to a definite volume (e.g., 50 *milliliters*) rather than a fraction of the unit volume discussed in relation to the above definition.[13] I will continue to use the definition as used in well logging literature, that is, bulk volume refers to a fraction of total volume. However, you should be aware of other usages of the term *bulk volume* such as described in the API procedures.

Traditional thought has held that actual resistivity measurements are preferred. However, an interesting study by Kaufman and Moore showed that, in one instance, better agreement of results among different laboratories analyzing samples from the same solution resulted when they used the calculated resistivities from the chemical analysis rather than from actual measured resistivities! Kaufman and Moore also stated that significant uncertainties in laboratory measurement of resistivities was probably the rule rather than the exception.[14]

Part of the problem of measuring resistivity of a sample and then extrapolating it to formation temperature arises because there are significant quantities of ions other than Na and Cl in the water. These other ions do not follow the same resistivity-versus-temperature relation as NaCl. Kaufman and Moore also cite specific recommendations for improving the quality of water resistivity measurements.[15]

1. Make resistivity measurements at the well site as soon as possible after sampling to prevent sample alteration by attacking bacteria.
2. Measure temperature of the sample in the resistivity measuring cell after reaching equilibrium.
3. Use a resistivity cell that permits temperature control of the fluid sample and precise temperature and resistivity readout.
4. Record the resistivity at three or more temperatures to improve accuracy of extrapolation to subsurface temperatures.
5. Do frequent recalibration of resistivity cells with standard NaCl solutions over a range of concentrations.

This may sound like overkill to assure valid resistivity data. However, this is a real, if somewhat overlooked, problem. Pickett and others have strived to develop log interpretation methods that are independent of just such uncertainties as those encountered with formation water resistivity measurements. The most successful log analysts are those who recognize that well log data quality problems are a fact of life and use error-resistant interpretation methods as well as statistical methods to identify errors and reduce their impact on log analysis. Well logging programs and well log interpretation still make up a small part of the total budget for well drilling and completion. In the competitive oil and gas markets of the near future, they are likely to make up an increasing part of the budget for the more successful oil and gas operators. Well logging service companies have already begun to place more emphasis on data quality and it is likely that quality control will play an increasingly more important role in the future.

Always remember to convert water sample resistivities (however derived) to the appropriate formation temperature when required by the equations used in your analysis. This seems to be a common lack of understanding among many people engaged in well log analysis.

How does one correct a water resistivity measurement to the correct formation temperature? Consider an example. Suppose we have a bottom hole temperature (BHT) measurement but would like to know the appropriate resistivity for a formation that is, say, 2,000 ft from the bottom in a 10,000 ft well. The common practice is to assume a linear temperature gradient with increasing depth. Assume a straight-line relation extending from the average annual temperature at the surface (corresponding to zero depth) to the measured BHT (corresponding to the maximum depth in the well). The BHT is usually recorded by a recording device on the logging tool that registers the maximum observed temperature in the borehole. Remember that the temperature we would most like to have is that temperature recorded as near as possible to the time the resistivity log measurements are recorded in the well. If we do not know the average annual temperature at the well site, we can usually make a reasonable estimate by allowing for seasonal variations from the recorded surface temperatures that will be on the well log heading information recorded at the time the well is logged.

FORMATION FACTOR

Formation factor (F) is defined as the ratio of the resistivity of a water saturated rock (R_o) to the resistivity of the water in the rock's pore system (R_w), where both are expressed in the same units.

$$F = R_o/R_w \qquad\qquad \textbf{(3–11)}$$

It is customary to express both in units of ohm-meters. The equation could also be expressed as the ratio of pore water conductivity to rock conductivity, with both expressed in units of millimhos per meter. This equation is one of Archie's key empirical relationships[16] that has stood the test of time and more recently has been developed from theoretical considerations by Sen.[17]

I stated earlier that most water saturation calculations are based on comparing the resistivity of a rock to the resistivity the same rock would have if it were 100% water saturated. One of the common ways of finding out what the rock's resistivity is when water saturated is to calculate the "would be" resistivity from the formation factor and a knowledge of the water resistivity R_w. Then we can compare the rock's measured resistivity to the calculated *wet* resistivity R_o. If the measured resistivity is higher than the wet resistivity, this implies the presence of hydrocarbon. The equation for calculating the amount of hydrocarbon saturation from these resistivity comparisons is presented below.

By rearranging the above equation, we can calculate R_o as the product of formation water resistivity R_w and formation factor. Assuming we can determine R_w from resistivity data from previously drilled wells or some other technique, we still need to know the formation factor of the rock to use the equation. In the past, this has sometimes been done directly from well log measurements that will be discussed later. The common method in use today is based on relating formation factor to the porosity of the rock.

Archie established such a relation from observation of measurements on core data.[18] As an aid to understanding these types of relationships, consider the following empirical development. Figure 3–5 shows two cubes, each with a volume of 1 m^3. The first cube contains formation water (from a

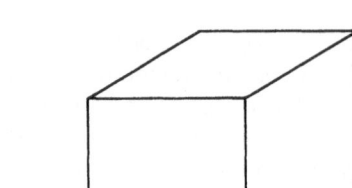 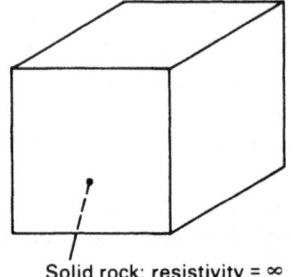

Water: resistivity = R_w Solid rock: resistivity = ∞

Figure 3–5. Cubes of water and rock, each with 1 m sides (volume $= 1$ m^3).

rock's pore system) of resistivity R_w. The second cube is assumed to be solid rock or rock without any pore space (zero porosity). The solid rock has infinite resistivity for practical purposes. Therefore, we should expect the resistance measured across the rock, say, from front to back to be infinite. If we measure the resistance across the cube of water, we should find from the relation for resistance (see Eq. 3–1) that the resistance is numerically equal to the resistivity R_w since the cross-sectional area is 1 m^2 and the length is 1 m.

Let us consider something more representative of a real rock with a pore system. Figure 3–6 represents a 1 m cube filled with water and spherical glass beads. For the moment, consider the glass beads to be of the same size and do not worry about how well packed they are. If we knew the total volume of glass beads included in the cube, we could readily find the porosity from

$$\phi = (1 \text{ m}^3 - \text{volume of glass beads in m}^3)/1.0 \text{ m}^3$$

It should be apparent that as the volume of glass beads increases, porosity decreases. Porosity is the storage space for fluids in the rock, so as porosity decreases we will find less and less of the cube in Figure 3–6 filled with water. There will be fewer current carriers, and the resistance offered to the flow of electrical current will increase until, when porosity approaches zero, the resistance will become infinite just as for the solid rock cube in Figure 3–5.

Now look at the parallel bundle of tubes filled with water of resistivity R_w running straight through the 1 m cube in Figure 3–7. Let the tubes represent pore space in the rock. The tubes will have a length of 1 m, the same as the dimension of the cube. If we denote the total cross-sectional area of the tubes by A', it should also be apparent that the porosity represented by these tubes is also equal to the ratio of this area A' to the total cross-sectional area of the cube.

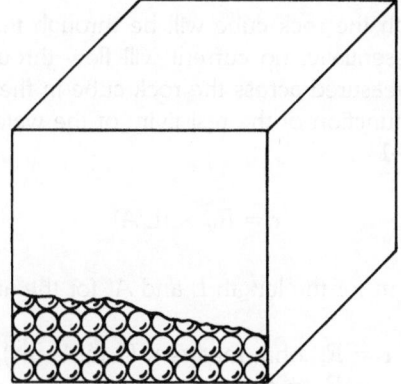

Figure 3–6. Block filled with spherical glass beads.

$$\phi = A'/1 \ \text{m}^2$$

This follows from the facts that porosity is the ratio of the pore volume to the total volume and that the tubes have the same linear dimension as the cube.

You can express area A' as

$$A' = \phi \ \text{m}^2$$

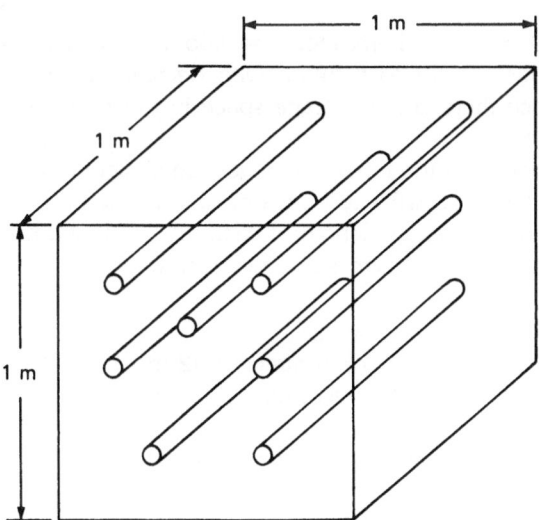

Figure 3–7. Bundle of parallel tubes porosity model.

Current flow through the rock cube will be through the water of resistivity R_w in the tubes. Essentially, no current will flow through the rock matrix. The resistance, r, measured across the rock cube in the direction parallel to the tubes will be a function of the resistivity of the water (R_w) in the tubes. According to Eq. 3–1

$$r = R_w \times (L/A)$$

Simply substitute 1 m for the length L and A' for the area A:

$$r = R_w \text{ ohm-meters} \times (1 \text{ m}/A' \text{ m}^2)$$
$$= R_w/\phi \text{ ohms}$$

The resistivity of the water saturated rock cube is

$$R_o = (A/L) \times r$$
$$= (1 \text{ m}^2/1 \text{ m}) \times (R_w/\phi \text{ ohms})$$
$$= R_w/\phi \text{ ohm-meters}$$

Formation factor is defined as R_o/R_w.

$$F = (R_w/\phi)/R_w$$
$$= 1/\phi$$
$$= \phi^{-1}$$

For $\phi = 0.10$, $F = 10$, whereas for $\phi = 0.05$, $F = 20$. Therefore, we should expect formation factor to increase with decreasing porosity. This would be logical since there is less storage space in the rock for current-carrying formation water.

Now consider that the tubes no longer run straight through the cube but follow some tortuous pathway from one side to the other. For the sake of discussion, assume that the length of each tube is now 2 m instead of 1 m. The resistance measured across the rock is now

$$r = R_w \times (L/A)$$
$$= R_w \text{ ohm-meters} \times (2 \text{ m}/A' \text{ m}^2)$$
$$= 2 R_w/\phi \text{ ohms}$$

The resistivity of the water saturated rock is now

$$R_o = (A/L) \times r$$
$$= (1 \text{ m}^2/1 \text{ m}) \times (2 R_w/\phi \text{ ohms})$$
$$= 2 R_w/\phi \text{ ohm-meters}$$

For this rock with more tortuous pore paths, formation factor is

$$F = R_o/R_w$$
$$= 2/\phi$$

For the same porosities as before, 0.10 and 0.05, the formation factor will be $F = 20$ and $F = 40$, respectively. Thus, formation factor will increase with increasing tortuosity of the pore connections.

As a final example, consider a bundle of straight tubes through another cube with 1 m sides. This time, let the total cross-sectional area of the tubes be 0.5 m². This is the same as letting porosity be 0.50. Suppose the water in the pores has a resistivity R_w of 0.10 ohm-meters. The resistance across the rock cube is

$$r = R_w \times (L/A)$$
$$= 0.10 \text{ ohm-meters} \times (1 \text{ m}/.50 \text{ m}^2)$$
$$= 0.20 \text{ ohms}$$

The resistivity of the rock cube is

$$R_o = (A/L) \times r$$
$$= (1 \text{ m}^2/1 \text{ m}) \times 0.20 \text{ ohms}$$
$$= 0.20 \text{ ohm-meters}$$

and the formation factor is

$$F = R_o/R_w$$
$$= 0.20 \text{ ohm-meters}/0.10 \text{ ohm-meters}$$
$$= 2$$

Now let half of the tubes be broken and displaced so that they are no longer continuous across the cube as shown in Figure 3–8. The porosity will still be 0.50, but the cross-sectional area for conduction is reduced to .25 m². Resistance measured across the tube is now

$$r = R_w \times (L/A)$$
$$= 0.10 \text{ ohm-meters} \times (1 \text{ m}/0.25 \text{ m}^2)$$
$$= 0.40 \text{ ohms}$$

and R_o for the rock must be 0.40 ohm-meters. Here $F = .4/.1 = 4$. If a relation such as

$$F = \phi^{-m}$$

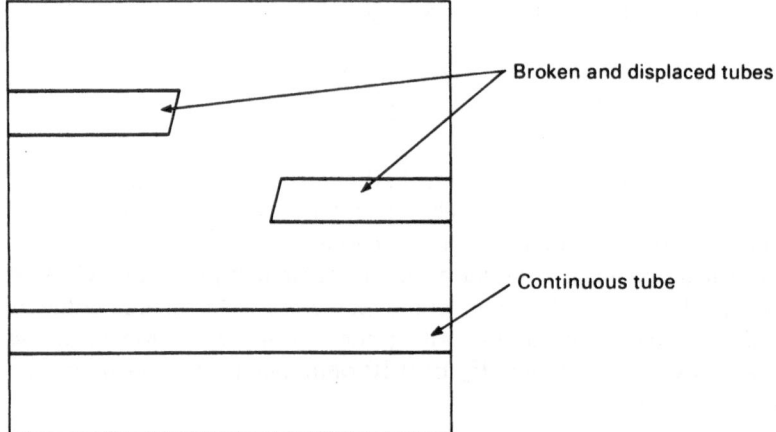

Figure 3–8. Bundle of parallel tubes with some broken tubes (cross-section).

is proposed to express formation factor in terms of porosity, for the first case above with $F = 2$, the exponent m must be equal to 1.0 if

$$0.5^{-m} = 2$$

For the second case with $F = 4$, the exponent m must be equal to 2.0 so that

$$0.5^{-m} = 4$$

Thus, when some pore spaces are not connected, such as in some vuggy type rocks, the exponent m will be higher. For these rocks, the increase in m is not related to tortuosity, although the exponent m is often referred to as the *tortuosity* exponent. It was more commonly called the *cementation* exponent in the past, although the term *tortuosity* exponent is more descriptive, with the exception of the effect of vugginess on m.

In real rocks you can expect the pore interconnection network to be very complex, with many different possible tortuous pathways. As Archie[19] and other researchers have found, the formation factor for rocks is best expressed as

$$F = \phi^{-m} \tag{3–12}$$

with porosity expressed as a fraction of total rock volume and the exponent m referred to as the *tortuosity* exponent. Note from this equation that, as porosity ϕ decreases, the formation factor F increases. From the idealized models we have just examined, we can also conclude that (1) as tortuosity increases, the exponent m increases and F also increases, and (2) as vuggi-ness increases, the exponent m also increases along with F.

It is revealing to examine Eq. 3–12 when it is plotted on a log-log grid as in Figure 3–9. In logarithmic form, Eq. 3–12 can be written as

$$-m \log (\phi) = \log(F) \qquad \textbf{(3–13)}$$

On the graph, the slope m is shown for values typical for reservoir rocks, ranging from 1.6 to 2.2. It is easy to see that formation factor decreases with increasing porosity to a limiting value of $F = 1$ when porosity is 100%. If this 100% porosity were possible, we would have only water of resistivity R_w left. By re-expressing Eq. 3–11

$$R_o = F R_{w'} \qquad \textbf{(3–14)}$$

Note that $R_o = R_w$ when $F = 1$. For convenience, I have labeled the porosity scale on the graph in porosity units rather than as a decimal fraction as required by Eq. 3–12. Note also that for very small porosities, moving to the right and down along a line of constant slope, the formation factor becomes very large.

This graph not only provides a conceptual look at the relation of formation to porosity, but it can also be used to solve Eq. 3–12 for formation factor in terms of porosity. For example, if we know $m = 1.6$, we can find the formation factor for a rock with porosity equal to 5% (0.05). We enter the graph of Figure 3–9 on the vertical porosity axis on the left and extend a horizontal line until we meet the line of slope 1.6. Then we drop vertically down to a formation factor F of approximately 121. If this rock contains water of resistivity R_w equal to 0.10 ohm-meters, we can find the rock resistivity from Eq. 3–14.

$$
\begin{aligned}
R_o &= (121) \times (0.10 \text{ ohm-meters}) \\
&= 12.1 \text{ ohm-meters}
\end{aligned}
$$

Figure 3–10 is the graph in Figure 3–9 with the example traced out.

Suppose $m = 2.2$. What is the formation factor for a rock with 30% porosity? What is the resistivity of the rock if the pore spaces contain forma-tion water with resistivity $R_w = 0.05$ ohm-meters? First, find the formation factor. Enter the vertical porosity scale at 30% and move horizontally to the right until you intersect the line with a slope of 2.2. Then drop vertically down to the formation factor F-axis and read $F = 14.5$. Now it is a simple matter to compute the rock resistivity.

$$
\begin{aligned}
R_o &= (14.5) \times (0.05 \text{ ohm-meters}) \\
&= 0.725 \text{ ohm-meters}
\end{aligned}
$$

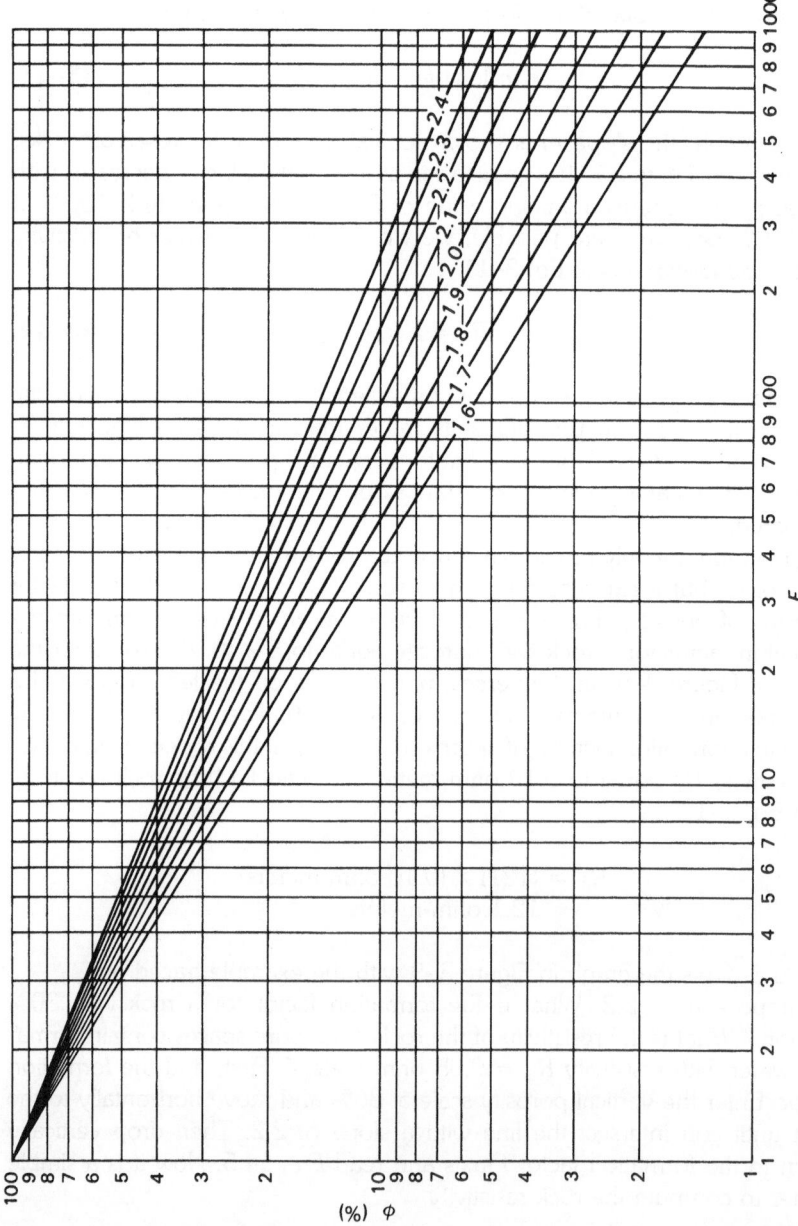

Figure 3-9. Porosity versus formation factor.

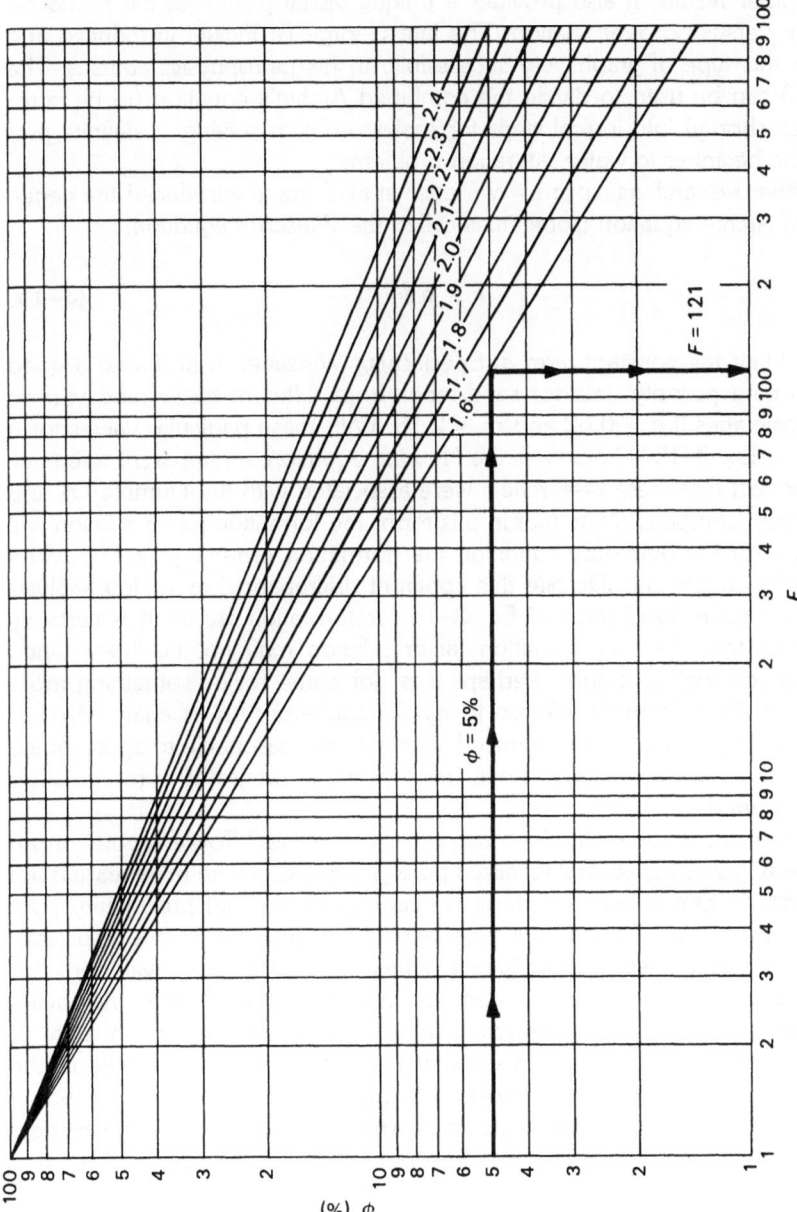

Figure 3–10. Example using porosity versus formation factor graph.

The graph of Figure 3–9 provides an easy way to check hand calculator or computer results. It also provides a unique visual picture of the formation factor versus porosity relation. The late George R. Pickett introduced and used this type of graph very successfully in his petrophysics courses. The graph can be used to divide the combined Archie's equation (to be introduced shortly) into logical steps for easier problem solving: a definite plus for the beginner to water saturation problems.

Other researchers, such as Winsauer et al.,[20] have introduced the *generalized* Archie equation (sometimes called the *Winsauer equation*)

$$F = a \, \phi^{-m} \qquad\qquad \textbf{(3–15)}$$

by adding the constant term *a*. Specifically, Winsauer et al. found a good fit to core porosity data for sandstones from different basins and various geologic ages if $a = 0.62$ and $m = 2.15$. With these particular values for *a* and *m*, Eq. 3–15 is known as the *Humble* equation; an apparent reference to the fact that these researchers were associated with the Humble Oil and Refining Company. Note that in this form, the formation factor relation will not satisfy the boundary condition that formation factor equals one when porosity equals one. Despite this apparent discrepancy, many log analysts use the generalized form of Eq. 3–15 for formation factor. It is certainly possible that the true formation factor relation may not be linear when plotted on log-log graphs. Perhaps it is, for some rocks, something more like the *Shell Formula* referred to by Schlumberger on its chart: Por-1.[21] Unless the curvature is pronounced, it still seems reasonable to approximate the curve with a straight line over most practical ranges of porosity values encountered.

Will *m* decrease or increase as porosity increases? Towle reports on an interesting study of several idealized pore geometries where *m* increases with porosity.[22] Etnyre confirms these results in a similar but preliminary pore geometry study.[23] Actually, we might expect the reverse to be true, especially as porosity becomes very nearly one. Whey should *m* increase with porosity? Taking one of the idealized pore systems studied by Towle, that being mutually perpendicular intersecting tubes of square cross section, note in the two-dimensional Figure 3–11 that the path the ions may take in the intersection widen out. Towle referred to this as the current *flare*.[24] Winsauer et al. also illustrate this widening of the current path as increasing the average length of the current paths through a pore system in an idealized packing of spherical grains.[25] Towle's work also confirms the earlier observations I made concerning the effect of vugginess.[26]

In any event, some workers have found it preferable to use the generalized form of the formation factor relation, where the intercept at 100% porosity ($\phi = 1.00$) will be $a R_w$ rather than just R_w.

$$R_o = FR_w$$
$$= a\phi^{-m}R_w$$
$$= a \times (1.00)^{-m} \times R_w$$

where 1 raised to any power is still 1. However, many log analysts still prefer the version of Eq. 3–12 that meets the boundary condition of $F = 1$ at 100% porosity. Jackson et al. feel that *tortuosity exponent* is a better term than cementation exponent and that m is reflective of particle shape, becoming smaller with grain sphericity. On the other hand, m becomes larger for plate-shaped or flatter grains. They also subscribed to the boundary condition that $F = 1$ when $\phi = 1$.[27] Ransom quotes other researchers as relating smaller values of m to grains with a minimum surface area to volume ratio.[28] Again, this would apply to the more spherical grains. Ransom also suggests that the parameter a should equal 1 when using dual porosity, dual water, or cation exchange capacity-type saturation equations for shaly sands that account for bound water conductivity, whereas values of a less than 1 would apply to shaly sand analysis using conventional (Archie type) single porosity, single water saturation equations.[29]

Bussian takes exception to the Humble formula to the extent that other equally good fits to the Winsauer data can be made with $a = 1$.[30] He also

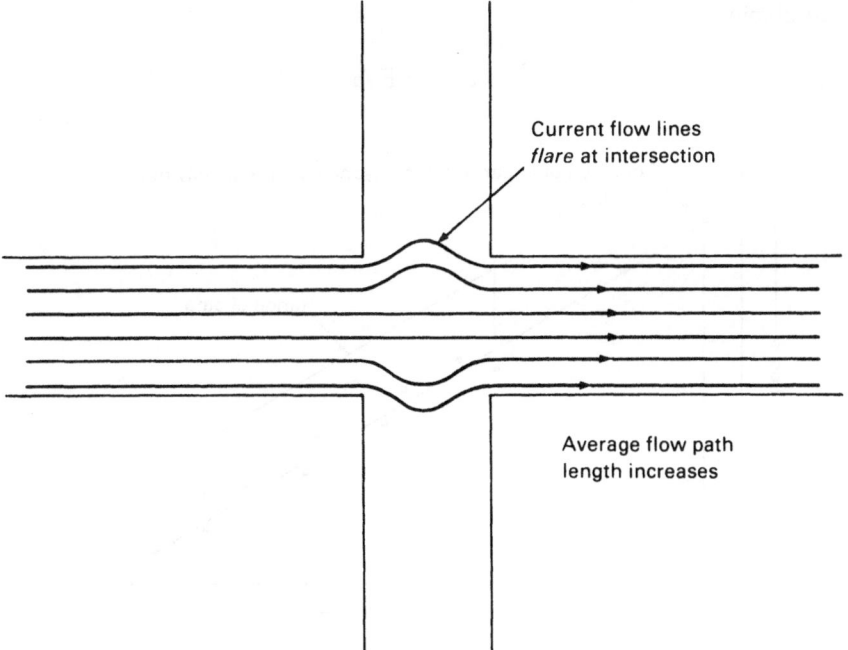

Current flow lines *flare* at intersection

Average flow path length increases

Figure 3–11. Tortuosity increase at tube pore system intersections.

states that a is an ad hoc parameter with no theoretical basis. I agree from previous statistical studies of the Winsauer data set that other equally valid fits to the data can be found, depending on the assumptions used in establishing the statistical model for fitting the data. Part of the problem in fitting data from core samples (or other well log data plots to be discussed later) is that we usually have only a limited range of data. Figure 3–12 illustrates the problem using the formation factor-versus-porosity graph in Figure 3–9. In laboratory measurements, F is actually measured (as the ratio of R_o to R_w) on the rocks for whatever porosity values are available. The problem is that the porosity values are often over such a limited range that it is really stretching things to say these data reliably establish a line. Etnyre also discusses the relevant statistical problems of data fitting with a limited range of data.[31] The late George R. Pickett tended to stick with $a = 1$ in most of his work, and keeping all this in mind, I propose to use $a = 1$ in most of the examples in this book. You may, on the other hand, one day find it useful to fit data with $a < 1$ or $a > 1$. There is certainly not universal agreement among the logging community on this subject of the ad hoc relational constant a in Archie's formation factor relation.

It is still possible to use the graph in Figure 3–9 to solve formation factor problems when a is not equal to 1. Denote the formation factor computed from the generalized formula (Eq. 3–15) by F'. Solving Eq. 3–15 for ϕ^{-m}, we obtain

$$\phi^{-m} = F'/a$$

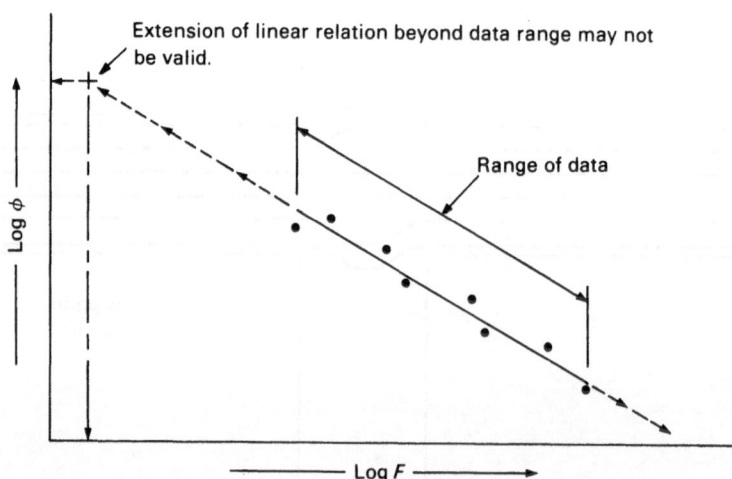

Figure 3–12. Limited range of formation factor versus porosity data. *(After Lee M. Etnyre, "Practical Application of Weighted Least Squares Methods to Formation Evaluation — Part I," The Log Analyst **25**, no. 1 [1984]: 13.)*

If we use the graph in Figure 3–9 to solve for F from ϕ^{-m}, we find

$$F = F'/a$$

and thus

$$F' = aF$$

Having found F, we calculate F', the formation factor from the generalized equation as the product of the constant a with F. For example, if $\phi = 0.10$, find the formation factor from the Humble equation. The Humble equation uses $a = 0.62$ and $m = 2.15$. First, solve for F assuming that Eq. 3–12 is valid (assume that $a = 1$ and use $m = 2.15$). Enter the chart (Fig. 3–9) on the left at $\phi = 0.10$ and extend a horizontal line over to a point halfway between the slopes (m) of 2.1 and 2.2. From there, extend a vertical line (either up or down) to the horizontal (F) axis at an F of approximately 141. Then multiply this value of F by $a = 0.62$ to obtain the formation factor $F' = 87$. Thus, to use the graph for formation factor when the constant a is not equal to 1, simply multiply the formation factor you find from assuming $a = 1$ by the actual value for a.

Typical values for the tortuosity exponent m are 2 for intergranular carbonates and well-cemented and consolidated sands, 1.6 to 1.8 for less consolidated sands (even less than 1.6 for some Gulf Coast sands), and > 2 for vuggy carbonates (values of 2.6 are not unusual).

RESISTIVITY INDEX

The resistivity index I is defined as the ratio of the resistivity of a rock R_t to the resistivity the rock has if it is 100% saturated with formation water, with resistivity $R_w : R_o$.

$$I = R_t/R_o \qquad\qquad (3\text{–}16)$$

where both resistivities are expressed in the same units, usually ohm-meters. Thus, a rock with all formation water in its pore system has a resistivity index of 1. A rock will also have a resistivity index greater than 1 if it contains hydrocarbons. Why is the resistivity index of a rock greater than or equal to 1? If the pores are saturated with formation water, the actual resistivity of the rock is then the same as R_o, by definition, and hence, $I = 1$. Next, imagine that in the parallel bundle of tubes in the cube of Figure 3–7 there are continuous strings of oil running through the middle of each tube such that the cross-sectional area of water in each tube is reduced by the presence of an oil stringer (centered in the tube) to a value less than the cross-sectional area of the tube. If no oil were present, the conductive cross-sectional area of the water would be the same as the cross-sectional area of the tube. Oil is an insulator and resists the flow of electrical current. Thus, the effective cross-sectional area of the conductor has been reduced, and by the relation for resistance with resistivity, we can see that the resistance across the cube must increase when oil occupies part of the pore (tubes, in this example) space. Thus, the resistivity of the rock with its pore system is now higher than it would be if there were only water in the pores.

Archie reviewed the work of other investigators and explained how their laboratory studies of the relationship of water saturation with resistivity index (R_t/R_o) should hold true for both oil and gas saturation in rocks underground.[32] The relation is

$$S_w = I^{-1/n} \qquad\qquad (3\text{--}17)$$

where I have taken the liberty of expressing the relation using the resistivity index I to represent the ratio of resistivities (R_t/R_o), and water saturation S_w is expressed as a *fraction of pore volume*. The exponent n is known as the *saturation exponent* and is often found to have a value of 2 or sometimes is assumed to be approximately equal to the tortuosity exponent m.

Both m and n can be measured on rock cores in a laboratory, but the measurement of the saturation exponent n is sometimes questioned on the grounds that the oil and gas in the pore systems of rocks in a relatively short-term laboratory experiment may not be distributed in the pore system in the same manner as the pore system of rocks that have been buried over geologic time.[33] Even though the relative saturations are the same, the fluids and gas can fill the pore spaces in a different manner.

Equation 3–17 can be expressed as a straight line on a log-log graph in the same fashion as the porosity-versus-formation factor relation (Eq. 3–12) is in Figure 3–9. Figure 3–13 is a graph of the logarithm of water saturation versus the logarithm of resistivity index.

$$-n \log (S_w) = \log(I) \qquad\qquad (3\text{--}18)$$

This time, the lines with various slopes from 1.6 to 2.2 represent different values for the saturation exponent n. This graph can be used in the same manner as that in Figure 3–9 to solve problems using Eq. 3–17. In the graph in Figure 3–9, we entered with a porosity value on the vertical axis to the left and extended horizontally and then down from the appropriate line with slope m to a value for F. With Figure 3–13, we enter with a resistivity index I on the horizontal axis, and extend vertically to the appropriate line with slope n and then to the vertical water saturation axis to read the water saturation.

This is just what we are looking for! If we know m and R_w, we can use the graph in Figure 3–9 to solve for R_o, the resistivity the rock should have

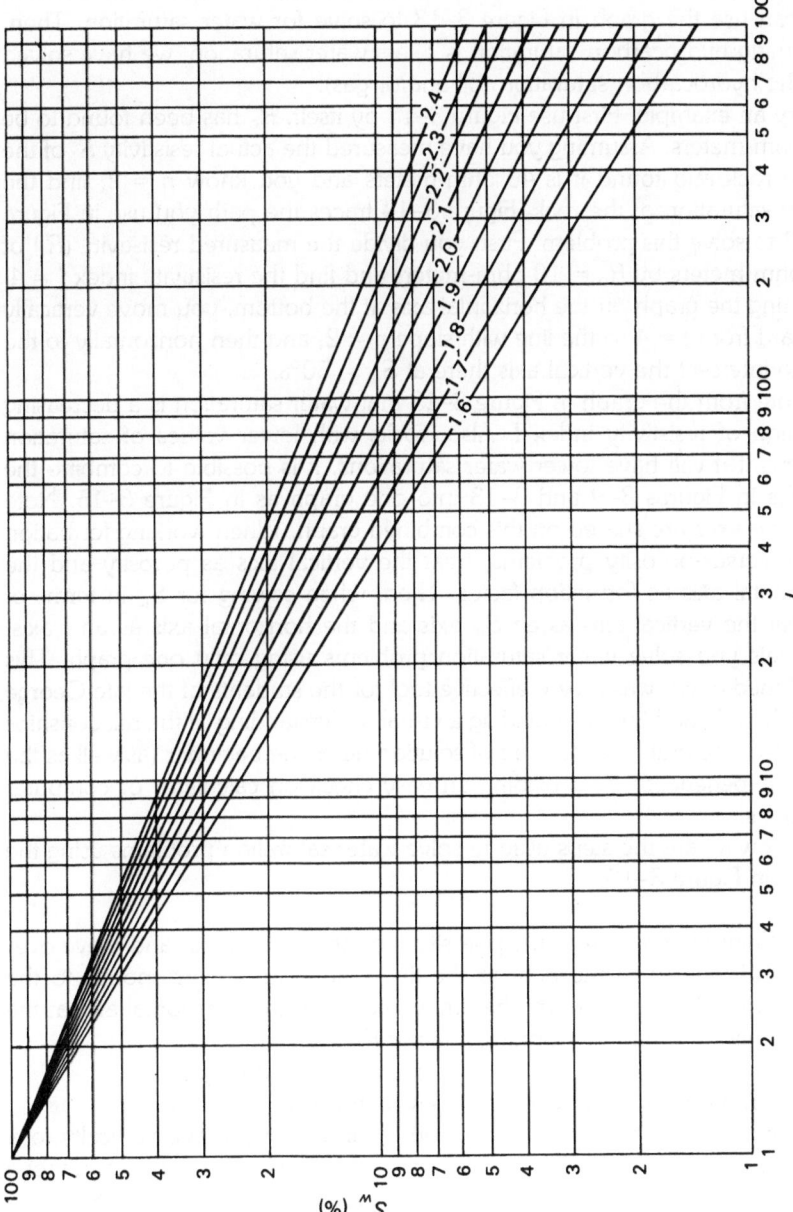

Figure 3–13. Water saturation versus resistivity index.

if it is 100% water saturated. Then, we can use the measured rock resistivity R_t and divide by R_o to obtain the resistivity index I. Now, if we know n, we can use the graph in Figure 3–13 to solve for water saturation. Then, assuming hydrocarbon saturation is 1 less water saturation, we have solved for the hydrocarbon saturation (oil and/or gas).

Try an example. First use Figure 3–13 by itself. R_o has been found to be 10 ohm-meters. Assuming you have measured the actual resistivity R_t of the same rock and found it is 40 ohm-meters and you know $n = 2$, find the water saturation of the rock. Figure 3–14 traces the path you use in Figure 3–13 to solve this problem. First, you divide the measured resistivity (R_t) of 40 ohm-meters by $R_o = 10$ ohm-meters and find the resistivity index $I = 4$. Entering the graph on the horizontal axis at the bottom, you move vertically upward from $I = 4$ to the line with slope $n = 2$, and then horizontally to the left to intersect the vertical axis there at $S_w = 50\%$.

Note from the graph in Figure 3–13 that water saturation is a decreasing function of resistivity index I. Also, rocks with lower values of saturation exponent n will have lower water saturations. It is possible to combine the graphs in Figures 3–9 and 3–13 into one graph as in Figure 3–15. Note how the axes are shared on this combined graph. When working formation factor-versus-porosity problems, treat the vertical axis as porosity and the horizontal axis as formation factor. Then, when solving for S_w in terms of I, treat the vertical axis as an S_w axis and the horizontal axis as an I axis. This lets you solve water saturation problems using only one graph. This combined graph was a very effective tool for the students of the late George R. Pickett. In addition to providing a visual understanding of the relationships involved, the graphical method of solution helps the beginner (as well as the more experienced) by providing an easy check on calculator or computer solutions.

Following are the steps used to solve water saturation problems using the graph in Figure 3–15.

1. Enter the graph with the porosity ϕ on the vertical axis and move over horizontally to the right to the line whose slope corresponds to the tortuosity exponent m, then move vertically to a horizontal axis at the bottom to find the formation factor F.
2. Multiply the formation water resistivity R_w by F to get R_o.
3. Divide R_t by R_o to get the resistivity index I.
4. Enter the graph with I on a horizontal axis and move vertically to a line whose slope corresponds to the saturation exponent n, then move horizontally across to a vertical axis to read water saturation S_w.

Now try an example calculation using this procedure. You need five values to solve for water saturation using this conventional approach: R_w, m, n, ϕ, and R_t. (If you use the generalized formation factor, you also need a).

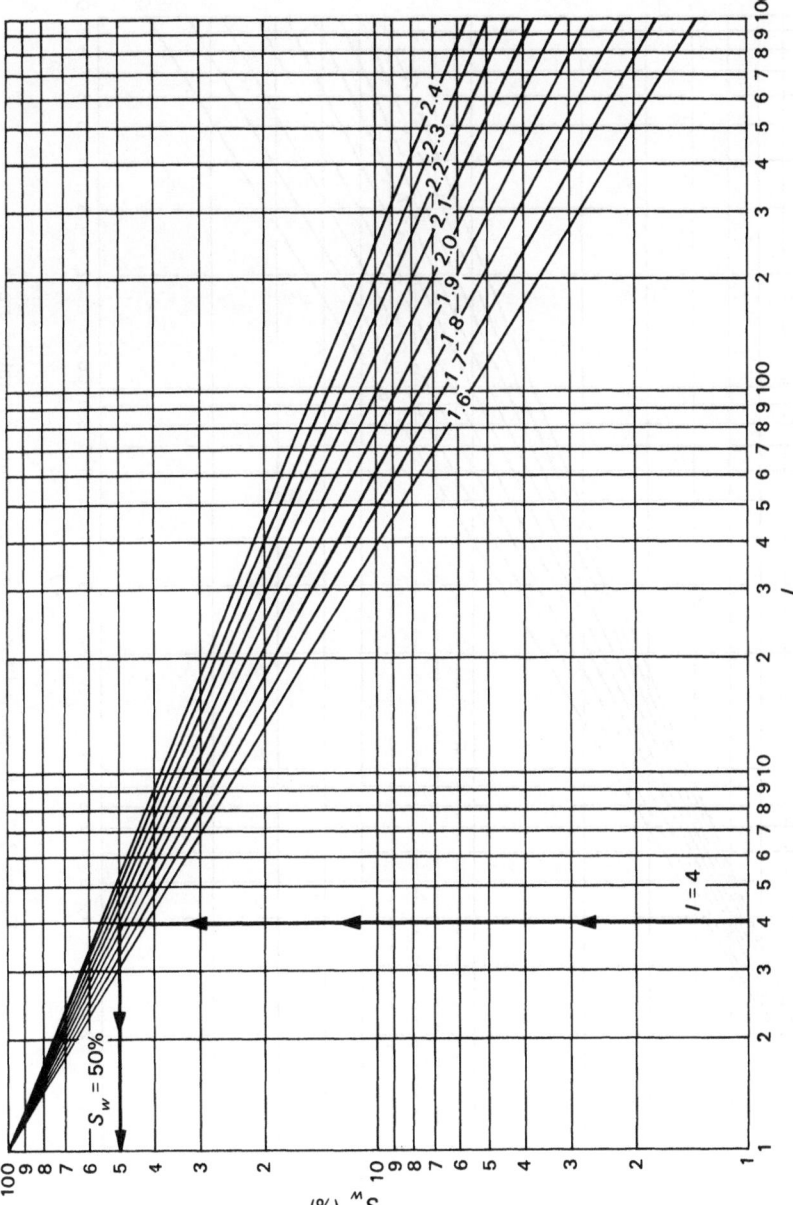

Figure 3–14. Using the water saturation versus resistivity index graph to find water saturation from the graph.

77

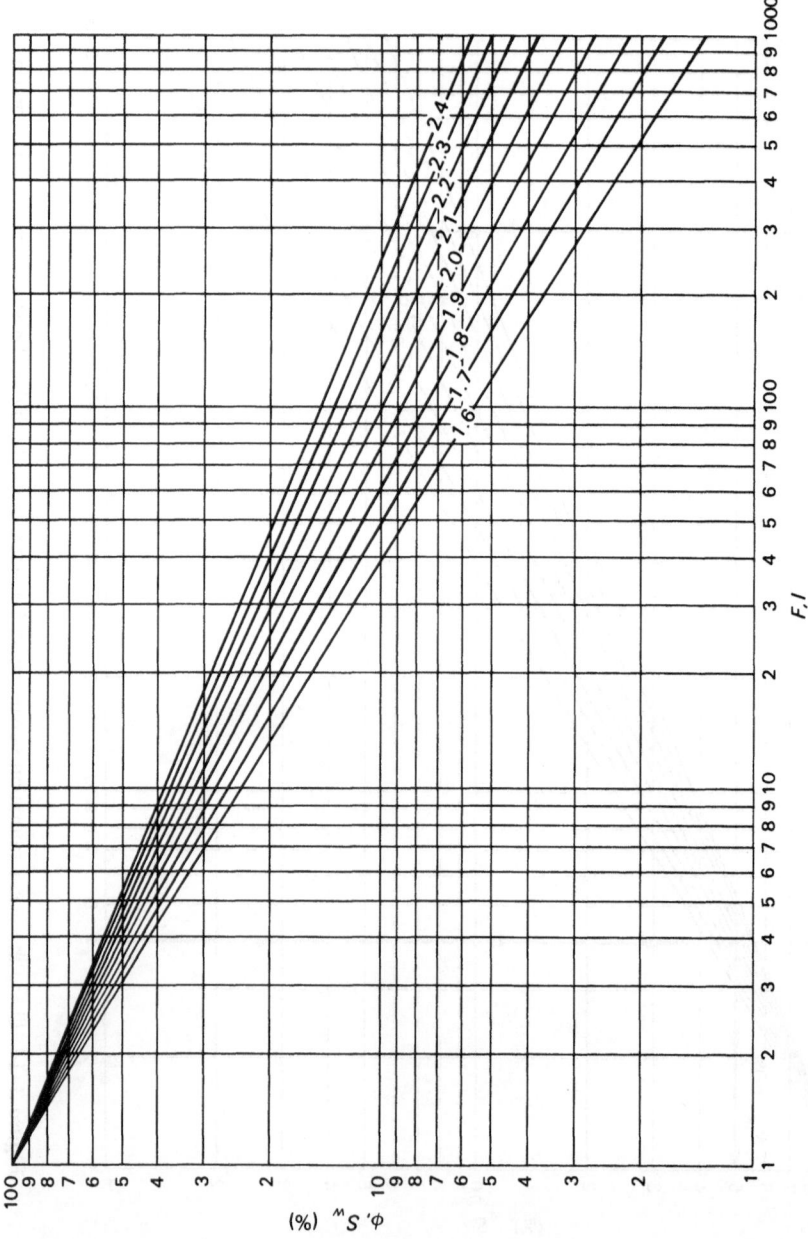

Figure 3–15. Combined formation factor–water saturation graph.

78

Suppose you measured the following values

$$R_t = 100 \text{ ohm-meters}$$
$$\phi = 25\%$$

and had a reliable measurement of R_w (maybe from produced water in a nearby well) that was 0.50 ohm-meters (at formation temperature). Further, assume $m = n = 1.8$. What is the water saturation for this zone?

First, enter the graph (Fig. 3–15) with the porosity of 25% and move to the line with slope (corresponding to m) equal to 1.8, then move down to a formation factor of $F = 12$. Multiply the water resistivity $R_w = 0.50$ ohm-meters by the formation factor of 12 to get $R_o = 6$ ohm-meters. This is the resistivity the rock would have if it were 100% saturated with formation water of resistivity $R_w = 0.50$ ohm-meters.

Next, divide the measured resistivity by this calculated R_o to get the resistivity index.

$$I = R_t/R_o$$
$$= 100 \text{ ohm-meters}/6 \text{ ohm-meters}$$
$$= 16.7$$

Go back to the horizontal axis on the graph in Figure 3–15 with this resistivity index and move up to the line with the appropriate slope corresponding this time to $n = 1.8$ and then move to the vertical axis to find the water saturation $S_w = 21$.

Figure 3–16 is a flow chart of the procedure outlined in the four steps above and that you used in the example.

Equations 3–14 and 3–17 may be combined by first substituting the expression for F from Eq. 3–15 into Eq. 3–14 and then substituting the expression for R_o of Eq. 3–14 into Eq. 3–16. Finally, use this latter expression for I in Eq. 3–17 to get

$$S_w = (R_t/a\phi^{-m}R_w)^{-1/n} \qquad \text{(3–19)}$$

Use this equation with the values for the five unknowns in the earlier example to see if you get the same answer you got with the graphical method. Remember that the example assumes $a = 1$. Try again if you do not get almost the same answer. People familiar with programmable hand-held calculators may find it easier to use the calculator, whereas those less familiar with calculators and mathematical manipulations may find the graphical approach more comfortable. Certainly, the graphical method provides a quick verification of calculator or computer results.

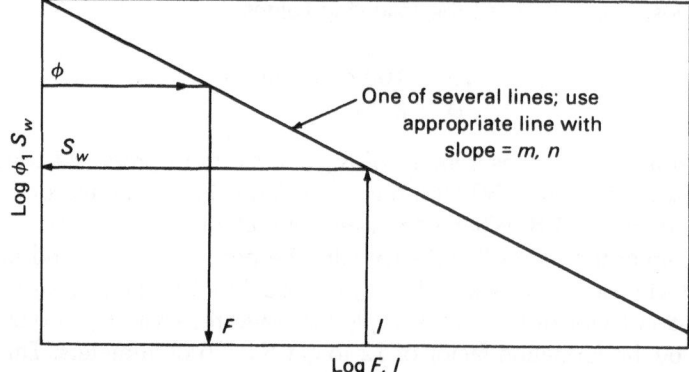

1. Enter with ϕ, move to desired m, find F.
2. Multiply R_w by F to get R_o.
3. Divide R_t by R_o to get I.
4. Enter with I, move to desired n, find S_w.

Note: m not necessarily equal to n; use appropriate line on graph

Figure 3–16. Using the formation factor–resistivity index graph to find water saturation.

From Eqs. 3–14, 3–16, and 3–17, you should easily verify the following variations of the fundamental relationships.

$$R_t = IR_o$$
$$= IFR_w$$

where

$$I = R_t/R_o$$
$$= S_w^{-n}$$

Also, incorporating Eqs. 3–15 and 3–19

$$S_w = (R_o/R_t)^{1/n}$$
$$= (FR_w/R_t)^{1/n}$$
$$= (a\phi^{-m}R_w/R_t)^{1/n}$$
$$= (aR_w/\phi^m R_t)^{1/n}$$

If you spend any time with log analysis literature, you will probably encounter these variations of the same basic relationships. You should verify their equivalence to your own satisfaction. I encourage you to develop proficiency in using any of the equations and their variations in your calculations.

Suppose we have a rock 100% saturated with water of resistivity R_w. From Eq. 3–17, we see that $I = 1$ for $S_w = 1$. If $I = 1$, from Eq. 3–16 we realize that $R_t = R_o$. Hence,

$$R_t = IFR_w$$
$$= FR_w$$

for a rock 100% saturated with water of resistivity R_w. This relation says that we can solve for the formation water resistivity from

$$R_w = R_t/F$$

where $R_t = R_o$ for water saturated rocks. It is customary to refer to this calculated R_w as an apparent water resistivity R_{wa} and define it by the equation

$$R_{wa} = R_t/F \qquad\qquad (3\text{–}20)$$

Also,

$$R_{wa} = R_t\,\phi^m \qquad\qquad (3\text{–}21)$$

or

$$R_{wa} = R_t\,\phi^m/a \qquad\qquad (3\text{–}22)$$

using the extra constant a in Eq. 3–22. If well log tool responses were perfect, the calculated R_{wa} should be equal to R_w for 100% water-bearing rocks. Of course, they are not perfect, so R_{wa} may not always be equal to R_w. For this reason and the possibility that there may be hydrocarbons present in the pore system, the R_w calculated from well log responses is known as an *apparent* water resistivity. In certain respects, it may be preferable to use a calculated R_{wa} instead of an R_w measured from a water sample because it reflects the water resistivity as seen downhole by the well log responses. The idea is that any errors in the logging tool responses may in part be compensated for by using the R_{wa}, which may be affected somewhat in the same fashion. By substituting R_{wa} from the right side of Eq. 3–22 into Eq. 3–19, we can easily verify that

$$S_w = (R_{wa}/R_w)^{-1/n} \qquad \qquad \textbf{(3–23)}$$

Note that this also implies

$$I = (R_{wa}/R_w)$$
$$= R_t/R_o$$

This rearrangement of Archie's relation has come to be known as the R_{wa} method or R_{wa} quick look method. By comparing the ratio of the calculated R_{wa} for a zone of interest to the known R_w (or, better yet, a calculated R_{wa} from a water-bearing zone), we can calculate S_w from Eq. 3–23. The only pitfall is that the zone of interest we are checking for hydrocarbons must have the same R_w as the wet zone from which we calculate R_w. Likewise, if we are using a measured R_w, say, from some produced water sample, the R_w of the suspected hydrocarbon-bearing zone must be the same as the zones that produced the formation water of resistivity R_w.

If there are hydrocarbons present in the pore system of a presumed water-bearing zone, the calculated R_{wa} will differ from R_w. Will it be too high or too low? From Eq. 3–17, I must be greater than 1 ($I > 1$) if the rock is less than 100% water saturated ($S_w < 1$). Thus, from Eq. 3–16, $R_t > R_o$ and the R_{wa} we would compute from any of the above equations would be *larger* than it would have been had R_t been equal to R_o (i.e., $R_t = R_o$ for water-bearing zones). The important thing to remember here is that the R_{wa} method is really nothing more than a rearrangement of Archie's relation that allows us to do the bookkeeping a little differently while we do the calculations. The R_{wa} method, when using a calculated R_w from a wet zone, thus requires one less parameter than the conventional method of calculating water saturation from Eq. 3–19. In fact, the R_{wa} method lets us calculate water saturation when *both* R_w and m (as well as a), the tortuosity exponent, are uncertain or unknown.

The *quick-look* application for the R_{wa} method is used when a logging service company presents a computed R_{wa} curve along with the other recorded curves on the well logs. It is then stated that when the R_{wa} for potential hydrocarbon-bearing zones is two to three times as large as the R_{wa} curve value occurring opposite possible water-bearing zones, this is an indication of producible hydrocarbons. From the graph in Figure 3–13, we can see that for $n = 2$ the ratio of $R_{wa}/R_w = R_t/R_o = 2$ would represent a water saturation of 71%, whereas a ratio of 3 would represent $S_w = 58\%$. If the ratio is greater than 3, the water saturation must be even less. What ratio would you like to see if the cutoff water saturation is 50% or less (assuming $n = 2$)? If you do not agree with $I = 4$, go back and review the relation of resistivity index and water saturation with the relational parameter n.

Another quick-look method similar to the R_{wa} approach is the presentation on the logs of an R_o curve. We have already seen R_o defined as the resistivity of a water saturated rock. R_o can be computed from a porosity tool reading with the use of Eqs. 3–14 and 3–15 or 3–12. We can then compare the actual resistivity R_t for a zone of interest and look for a suitable ratio for I as in the R_{wa} method. This method is also a rearrangement of Archie's relationships. We can use this method with either an R_w or a calculated R_{wa} from water-bearing zones.

Methods for calculating water saturation other than Eq. 3–19 are also available. Several R_{xo} methods depend on resistivity readings from the *flushed* zone adjacent to the well bore. Most of the R_{xo} methods require more measurements for their solution than Eq. 3–19. Usually, to use them, a measurement of R_{xo} and R_t is required along with a knowledge of the resistivity of the mud filtrate water (R_{mf}) from the mud system as well as R_w. This mud filtrate water flushes the original formation water from the pore system next to the well bore. This will be discussed in detail in Chapter 5.

According to Pickett, another parameter method, known as F_r/F_s, turns out to be a rearrangement of the bookkeeping for Eq. 3–19.[33] In this method, we use the ratio of two calculated formation factors. One is calculated from a resistivity tool measurement as

$$F_r = R_t/R_w$$

Note that for water-bearing rocks, $R_t = R_o$, and this expression then is the formation factor.

The other formation factor is calculated from an acoustic or sonic log (which is calibrated to yield a porosity measurement)

$$F_s = \phi_s^{-m}$$

where the subscript s denotes that porosity ϕ came from a sonic or acoustic log. You should recognize this as being the same formation factor we calculated from Eq. 3–12 or by using the graph in Figure 3–9. The ratio F_r/F_s is seen to be another arrangement of Eq. 3–19 with $a = 1$.

$$(F_r/F_s)^{-1/n} = (R_t\phi^m/R_w)^{-1/n}$$

where ϕ_s is replaced with ϕ and the quantity F_r/F_s is raised to the $-1/n$ power. Thus, a ratio of $F_r/F_s \geq 4$ would indicate, for example, that $S_w \leq 0.5$.

A similar parameter method is the presentation of a calculated porosity curve on the logs. This calculated porosity is based on the actual resistivity reading and is made from another re-expression of Archie's relationships. Visualize

$$R_t = IFR_w$$

from an earlier paragraph. For water wet zones, R_t becomes R_o and $I = 1$. Now use Eq. 3–12 and we have

$$R_o = \phi^{-m}R_w$$

Solving for ϕ and using a prime to indicate this particular calculated value

$$\phi' = (R_w/R_o)^{1/m}$$

For wet zones where $I = 1$, ϕ' should equal the actual porosity of the rock (as long as we have a correct value for R_w). When hydrocarbons are present, we will divide by $R_t = IR_o$ instead of just R_o in the above equation. Since this value is larger than R_o, the calculated porosity will be too low. We will actually be calculating

$$\phi' = (R_w/IR_o)^{1/m}$$

Since the actual porosity, as reflected by a porosity tool reading, should be

$$\phi = (R_w/R_o)^{1/m}$$

we can take the ratio of ϕ to ϕ' and raise everything to the m power.

$$\phi/\phi' = I^{1/m}$$

Rearranging

$$I = (\phi/\phi')^m$$

Thus, we are back to another bookkeeping method for Archie's equation. For $S_w \leq 0.5$ and $m = 2$, we would like to see the porosity tool reading ϕ at least twice as high as the porosity calculated from an R_t reading using the rearrangement of Eq. 3–12 above.

In actual practice, where will we get the values for all the unknowns in Archie's equation? Figure 3–17 is a flow chart for typical sources of data for the various parameters used in solving Archie's equation.

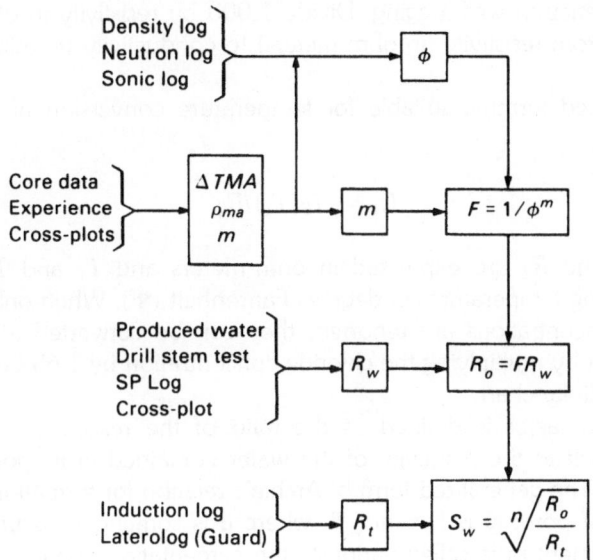

Figure 3–17. Flow chart for sources of data for porosity and saturation.

SUMMARY

Electrical current flows in rocks primarily through the formation water present in the pore system. The fundamental rock property that determines the resistance a rock offers to the flow of electric current is called *resistivity*.

The resistivity of an electrolyte solution, and hence, a rock, is dependent on the temperature of the solution. The *temperature* of any resistivity measurement on rocks or electrolyte solutions must be quoted along with the measured resistivity. A formation water or mud filtrate resistivity must be converted to the appropriate formation temperature before using the number in any log analysis formulas involving formation resistivities.

The resistivity of electrolyte solutions decreases with increasing temperature because of increased ion mobility. Since temperature increases with depth in a well, the resistivity of formation water and mud filtrate is smaller at depth for the same ionic concentration.

Resistivity determines the resistance offered to the flow of current. Resistance is usually expressed in units called *ohms*. Resistance is increased by longer current paths and smaller cross-sectional areas perpendicular to the flow of current. Resistivity is usually expressed in units of ohm-meters (the more formal ohm-meters^2/meter is sometimes used). Conductivity is the reciprocal of resistivity but is usually expressed in units of millimhos per meter for convenience in well logging. Divide 1,000 by resistivity in ohm-meters to convert from resistivity (in ohm-meters) to conductivity (in millimhos per meter).

A simplified formula suitable for temperature conversion of electrolyte solutions is

$$R_1 = (T_2/T_1)/R_2$$

where R_1 and R_2 are expressed in ohm-meters and T_1 and T_2 are the corresponding temperatures in degrees Fahrenheit (°F). When only chloride (Cl) ion concentrations are reported, they can be converted to parts per million NaCl by multiplying the chloride concentration by 1.65 before using the NaCl salinity chart.

Formation factor is defined as the ratio of the resistivity of a water-saturated rock to the resistivity of the water contained in its pore system: $F = R_o/R_w$. The generalized form of Archie's relation for formation factor as a function of porosity is $F = a\phi^{-m}$, where a is sometimes assumed to be 1. The exponent m is called variously the cementation exponent, porosity exponent, or tortuosity exponent of the formation factor relation. It increases as the tortuosity or vugginess of a rock's pore system increases. Typical values for m are 2 for carbonates and well-cemented rocks, 2.6 for vuggy carbonates, and 1.8 for some less consolidated sandstones. In loosely consolidated Eocene or Miocene sands of the Gulf Coast m can be as small as 1.3 to 1.6.

Resistivity index is defined as the ratio of a rock's resistivity to the resistivity it would have if 100% saturated with formation water: $I = R_t/R_o = S_w^{-n}$, where n is the saturation exponent. It is usually assumed to be the same as m or else always equal to 2. Saturation exponent n is more difficult to measure accurately than m, although both exponents are probably more uncertain in practice than some would believe. Water saturation may be solved directly from Archie's relation $S_w = (R_t/a\phi^{-m}R_w)^{-1/n}$ or from the sequential approach using the charts in Figures 3–9 and 3–13.

1. Use the graph in Figure 3–9 with a porosity measurement and m to find formation factor F. If using the generalized form, multiply the resultant F value from the graph by a to find the generalized formation factor F'.
2. Multiply R_w by F (or F') to find R_o, the wet rock resistivity.
3. Divide the measured rock resistivity by R_o to get the resistivity index I.
4. Use the graph in Figure 3–13 with a value for n to solve for S_w. Then, hydrocarbon saturation S_{hc} may be found from $S_{hc} = 1 - S_w$.

Several parameter methods for solving water-saturation problems are also available. However, they amount to nothing more than a rearrangement of

Archie's relation for S_w. One of the commonly used forms is the R_{wa} method. It is useful for obtaining R_w values from known water-bearing formations for use in Archie's equation when there are no actual measurements of formation water available. In a sense, R_{wa} is an in situ measurement of R_w as seen by the logging tools in water-bearing zones. R_{wa} is defined as the ratio R_t/F. For water-bearing zones, $R_t = R_o$ and $R_{wa} = R_w$.

PROBLEMS

3–1.   If formation water resistivity is 0.20 ohm-meters at 75°F, what is it at 210°F? What concentration in parts per million NaCl does this represent?

3–2.   You have been given a water sample analysis that reports only chloride concentration of 50,000 ppm. What is the resistivity of this sample at 120°F? What conductivity is this in millimhos per meter?

3–3.   Use Eq. 3–4 to solve the first part of problem 3–1 and compare the results.

3–4.   Calculate the resistivity at 68°F for the following formation water sample.

	Cations	Anions	
	7,208 mg/l Na	12,152 mg/l Cl	
	1,323 mg/l Ca	1,889 mg/l SO$_4$	
	122 mg/l Mg	464 mg/l HCO$_3$	

3–5(a).   If $m = 1.8$ (assume $a = 1$), what is F for $\phi = 0.12$?

3–5(b).   If $a = 0.5$, what is F for $\phi = 0.12$?

3–5(c).   If $m = 2.2$ and $F = 400$, what porosity is represented if $a = 1$?

3–5(d).   If $m = 1.6$, $F = 200$, and $a = 0.5$, what porosity is implied?

3–6(a).   If $R_t = 20$ ohm-meters and $R_o = 4$ ohm-meters with $n = 2$, what is S_w?

3–6(b).   If $R_t = 200$ ohm-meters and $R_o = 80$ ohm-meters with $n = 1.9$, what is S_w?

3–7(a).   If $R_w = 0.20$ ohm-meters, $\phi = 14\%$, and $m = 1.9$, what is R_o?

3–7(b).   What value of R_t is required for hydrocarbon saturation to be at least 50%?

3–7(c).   What value of R_t will it take to ensure that hydrocarbon saturation is greater than 50% if $R_w = 2$ ohm-meters?

3–7(d).   What value of R_t will it take if $R_w = 0.02$ ohm-meters?

3–8(a).   Suppose you are evaluating a formation where R_w, porosity, and m all remain constant. The lower part of the formation is usually water wet so that you can assume that $S_w = 100\%$ in the lower part. If the lower part of the formation has a resistivity of 5 ohm-meters and the upper part has a resistivity of 11 ohm-meters, what is the hydrocarbon saturation in the upper part of the formation if $n = 1.7$?

3–8(b).   Suppose n is known to vary from 1.6 to 2.2. What would be the possible range for hydrocarbon saturation in the upper part of the formation?

3–8(c).   Consider the practical situation where porosity can vary from 15% to 20% and $m = 1.8$ (where n can still vary from 1.6 to 2.2). What is the possible range of hydrocarbon saturation in the upper part of the formation?

3–9(a).   If $R_w = 0.5$ ohm-meters at 75°F and $R_o = 10$ ohm-meters at 75°F, what is the formation factor?

3–9(b).   What porosity does this represent if $a = 1$ and $m = 2$?

3–9(c).   If the measured rock resistivity is 50 ohm-meters at 150°F, what porosity would this represent for the same $a = 1$ and $m = 2$?

3–9(d).   If the rock's pore system contains hydrocarbons, will the porosity calculated in (c) be too high or too low?

3–9(e).   If $n = m = 2$, what is the hydrocarbon saturation if the actual rock porosity is 28.4% in part (c)?

REFERENCES

1. D. R. Crow, *Principles and Applications of Electrochemistry*, 2nd ed. (Chapman and Hall, London, 1979), p. 52.
2. Author's class notes from course on well log interpretation presented by George R. Pickett at the Colorado School of Mines, Goldon, Colorado, 1976.
3. Sylvain J. Pirson, *Handbook of Well Log Analysis for Oil and Gas Formation Evaluation* (Prentice-Hall, Englewood Cliffs, N.J., 1963), p. 36.

4. Dresser Industries, *Log Interpretation Charts* (Dresser Industries, Inc., Houston, Tex., 1979), p. 6.
5. Schlumberger, *Log Interpretation Charts* (Schlumberger Ltd, New York, 1985), p. 5.
6. Douglas W. Hilchie, "A New Water Resistivity versus Temperature Equation," *The Log Analyst* **25**, no. 4 (1984): p20, 21.
7. Dresser Industries, *Log Interpretation Charts*, p. 6.
8. R. I. Martin, *Fundamentals of Electric Logging* (Petroleum Publishing Company, Tulsa, Okla., 1955), p. 3.
9. Author's class notes, 1976.
10. Dresser Industries, *Log Interpretation Charts*, p. 6.
11. Schlumberger, *Log Interpretation Charts* (Schlumberger Ltd, New York, 1979), p. 89.
12. "Core Water Salinity Determination," para. 6.2 in *API Recommended Practice for Core-Analysis Procedure*, API RP40 (The American Petroleum Institute, Dallas, Tex., 1960), p. 15.
13. Ibid., p. 15.
14. R. L. Kaufman and C. V. Moore, "Resistivity Techniques Can Cause Unsuspected Problems," *World Oil* (February 1, 1983).
15. Ibid.
16. G. E. Archie, "The Electrical Resistivity Log as an Aid in Determining Some Reservoir Characteristics," *Transactions of the AIME* 146: 541–562.
17. P. N. Sen, "The Dielectric and Conductivity Response of Sedimentary Rocks," paper presented at the 55th Annual Fall Technical Conference and Exhibition of the SPE of AIME, Dallas, September 1980.
18. Archie, "Electrical Resistivity Log."
19. Ibid.
20. W. O. Winsauer, A. M. Shearin, Jr., P. H. Mason, and M. Williams, "Resistivity of Brine-saturated Sands in Relation to Pore Geometry," *Bulletin of the American Association of Petroleum Geologists* **36**, no. 2 (1952): 263–277.
21. Schlumberger, *Log Interpretation Charts*, 1979, p. 12.
22. Guy Towle, "An Analysis of the Formation Resistivity Factor–Porosity Relationship of Some Assumed Pore Geometries," paper presented at the 3rd Annual Logging Symposium of the Society of Professional Well Log Analysts, 1962.
23. Lee M. Etnyre, Unpublished research on computer-aided study of pore geometries.
24. Towle, "Analysis of Formation Resistivity."
25. Winsauer et al., "Resistivity of Brine-saturated Sands."
26. Towle, "Analysis of Formation Resistivity."
27. P. D. Jackson, D. Taylor-Smith, and P. N. Stanford, "Resistivity-Porosity-Particle Shape Relationships for Marine Sands," *Geophysics* **43**, no. 6 (1978): 1250–1268.
28. R. C. Ransom, "A Contribution Toward a Better Understanding of the Modified Archie Formation Resistivity Factor Relationship," *The Log Analyst* **25**, no. 2 (1984): 7–11.
29. Ibid., p. 7–11.
30. A. E. Bussian, "Electrical Conductance in a Porous Medium," *Geophysics* **48**, no. 9 (1983): 1258–1268.
31. Lee M. Etnyre, "Practical Application of Weighted Least Squares Methods to Formation Evaluation—Part I," *The Log Analyst* **25**, no. 1 (1984): 11–21.
32. Archie, "Electrical Resistivity Log."
33. Author's class notes, 1976.
34. Ibid.

CHAPTER 4

Borehole Acoustic
Waveform Logging

Acoustic properties are one of three general categories of properties that are commonly logged. The other two—electrical and nuclear—are covered in Chapters 5 and 7, respectively. Acoustic properties of rock are generally ascertained from recordings of acoustic or sound waveforms in the borehole. From these waveforms we can measure amplitude, attenuation, travel time, and apparent frequency for several component wave types. Travel time is the most commonly used measurement, but the others are becoming more important in modern well logging.

Borehole acoustic waves are also sometimes referred to as *elastic waves* in reference to the mathematical theory of elasticity, which is often used for explaining the deformation of elastic bodies (those obeying Hooke's law).

The first logging tool developed for measuring porosity in a well bore was the acoustic or *sonic* log. Although the sonic log was actually developed to replace seismic geophone surveys, it was soon seen that it could be used as a porosity tool. The principle is relatively simple: measure the travel time of sound in a rock, which is related to the porosity of the rock. In practice, however, relating travel time of sound in a rock to porosity is complicated by uncertainties in the calibration parameters needed to relate the rock porosity to the acoustic log response. Yet, there are many areas where the sonic log is still used as the preferred porosity tool. I use the terms *sonic log* and *acoustic log* interchangeably since they refer to measurement of the same rock property. The only difference you will find in practice is that a service company may prefer one or the other of the terms for its own commercial designation.

In modern logging the sonic log has been replaced as the porosity log of choice by neutron-density logs. Nevertheless, the sonic log is still useful for obtaining porosity data from many rock types. For example, when grain densities are variable, sonic-log-derived porosity may be more reliable than density-log-derived porosity. Also, if proper use is made of the sonic log response relations, they can be used as a powerful rock typing tool, even when the sonic log is the only porosity tool run in a well.

The sonic log is probably used in more different applications to formation evaluation than most other modern logging tools. It also provides us with an excellent illustration of the general concept of response relations. Some of the many applications for acoustic or sonic logs in modern formation evaluation are:

• porosity measurement

• cement bond evaluation

• fracture detection

• lithology determination

• mechanical rock properties measurement

• borewall and casing inspection

• seismic calibration

• abnormal formation pressure detection

• gas-bearing formations identification

• shale indicator

SONIC LOG OPERATION

Commonly used sonic logs require fluid in the borehole to operate. Specialized tools have been designed to operate without fluid by pressing the acoustic transducers against the side of the borehole (*sidewall* tools). However, logging with these tools is a slow process, thus, they are limited to special applications where logging speed is unimportant. In petroleum applications, it is important to record the logs as quickly as possible, consistent with obtaining good data.

Most sonic log tools generate signals in the 20 to 30 kilohertz range. Specialty tools used for cement-bond evaluation and casing inspection operate at higher frequencies. Equations that relate acoustic log behavior to porosity are based on Hooke's law of elastic behavior: stress is proportional to strain. In actual rocks this may not always be true.[1]

Sonic logging tools have bow spring centralizer devices at one end and a three-arm caliper at the other to keep the transducers centered in the borehole. Otherwise, the received acoustic signals could be distorted and yield unreliable travel time data. With modern logging tools, it is possible to run the sonic log in tandem with a resistivity log so that both logs may be recorded during the same logging pass downhole. Of course, more *rat hole* must be drilled to accommodate the additional length of the combined tool so that both measurements can be recorded opposite the deepest desired depth.

In Figure 4–1, the arrows illustrate the ray path for the acoustic signal of interest. This same path is taken circumferentially all around the borewall but is shown here for only one side in the profile drawing. Although acoustic waves move out in all directions from the transmitting transducer labeled *TR*, only those striking the borewall at the critical angle for refraction, α, are responsible for the formation signal of interest: the signal recorded by the two receivers, R_1 and R_2. Other signals from the transmitter will either be

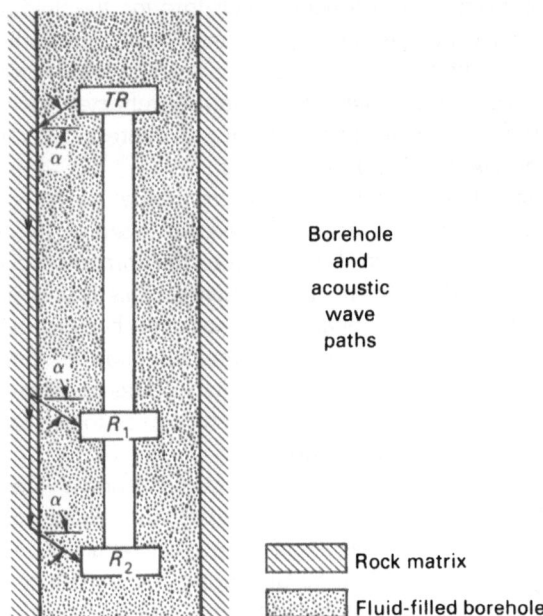

Borehole
and
acoustic
wave
paths

Rock matrix

Fluid-filled borehole

Figure 4–1. Schematic of sonic log tool in a borehole.

reflected at the borewall or refracted into the formation at an angle that will result in their being radiated away from the borewall where they cannot be detected.

The transducers are piezoelectric or magnetostrictive devices that change an applied voltage or current into a physical deformation of the transducer crystal or device. The motion of the deformation is an alternately expanding and contracting motion corresponding to the variations in the input signal. These are the *transmitter* types of transducers. Transducers also operate in the reciprocal mode, that is, they can produce an alternating voltage or current in response to an externally imposed deformation. These are *receiving* transducers.

The critical angle for refraction can be determined from Snell's law if we know the velocity of sound in both the fluid in the borehole and the formation. There is no need to be too concerned about calculating this angle now. There are actually two angles of refraction that are of interest, although we will only be concerned with the one that results in the transmission of the fastest wave along the borewall: the compressional wave. The other angle corresponds to conversion of the compressional wave in the fluid to a shear wave at the borewall–fluid interface. For most applications, the travel time of the compressional wave is used to calculate porosity, although shear wave travel time is also related to porosity. Shear wave travel time is more difficult to detect, though, since it arrives after the compressional wave.

After a signal from the transducer travels through the fluid and strikes the borewall at the critical angle, a compressional wave is transmitted down the borewall in the direction of the receivers. As it propagates along the borewall toward the receivers, fluid waves are continually refracted back into the fluid. However, only two refracted ray paths are of interest: the ray paths taken back toward the two receiving transducers.

The rays that follow the path from the transmitter to the receivers shown by the arrows in Figure 4–1 represent the path taken by the acoustic wave that will travel fastest through the fluid, into the formation down the bore-wall, and be refracted back into the receivers. This first arrival at each of the receivers is the compressional wave, so-called because the direction of wave vibration or oscillation is in the same direction as that of the wave propagation. When the wave is in the fluid traveling from transmitter to borewall or in the fluid traveling from borewall back to the receivers, it also propagates in the compressional mode.

The spacing between the two receivers is a known constant and will correspond to a distance along the borewall and slightly above the two receivers as determined by the angle for critical refraction. As long as the sound velocity in the rock is constant along the corresponding distance, this corresponding distance will be identical to the receiver spacing. If the rock sound velocity changes along this distance, the distance traveled by the compressional wave in the rock adjacent to the borewall may be slightly different than the receiver spacing due to slight differences in the critical angle for refraction opposite the two receivers.

Therefore, if we measure the difference in time of arrival of the compressional wave at the two receivers, we can determine the compressional wave travel time through the rock formation opposite the two receivers. The measurement is assumed to be taken at the midpoint between the two receivers. This interval travel time, usually measured in microseconds per foot (μsec/ft) or microseconds per meter (μsec/m), corresponds to the velocity of sound in the rock. The transmitter-to-first-receiver spacing is from 3 ft to 10 ft. The signal received at a receiver closer to the transmitter will, of course, be stronger than that for a receiver spaced further away. Receiver spacing (the spacing between the two receivers) for most modern tools is 2 ft, although older tools and some modern tools use other spacings such as 1 ft or 3 ft. We will also see that some tools use two transmitters to achieve compensation for changes in borehole size.

Typical sonic log tools have diameters of 3 3/8 in. to 3 5/8 in., although *slim hole* tools exist. Slim hole sonic or acoustic tools are only 1 1/16 in. in diameter.

ACOUSTIC WAVEFORMS AND BOREHOLE EVENTS

Figure 4–2 illustrates various paths that acoustic waves may take in the borehole. Waves received along these different paths may destructively interfere with each other. When this happens, any of the various acoustic *events* on the wave train may have too small an amplitude to be detected. The first arrival would actually be a direct wave through the tool, but this arrival is suppressed by including slots in the tool housing between the transmitter and the two receivers. These slots or acoustic isolators result in a great reduction in amplitude of this direct tool signal relative to the desired signals.

The direct fluid wave, labeled f in Figure 4–2, will be one of the later arrivals because the speed of sound in the fluid is slow compared to the formation speeds for most modes of propagation. Note that the transmitter (Tx) in this diagram is shown below the two receivers. In practice, it can be above or below the receivers, and in borehole-compensated tools discussed later, there are two transmitters: one above and one below the receivers.

The desired signals arrive at the receivers along the refracted path labeled r in Figure 4–2. Signals may also be reflected from vertical acoustic discontinuities such as fractures or a vertical boundary between rocks of widely differing sound velocity. These rays will follow the path labeled R in Figure 4–2. Signals following such paths will arrive later than the desired refracted rays and may obscure the arrival of slower components of the wave train such as the shear wave. Reflected waves may also propagate from horizontal

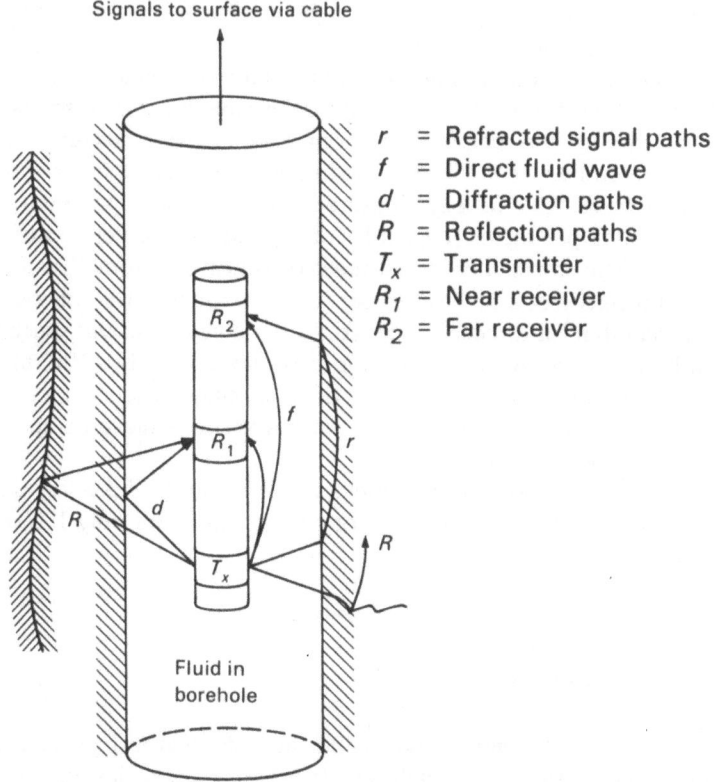

Signals to surface via cable

r = Refracted signal paths
f = Direct fluid wave
d = Diffraction paths
R = Reflection paths
T_x = Transmitter
R_1 = Near receiver
R_2 = Far receiver

Fluid in
borehole

Figure 4–2. Borehole acoustic wave paths.

acoustic discontinuities such as shown in Figure 4–2. Such a discontinuity occurring between the transmitter and receivers may result in severe attenuation of the received signals and make detection at the receivers difficult.

It is also likely that acoustic waves may be diffracted at irregular or *rugose* borewall surfaces and follow paths such as that labeled d in Figure 4–2. The usual effect of acoustic waves following all these pathways other than that of the refracted wave is to create interference in the received wave train. However, these interference events sometimes create waveform patterns that provide useful information.

There are four measurable characteristics of acoustic waves in the borehole: velocity, amplitude, attenuation, and apparent period (or reciprocal of frequency). We have already noted that velocity (or reciprocal velocity, the interval travel time) can be related to porosity. In Figure 4–3 the velocity would be measured by the difference in the time of arrival of the event at the two receivers ($t_{2c} - t_{1c}$ for the compressional wave in the figure). Subscript 1 refers to the near receiver, closest to the transmitter and subscript 2 refers to the far receiver, most distant from the transmitter. Subscript c refers to the compressional wave arrival, whereas s refers to the shear wave arrival.

In Figure 4–3 we note that amplitude is measured between the maximum positive and the maximum negative excursion of the wave. Attenuation is the change in amplitude that can be measured by the decrease in signal amplitude at the second receiver as compared to the amplitude at the receiver closest to the transmitter. The apparent period (p in Figure 4–3) is the time between two adjacent positive or negative peaks.

Another interesting feature of the acoustic waveforms at the two receivers illustrated in Figure 4–3 is the *moveout* or increased separation of corresponding acoustic events on the two waveforms. Note, for example, that the arrival of the shear wave at the far receiver (number 2) in the diagram is separated further from the shear wave arrival at the near receiver (number 1) than the compressional arrival at the far receiver is separated from the corresponding compressional wave arrival at the near receiver. This is to be expected since the compressional wave travels faster and takes less time to move between the two receivers.

Figure 4–4 illustrates the various events on the acoustic waveform. *Event* refers to the arrival of a waveform of a particular mode of propagation. The first arrival is the compressional wave (labeled c in Fig. 4–4). Next is the shear wave (labeled s in Fig. 4–4). Its velocity is limited to about 70% of the compressional velocity (or about 1.4 times the compressional wave travel time) for waves in elastic solids that obey Hooke's law. It may be as slow as the direct fluid wave travel time at the other extreme. The shear wave often has a much larger amplitude than the compressional wave. Yet, sometimes no shear wave is received.

Figure 4–5 illustrates the marked difference in the mode of propagation of compressional waves and shear waves. The compressional wave motion has particle movement in the direction of propagation, whereas the shear wave motion results in particle motion perpendicular to the direction of propagation. Other arriving events may have more complicated modes of particle motion with the exception of the direct fluid wave, which is a compressional wave.

The direct fluid wave travels slower than compressional and shear waves and arrives after them. Its amplitude is variable, but sometimes it is larger than the other waveforms. After the fluid wave, another waveform known as the Stonely wave sometimes arrives. It propagates along the borewall-fluid interface. Another waveform, the borehole analogy of the Rayleigh wave, travels along the borewall and arrives at nearly the same time as the shear wave. It is often this wave that we actually see rather than the shear wave, and its time of arrival is taken as equivalent to the shear wave arrival although it may be somewhat slower (90% or less of shear wave velocity according

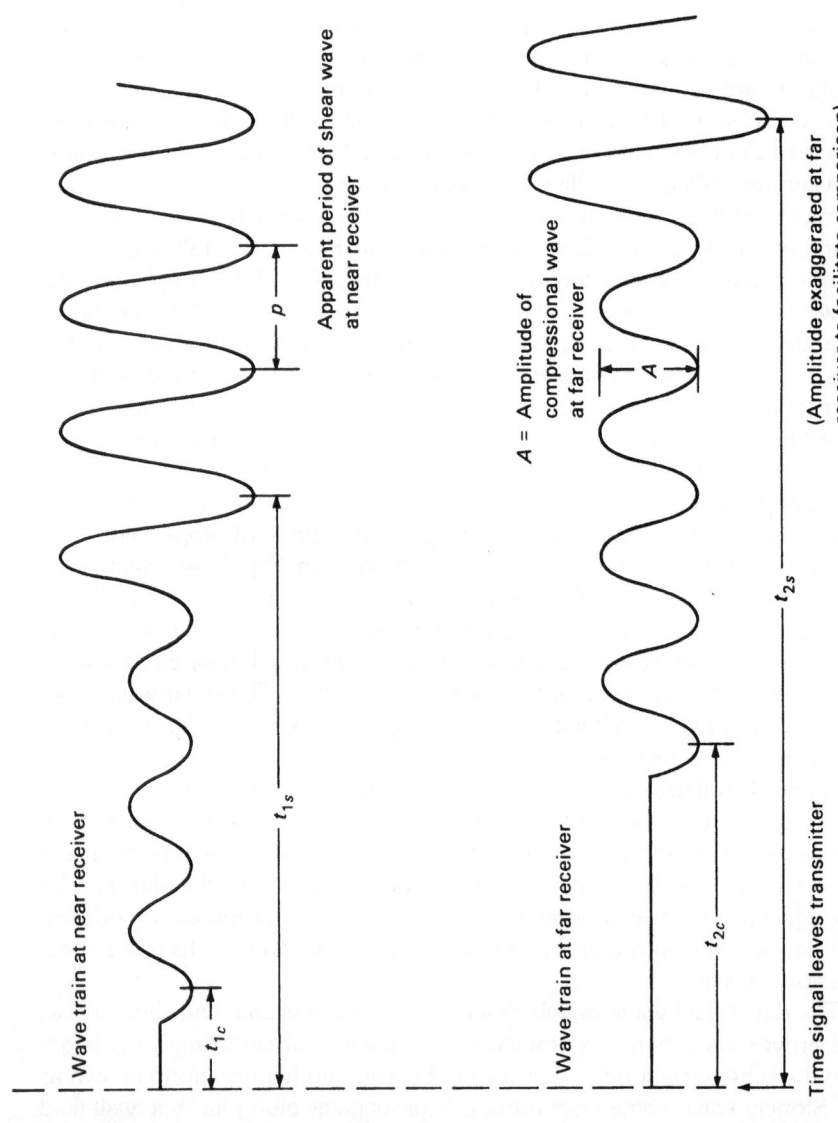

Figure 4-3. Measurable characteristics of sound waves.

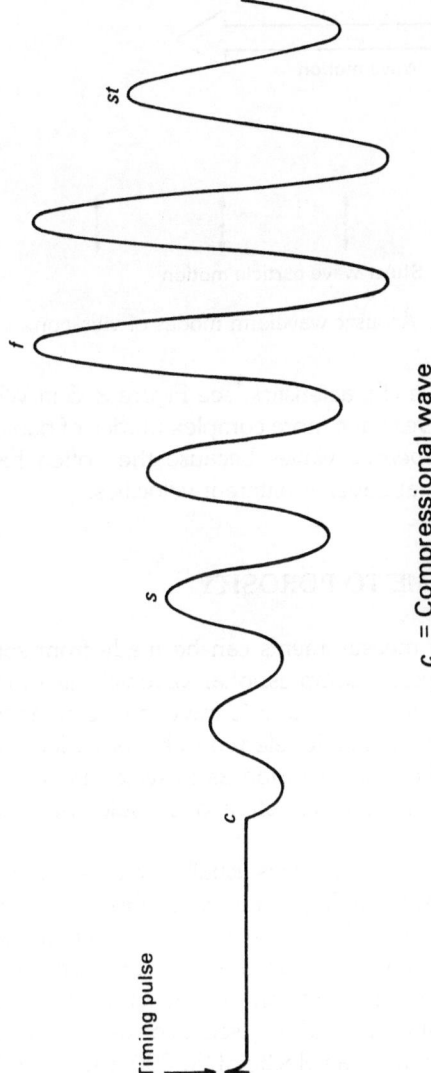

c = Compressional wave
s = Shear wave
f = Fluid wave
st = Stonely wave

Figure 4–4. Borehole acoustic wave.

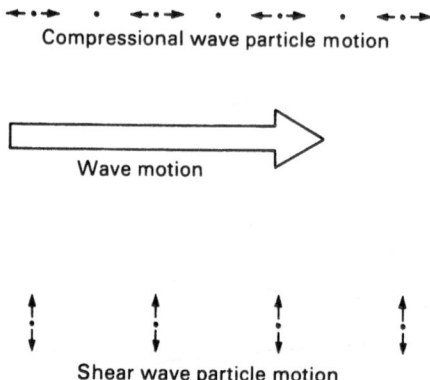

Figure 4–5. Acoustic waveform modes of vibration.

to a graph of its dispersive characteristics [see Figure 2–5 in White[2]]). Both Rayleigh and Stonely waves have more complex modes of particle vibration. They are also called *dispersive* waves because they often have different frequency components that travel at different velocities.

RELATING TRAVEL TIME TO POROSITY

Although many different measurements can be made from sonic logs, the most common is the reciprocal compressional wave velocity or interval travel time, which is the time it takes for sound to travel between the two receivers in the formation. It is also frequently referred to by its symbol: *delta T* (ΔT). When there is likely to be any confusion as to which travel time is meant (compressional wave or shear wave, etc.), one could use a subscript, for example, ΔT_c.

The compressional wave travel time is usually measured in units of microseconds per foot or microseconds per meter. In this book I use the English system of units, but if you work with well logs on international standards, the metric system will commonly be used and you should be able to work with either system and convert from one to the other. For example, 1 m is 3.28 ft. So, a reciprocal velocity of 50 μsec/ft corresponds to 164 μsec/m. This same figure corresponds to a velocity of 20,000 ft/sec or 6,097.6 m/sec.

Sonic Log Response

Although we can measure the travel time of sound in a rock, what we really would like to know is the rock's porosity. Given that we have the travel time, we could hypothesize some linear relation of the form

$$\Delta T = A + B\phi \qquad\qquad \textbf{(4-1)}$$

which we saw in Chapter 1. There is no particular magic about linear relations. They are easy to work with, and we can at least approximate the relationship between porosity and ΔT with a straight-line relationship over a practical range of porosities. Of course, if a linear relation is not adequate, we must resort to something else, such as quadratic or maybe even an exponential relationship. In fact, we examine a recently proposed nonlinear relationship in this chapter, although the linear relation of Eq. 4–1 will prove more than adequate for many practical applications. If porosity ϕ is expressed as a percent of the total rock volume and ΔT is stated in microseconds per foot, then A must be in units of microseconds per foot and B in units of microseconds per foot per percent porosity (μsec/ft/%).

At this point, I have simply postulated that some linear relation exists between the travel time for sound in a rock and the porosity of that rock. It would be nice if we could go one step further and ascribe some physical meaning to the constants A and B. However, this would actually be unnecessary if we had, for example, made some measurements on a rock core of both travel time and porosity and fitted a line (visually or by some mathematical procedure) to the data. As long as we were dealing with the same rock type, we would logically extend the relation to measurements of travel time on other similar rock samples to estimate their porosity as well. The idea is to establish some known and predictable correspondence between the well log measurement that we make (interval travel time) and the property that we want to know (porosity). Although we have not been able to ascribe physical significance to the constants A and B, they still serve as *calibration* constants.

One of the early studies on acoustic velocities in reservoir rocks showed that the velocity of sound increased significantly with an increase in effective stress or differential pressure.[3] Differential pressure is the difference between the overburden pressure of the rocks above the rock layer of interest and the internal pressure of the fluids in the rock's pore system. The pore pressure will be approximately that of a fluid column extending from the surface to the depth of interest. It is approximately 0.433 lb/in.²/ft of depth. Some formations, however, may have abnormal pressures, either above or below this value. The overburden pressure will be determined by the weight of the rocks above. Figure 4–6 illustrates effective stress or differential pressure. The overburden pressure appears at the grain contacts and attempts to press the grain together, whereas the pore pressure from the fluids in the pores resists this overburden pressure. As depth increases, the differential pressure will increase so that the interval travel time can be expected to decrease for the same rock type as depth increases.

This property of increase in velocity with depth has been used to identify abnormally high-pressured zones in reservoirs before they were drilled

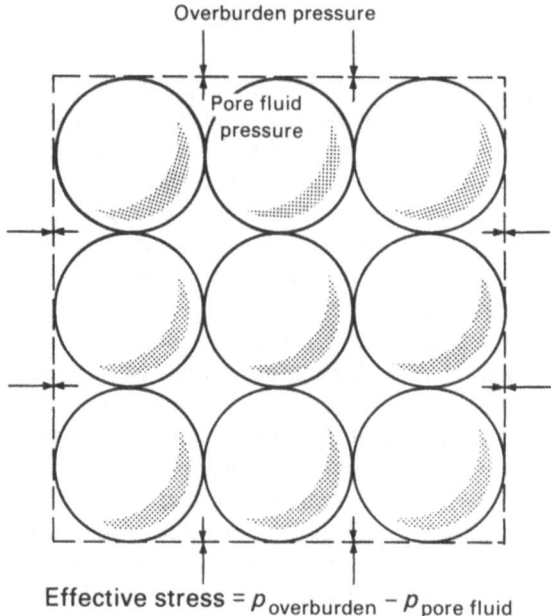

Overburden pressure

Pore fluid pressure

$$\text{Effective stress} = P_{\text{overburden}} - P_{\text{pore fluid}}$$

Figure 4–6.   Effective stress.

and caused a hazard to personnel. The travel times in shales will normally decrease with depth in the well until a shale overlying the abnormally high-pressured zone is encountered. This shale will exhibit a slower than normal (longer) travel time, as if it were at a shallower depth. Abnormal pressures are created when a sealing barrier has prevented the normal escape of water from the formations during the compaction process after burial.[4] The fluid pressures in such sands can approach that due to the weight of overburden. The associated overpressured shales are undercompacted due to the excess of water, have lower densities, and correspondingly larger acoustic travel times. This effect in the shales can occur as much as 1,000 ft above the overpressured sands.

Figure 4–7 illustrates an idealized picture of wave propagation through a rock with spherical grains. Note that the pore fluid is slightly compressible. As the grains are alternately compressed and then allowed to expand by the passing compressional wave, the grain contact points would become somewhat flattened and then alternately allowed to resume a more relaxed tangential contact. Hicks and Berry also verified different theoretical relationships that showed that the sound velocity decreased with increasing compressibility of the pore fluids.[5] For oil, the decrease is 15% to 20%. The decrease may be even more severe for gas in the pore system. Some ad hoc corrections used in the past advocated a 42% decrease in velocity for

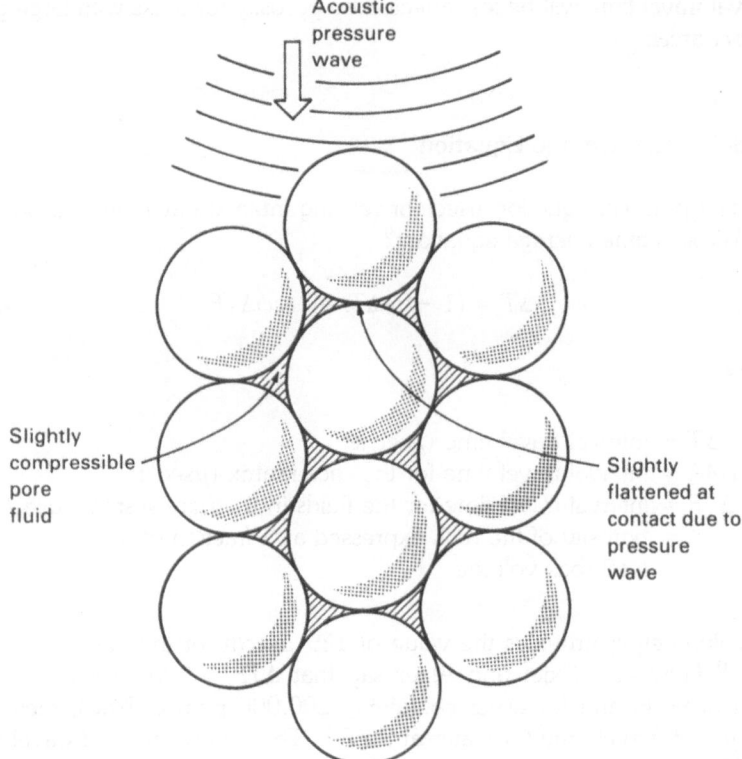

Figure 4–7. Sound wave propagation and compressibility. Wave velocity is determined by fluid, grain compressibilities, grain contact area, and effective stress.

gas-saturated rocks. However, this may vary depending on the actual rock involved. Hicks and Berry also verified the decrease in velocity with increasing porosity in the rocks.[6]

Geertsma and Smit showed that Biot's equations[7] for propagation of dilatational waves in fluid-saturated porous solids could be expressed in terms of compressibilities and porosity.[8] The compressibility of the rock matrix as well as the pore fluids affected the acoustic velocity.

Thus far we have seen that acoustic travel time in rocks will vary according to differential pressure, pore fluid and rock compressibility, and porosity. If we are to relate porosity to acoustic travel time, we will somehow have to account for these additional effects on travel time. Examining Figure 4–7, we see that for real rocks the grain structure, primarily the grain contact areas, will likely affect acoustic wave propagation. For large grain contact areas we might anticipate that the pressure of the advancing acoustic wave is not transmitted as much through the fluid. Thus, the acoustic velocity or interval travel time will be less affected by porosity for rocks with large grain contact areas.

Wyllie's Time-average Equation

The first practical equation used for relating interval travel time to porosity was Wyllie's *time-average* equation:[9]

$$\Delta T = (1 - \phi)\Delta TMA + \phi\Delta TF \qquad\qquad (4\text{--}2)$$

where

ΔT = interval travel time (μsec/ft)
ΔTMA = interval travel time for the rock matrix (μsec/ft)
ΔTF = interval travel time for the fluids in the pore system (μsec/ft)
ϕ = porosity of the rock expressed as a fraction of the total rock volume

Schlumberger provides the value of 189 μsec/ft for the fluid travel time ΔTF.[10] However, Tixier and Alger say that ΔTF = 200 μsec/ft for fresh water at 90°F, and for water containing 200,000 ppm of NaCl, they give 176 μsec/ft travel time for water at 90°F.[11] They show the fluid travel time of 189 μsec/ft as corresponding to 100,000 ppm NaCl at 90°F. The value of 189 μsec/ft probably is valid for many applications, but you might want to consider a shorter travel time for very salty fluids or a longer travel time for fresh fluids.

Geertsma and Smit took exception to the time-average formula on the grounds that it neglected the effects of pressure-dependent bulk deformation properties of the rock material.[12] They feared that the time-average model, which views rock as alternating, sandwich-style layers of porosity and rock matrix, was an oversimplification. They felt its frequent use would surround it with an aura of scientific truth that was unjustified since it did not account for the bulk deformation properties of the rock.

The simplicity of the time-average relation can be seen in Figure 4–8 where the rock and its accompanying pore volume have been separated such that we consider the travel times through the rock solid matrix and that through the pore fluids in the rock to be additive. This equation has worked well in nonshaly sandstones at depths of 5,000 ft or more. It has not worked well in carbonates or sandstones buried at shallower depths. Some ad hoc corrections have been employed to improve the usefulness of the relation. Despite its continued popularity, it entails other problems that we examine

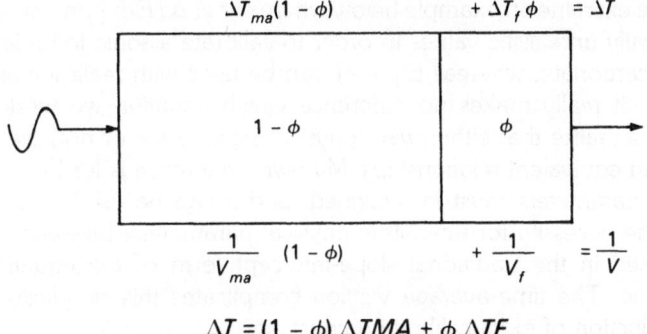

$$\Delta T_{ma}(1 - \phi) \qquad + \Delta T_f \phi = \Delta T$$

$$1 - \phi \qquad \phi$$

$$\frac{1}{V_{ma}}(1 - \phi) \qquad + \frac{1}{V_f} = \frac{1}{V}$$

$$\Delta T = (1 - \phi)\, \Delta TMA + \phi\, \Delta TF$$

Figure 4–8. Wyllie's time-average equation for porosity.

in more detail. Pirson described the equation as having been "extrapolated beyond reason."[13]

However, it is well to understand how to use this equation if you hope to discuss results, which may be based on a more realistic approach, with your peers. Equation 4–2 can be solved for porosity and a correction term can be added to account for rocks whose velocity does not conform exactly to the model.

$$\phi = (\Delta T - \Delta TMA)/c_p \cdot (\Delta TF - \Delta TMA) \qquad \textbf{(4–3)}$$

where the units are the same as for Eq. 4–2, and the compaction correction c_p is an extra, dimensionless correction term (sometimes called *lack of compaction correction*). The compaction correction can be selected from acoustic interval travel times in adjacent shales: [14]

$$c_p = 100/\Delta T_{sh} \qquad \textbf{(4–4)}$$

where ΔT_{sh} is the interval travel time in the adjacent shale expressed in microseconds per foot.

It turns out that Eq. 4–2 is easily expressed as a variation of the more general Eq. 4–1: let $A = \Delta TMA$ and $B = c_p \cdot (\Delta TF - \Delta TMA)$, and the equivalence is apparent. The only difference, in fact, is that the time-average equation requires three parameters for its solution (ΔTMA, ΔTF, and c_p), whereas the more general equation, advocated for practical use by Pickett[15] as well as other researchers, requires only two parameters: A and B. From the standpoint of finding a sonic log response relation, Eq. 4–1 is simpler. One could assert that Eq. 4–2, with its *compaction correction*, along with ΔTMA and ΔTF, has a physical significance that is not apparent in Eq. 4–1 with A and B. However, physical significance can also be attributed to A

and B. We examine an example below where either ΔTF or c_p must be assigned physically unrealistic values in order to calibrate a sonic log relation in a granular carbonate, whereas Eq. 4–1 can be used with realistic values.

In fact, it really makes no difference which equation we work with as long as we realize that either way (physical significance or not) we can find usable and equivalent relationships. My own preference is for Eq. 4–1 since only two parameters must be assigned, and it can be used in carbonates without the necessity for unrealistic physical parameters. Likewise, Eq. 4–1 is expressed in the traditional slope-intercept form of the equation for a straight line. The time-average version complicates this simplicity through the introduction of extra calibration constants.

Pickett's Relation for Sonic Log Porosity

Pickett strongly recommended using Eq. 4–1 after considering theoretical and laboratory results of several investigators (Geertsma[16] and Biot,[17] in particular).[18] The constant A is lithology dependent and equal to the matrix travel time (ΔTMA) for the given lithology. The constant B depends on lithology, effective stress (or differential pressure), and grain structure or grain contact area. B is larger for rocks with smaller grain contact areas as in finer-grained rocks, and smaller for rocks with larger grain contact areas (it is smallest for vuggy rocks).

In some interpretation methods employing the time-average equation, the sonic log is said to *ignore* vuggy type porosity. If we use Eq. 4–1 with appropriate small values for B (found in vuggy rocks):

$$\phi = (\Delta T - A)/B \qquad (4\text{–}5)$$

we find larger calculated porosities than can be obtained from the time-average formula with physically realistic parameters. We can use either values of compaction factor (c_p) less than 1 or ΔTF values less than 189 μsec/ft when using the time-average formula. Although both are physically unrealistic, they still give satisfactory results in vuggy rock types. Thus, it is indeed a matter of interpretation as to whether the sonic log really *ignores* vuggy porosity, or whether we have extrapolated the time-average formula beyond physical reality.

Pickett found from laboratory studies on rock cores representing many different geologic provinces that A (or ΔTMA) was 50–55 μsec/ft for sands, 45–50 μsec/ft for limestones, and 42–48 μsec/ft for dolomites.[19] The parameter B was 130–300 μsec/ft/fractional porosity with an average of 150 μsec/ft/fractional porosity for sands, and 60–120 μsec/ft/fractional porosity with an average of 100 μsec/ft/fractional porosity for carbonates. B normally

Table 4–1. Constants for Acoustic Log Calibration $\Delta T = A + B\phi$

Lithology	A (μsec/ft)	B (μsec/ft/fractional porosity)
Sands	50–55	130–300
Limestones	45–50	60–120
Dolomite	42-48	60–120
Anhydrite	50	
Salt	67	
Coal	110–170	
Shales	(similar to the high values for sands)	
Steel Pipe	57	

increases with shaliness. Table 4–1 lists these values for sands and carbonates along with some ΔTMA or A values for several other substances. If porosity is expressed as a percent of pore volume, the B values in Table 4–1 must be divided by 100. Note that the *pipe* value listed in Table 4–1 can be used as a calibration check behind casing.

Example of Parameter Determination
for Sonic Porosity

An interesting illustration of parameter selection for a practical problem comes from Hilchie.[20] In order to fit the time-average relation to a graph of sonic log compressional wave travel time (ΔT) versus porosity taken from core data in the Nisku formation (in the vicinity of Alberta, Canada), he finds that $\Delta TF = 166\mu$sec/ft must be used along with $\Delta TMA = 46\mu$sec/ft and a compaction correction of 1.00. The sonic log porosity relation with these parameters looks like

$$\phi = (\Delta T - 46)/(166 - 46)$$
$$= (\Delta T - 46)/120$$

It is easy to see that this corresponds to using Pickett's more general sonic relation with $B = 120$. We note from Table 4–1 that this value for B corresponds to a typical upper value for a granular carbonate rock.

There is no problem with using this relation to establish porosity, unless we fitted the data to the time-average form of the porosity relation and had difficulty accepting the fluid velocity value of 166 μsec/ft since it does not correspond to reality. What is the alternative? The only thing left in the denominator of the time-average relation would be the compaction factor. If we are to let fluid travel time equal the fastest that might be expected from a very saline water, we have to adjust the compaction factor. For very saline water, the travel time is 176 μsec/ft. If the denominator of the sonic-versus-porosity relation is to equal 120

$$c_p = 120/130 = 0.923$$

Once again, we are left with another physically impossible number to contend with. However, we obtain the same result whether we use the time-average relation with calibration constants that cannot be related to reality or the more general form of sonic porosity relation (Eq. 4–5) with the appropriate calibration constants. The only problems with the time-average relation is that it sometimes requires us to accept physically unrealizable values for calibration constants and requires three constants, instead of two, to establish the straight line relationship.

Raymer-Hunt-Gardner Transform for Sonic Porosity

Raymer, Hunt, and Gardner have proposed a new sonic-log-versus-porosity response relation in the form of a transform that is expressed as several different and distinct relations, depending on the porosity range.[21] The use of this transform is proposed to provide better porosity values over the entire porosity range of 0% to 100% without the use of special corrections such as the "lack of compaction correction factor" we saw in the time-average relation. The transform can be used for three distinct porosity ranges, each with its own porosity-versus-travel-time relation. Two of the ranges have an alternative or additional relation that can be used. These relations can be programmed into a calculator or computer, but the easiest method for using the transform is to use the figure proposed by the authors. As another alternative, the transform could be approximated by the equation given by Dewan.[22]

$$\phi = .63(1 - \Delta TMA/\Delta T) \qquad \text{(4–6)}$$

The transform may have application where parameters A or B are unknown (note that rock type, e.g., sand, lime, or dolomite would still need to be known) and a ballpark estimate of porosity is needed. It is based on an empirical fit to data over a wide range of porosity, geography, and geology. The transform should not be expected to perform better than Eq. 4–1 with appropriate calibration constants.

Raymer, Hunt, and Gardner have not proposed the transform in terms of the theory of acoustic wave propagation in a porous medium and accordingly do not attempt to explain or justify it on that basis.[23] It seems to provide reasonable results but may not see wide usage because of its rather complex implementation. Simplifications, such as that of Dewan above, may make it more accessible for ballpark estimates of porosity, but the simplification departs significantly from what appear to be more accurate values. From the example of the previous section, the Nisku formation exhibits an average travel time of 58 μsec/ft for 10% porosity. Dewan's approximation of the transform, on the other hand, gives 13% as the porosity for 58 μsec/ft travel time. Note that the range of core porosity measurements for a 58 μsec/ft travel time in Hilchie's example was from a low of about 7% to a high of about 13%. Thus, the approximation to the transform falls at the extreme upper limit of this range of values. Using Figure 9 from Raymer et al., we would obtain 9% porosity, which is closer to reality.[24]

SONIC LOG PRESENTATION

Much of the data examined in this section is common to most types of well logs and is not repeated in later chapters. Therefore, students who are not familiar with well log presentation should pay particular attention to this section. Those who are fairly conversant with well logs will probably find this material on log presentation repetitious.

A typical sonic log is shown in Figure 4–9. In track 1 (the usual term for the left side of the log) a gamma ray log is recorded on a scale from 0 to 100 American Petroleum Institute standard units (API units) for gamma ray measurements. Note at the top of the log that a scale change was made at 8,300 ft. Above that depth the scale was from 0 to 200 API units. Such scale changes are often made when significant rock type changes are encountered during logging.

A caliper curve is recorded as a dashed line. Its scales are not shown on the figure but can be ascertained from the log scale headings that are illustrated later. The sonic log caliper is a three-arm device that registers hole size. It can be used as an aid to evaluating log quality. Note that a sonic log is probably the most reliable porosity device to use in an enlarged hole where radioactivity type porosity devices develop most of their response from the washed out borehole rather than the formation.

A sonic log scale change is noted at 8,350 ft on the sonic log portion on the right side of the log. That side of the log has two tracks: track 2 and track 3. Above 8,350 ft, the sonic log scale in track 2 is 140 μsec/ft on the left side and 90 μsec/ft on the right. The scale is continued on track 3 with 90 μsec/ft on the left to 40 μsec/ft on the right. Sometimes data is recorded separately in tracks 2 and 3, or they are combined as in this example. The scale changed from 140 μsec/ft on the left and 40 μsec/ft on the right to 80 μsec/ft on the left and 40 μsec/ft on the right. Also, there is a corresponding

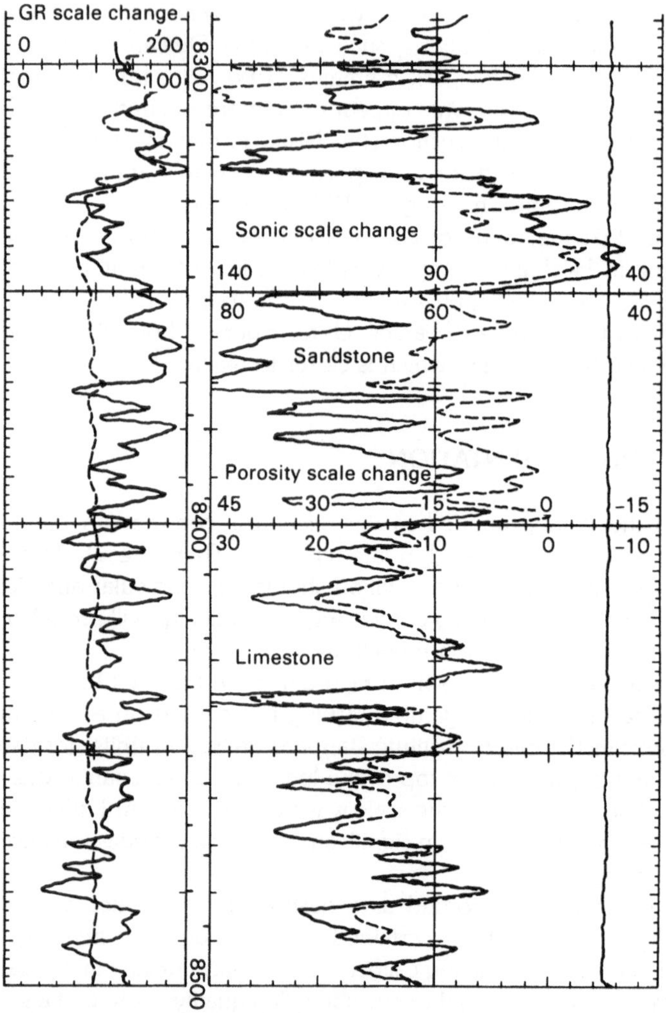

Figure 4–9. Typical sonic log.

change in the porosity scale shown at 8,400 ft. The latter change is from 45% on the left and −15% porosity on the right to 30% porosity on the left and −10% porosity on the right. The abbreviated sonic log scale is used to enhance the response opposite the carbonates (below 8,350 ft), where travel times are normally much faster (ΔT is smaller).

The logging engineer has noted on the log that the scale changes correspond to a lithologic change from a limestone below to a sandstone above. Note that the interval travel time is denoted by a solid line on the log and that the corresponding porosity values are denoted by a dashed line. These are both recorded in tracks 2 and 3, combined. On this presentation the solid curve is the actual recorded data: the interval travel time. The dashed curve in this case is a *derived* or calculated curve.

A tension curve also is recorded on this log in track 3. It is between the second and the third of twenty divisions, measured from the right side of track 3. It remains almost unchanged on this portion of the log. The scale values for this curve are not shown. Sometimes a tool sticks in the hole, and when it is pulled free, it suddenly jumps several feet up the hole. Any log values recorded in this jumped interval where the tool moves rapidly up-hole will generally be invalid. These intervals are identified by increased cable tension.

This portion of the log has been recorded on a vertical scale of 5 in. per 100 ft, or 1:240, which is the standard American logging scale. Other vertical scales may be used, for example, 1:200.

Note that the depth is recorded in the empty space separating tracks 1 and 2. No space separates the juxtaposed tracks 2 and 3, however. There are also small, horizontal tick marks, both to the right of track 1 and to the left of track 2. Those to the left of track 2 are integrated travel time, and those to the right of track 1 are integrated hole volume. The integrated travel time is used as an aid in seismic section calibration, and the integrated hole size is used by the drilling engineer to calculate the volumes of cement needed downhole to cement the casing to the borewall, for example.

Figure 4–10 is the main log heading. One of these is prepared for each principal log type. Gamma ray logs, tension curves, and caliper curves may be recorded simultaneously and presented on the same log recording with the sonic log. The heading provides well information at the top: operating company, well name and location, and field (if the well is drilled in a known field). The block labeled *Other Services* tells which other well logs were run in this well. They are usually listed using unique service company abbreviations. Those listed for this well are DLL for dual laterolog, μsfl for microspherically focused log, GR for gamma ray, FDC for compensated formation density log, CNL for compensated neutron log. The *Cyberlook*, a Schlumberger trademark, is a computed log based on all the other recorded logs and can be used for evaluation.

Other data on the log heading include the elevations measured from the Kelly bushing (K.B.), drilling floor (D.F.), or ground level (G.L.) as well as other pertinent data. Most of the data listed in the first column is self-explanatory, but you should note several items.

Wells are often logged before drilling is complete to the final total depth. These runs are denoted as *run one, run two*, and so forth, culminating with the final run number for the log run down to total depth. In this well, only one run was made, thus it is run one. The date of the run is also recorded.

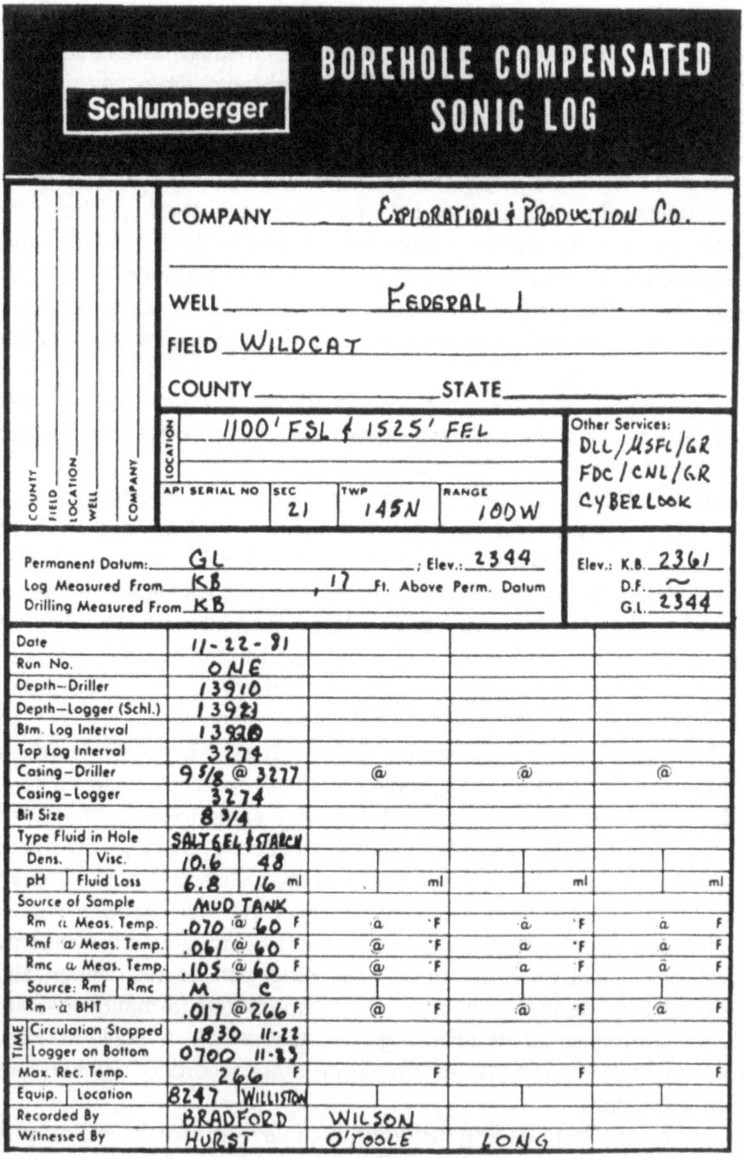
Schlumberger

BOREHOLE COMPENSATED SONIC LOG

COUNTY / FIELD / LOCATION / WELL / COMPANY

COMPANY EXPLORATION & PRODUCTION CO.

WELL FEDERAL 1

FIELD WILDCAT

COUNTY ______ STATE ______

LOCATION 1100' FSL & 1525' FEL

API SERIAL NO	SEC	TWP	RANGE
	21	145N	100W

Other Services: DLL/MSFL/GR, FDC/CNL/GR, CYBERLOOK

Permanent Datum: GL ; Elev.: 2344

Log Measured From KB , 17 Ft. Above Perm. Datum

Drilling Measured From KB

Elev.: K.B. 2361, D.F. ~, G.L. 2344

Date	11-22-81			
Run No.	ONE			
Depth—Driller	13910			
Depth—Logger (Schl.)	13923			
Btm. Log Interval	13920			
Top Log Interval	3274			
Casing—Driller	9 5/8 @ 3277	@	@	@
Casing—Logger	3274			
Bit Size	8 3/4			
Type Fluid in Hole	SALT GEL & STARCH			
Dens. / Visc.	10.6 / 48			
pH / Fluid Loss	6.8 / 16 ml	ml	ml	ml
Source of Sample	MUD TANK			
Rm @ Meas. Temp.	.070 @ 60 °F	@ °F	@ °F	@ °F
Rmf @ Meas. Temp.	.061 @ 60 °F	@ °F	@ °F	@ °F
Rmc @ Meas. Temp.	.105 @ 60 °F	@ °F	@ °F	@ °F
Source: Rmf / Rmc	M / C			
Rm @ BHT	.017 @ 266 °F	@ °F	@ °F	@ °F
TIME Circulation Stopped	1830 11-22			
TIME Logger on Bottom	0700 11-23			
Max. Rec. Temp.	266 °F	°F	°F	°F
Equip. / Location	8247 / WILLISTON			
Recorded By	BRADFORD	WILSON		
Witnessed By	HURST	O'TOOLE	LONG	

Figure 4–10. Typical well log heading.

The depth information, both driller's and logger's, is recorded next. These may not always agree to the exact foot, and significant differences, say, more than a few feet, should be investigated. In this example there was a 13-ft difference. It might have been caused by an error in estimating the logging cable stretch at depth.

The bottom (Btm.) log interval refers to the deepest depth that any valid logging measurement is made. This is determined by which sensor or sonde is placed at the lower end of the logging tool. The top log interval is the shallowest depth for which any log was recorded on this particular run. Another run may be made in the hole with a different tool that may have several sensors on it. Several different measurements may be made with this sonde, and conceivably, all the depths could be different with the exception of the driller's depth. Normally, the logger's depth will be the same as for other tools run in the hole as part of the same logging suite.

Casing and bit sizes are given (usually in inches in the United States). The casing depth, which is the deepest point in the well where there is casing, is given. Most logging tools cannot record valid measurements above this depth. We have seen that a sonic log should record the ΔT value for the casing once inside the casing: 57 μsec/ft.

Much data are recorded about the drilling fluids or mud system: type of fluid, its density, viscosity, pH, and fluid loss. *pH* refers to the relative acidity or alkalinity of the drilling fluid. A pH of 7.0 is neutral. The fluid loss of the mud system will affect the buildup of mudcake on the borewall.

The source of the mud or drilling fluid sample is very important. It should come from the fluids circulating from the well bore. Note that in Figure 4–10 the sample comes from the mud tank. The problem with mud system measurements taken from anything but the circulated fluids (such as the mud pit) is that they may not reflect the conditions downhole nearest the time the logging tools are recording data. Some interpretation techniques require accurate knowledge of the mud system properties. Therefore, the mud system properties should be checked at the time the logging company is making the mud system measurements on the logging truck.

Next the three mud system measurements and the temperatures of measurement are recorded. The temperature of measurement is very important to fluid resistivity measurements and salinity determinations as pointed out in previous chapters. The resistivity of the mud (R_m), mudcake (R_{mc}), and mud filtrate (R_{mf}) are all recorded.

The mudcake is formed on the borewall opposite porous and permeable rocks where the water from the mud system (mud filtrate) can migrate into the pore system of the rock. As the fluid migrates into the formation, a filter cake is left deposited on the borewall.

The source of these mud measurements is often given as measured (M) or calculated (C). Only measured values are valid, but even they may not be totally reliable. These measurements are made at the logging truck after placing the drilling fluid sample in a suitable pressure chamber with a filter paper to simulate the mudcake and mud filtrate formation downhole. The measurements may be extrapolated to temperatures other than that of measurement, as we have already learned to do. Remember that conversion using temperature charts can be faulty for some types of ions.

The mud system measurements can be, and sometimes are, inferred from suitable mud system charts. However, these charts are based on *average properties* and may be misleading. Sometimes an engineer in the field may elect to alter the mud system measurements to better agree with values from charts for mud systems, but this practice should be discouraged.

The time that circulation of fluids was stopped in the well is recorded as well as the time that the logger reached bottom depth with the logging tool. Dates are recorded with the times.

Frequently, thermometers designed to record the maximum observed temperature are attached to the logging tool. If so, this data is recorded. It is important to record the bottom hole temperature (BHT) for conversion of mud system resistivities to formation temperature. Mud system measurements are normally used in conjunction with resistivity measurements, so the maximum recorded temperature should be taken from the resistivity log. In some applications, the temperature measurement recorded on each logging tool run in the hole might be used to plot temperature as a function of time to gain information about the rock's thermal properties.

Finally, the names of the logging engineers are listed opposite *Recorded by* and the operating company representatives' names are recorded opposite *Witnessed by*. This log heading is almost universal, although each service company may introduce some format variations. Nevertheless, most of the information is common to most log headings.

Figure 4–11 is a continuation of the well log heading. This part usually appears on the second folded portion of the continuous well log paper. At the top left is the run number, service order number, and a statement about the fluid level in the well. This is followed by equipment data such as tool serial numbers.

Then follows significant information about centralizers and standoffs. Sonic logs use centralizing devices to center them in the well bore. Without such devices, the recorded travel times would be unreliable and probably not repeatable.

The logging speed in feet per minute (FPM) is listed. However, a better check on the logging speed can be made from the 1 min timing marks on the recorded well log, rather than using this heading information. The timing marks are usually made by momentarily interrupting the recording of the left margin of track 1 on the log each minute, leaving a small, readily observable gap. The log feet separating adjacent marks then gives the logging speed in feet per minute. These marks can be seen on a later figure.

The speed of 60 ft/min is about the maximum recording speed for a sonic log. This is actually a little fast for recording a quality gamma ray log. We see from the logging engineers remarks at the lower right part of

			SCALE CHANGES			
Run No.	ONE		Type Log	Depth	Scale Up Hole	Scale Down Hole
Service Order No.	250962		GR	8300	0 ~ 200	0 ~ 100
Fluid Level	FULL		Øs	8400	45 ~ -15	30 ~ -10
			Δt	8350	140 ~ 40	80 ~ 40

EQUIPMENT DATA

Sonic Panel No.	SLM	DA	1311
Sonic Cart No.	SLC	FA	1256
Sonic Sonde No.	SLS	RA	2156
Mem. Panel No.	CPU	CBS	1909
G.R. Cart. No.	SGC	dc	3204
G.R. Panel No.	NLM	BC	1128
Caliper No.	MLD	DA	1378
TTR No.	MTU	D	4912

LOGGING DATA

		Porosity Selectors		Depth	
Δm	Δt	Cp	Ø Scale	From	To
55.6	189	100	.30 ~ -.10	T D	13690
47.6	189	100	.30 ~ -.10	13690	8400
55.6	189	100	.45 ~ -.15	8400	L R

Centralizers: No.	ONE
Type	CMEZ
Standoffs: No.	—
Type	
time constant D.C. FT.	1
Speed - F.P.M.	60

Type: ONE / CALIPER

REMARKS:
TENSION MEASURE POINT IS: + 4"

GR CURVE IS MERGED FROM DLL/USR LOG.

CALIPER MAX AT 15.25"

Velocity (ft./sec.) = 1,000,000 / Interval Transit Time (microseconds per foot)

CALIBRATION DATA

GR	BKG. CPS	33	
	Source CPS	183	
	DC Req. FT	1	

All interpretations are opinions based on inferences from electrical or other measurements and we cannot and do not guarantee the accuracy or correctness of any interpretations, and we shall not, except in the case of gross or willful negligence on our part be liable or responsible for any loss, cost, damages or expense incurred or sustained by anyone resulting from any interpretation made by any of our officers, agents or employees. These interpretations are also subject to our General Terms and Conditions as set out in our current Price Schedule.

CHANGED PARAMETERS

NAME	VALUE	UNIT		NAME	VALUE	UNIT
DTM	55.60	US/F				

6.000 _ CAL(SIN _) 16.00   TENS(LB) 0.0

DTM	55.60		10000.
GR (GAPI)		DT (US/F)	
0.0	200.0	140.0	40.00

Figure 4-11. Typical well log heading (continued).

the log heading that the gamma ray curve has been merged from the dual laterolog/microspherically focused log run. The speed of that log will then determine the quality of the gamma ray log. Other pertinent remarks may be recorded, including information about equipment failure or limitations. For example, on the log heading in Figure 4–11, the engineer has noted that the maximum caliper reading is 15.25 in. This might be important if the hole is washed out larger than a diameter of 15.25 in. In that case, calculations of cement volumes would be in error since the tool caliper did not *see* the full hole volume. As a matter of interest, the caliper run with a density log or dip meter would be better. These tools are run slower in a hole than a sonic log and provide better *resolution* of rapid hole size changes, whereas the sonic log caliper may smooth out some of the rapidly occurring changes. The *Scale Changes* section of the heading information, also shown in Figure 4–11, can be very important to successful interpretation. In this section for this example log heading, you see the same changes already discussed relative to Figure 4–9. This heading is followed by *Logging Data*, which is also important to successful interpretation. For example, I mentioned that the solid line in tracks 2 and 3 of Figure 4–9 denotes the *raw* data recorded: interval travel time. The dashed line was the sonic-derived porosity with scales as indicated. The question answered in the logging data section is "what parameters were used to relate porosity to the travel time measurement?" Here we see the traditional ΔTMA (listed as Δt_m), ΔTF (listed as Δt_f), and C_p, compaction correction or *lack of compaction correction*, as it is sometimes called. Also note the abbreviation *LR* used to refer to the depth for the last reading taken uphole.

The parameters used to relate porosity to measured travel time are important. They should be checked before they are accepted as valid. Do the parameters agree with what is known about the rock types? As we have already seen, both sands and carbonates can have varying matrix travel times. Different pore structures and depth of burial will also play their role in influencing the observed travel time. They must be properly accounted for if the porosity relationship is to be usable.

On the lower part of Figure 4–11 we see the top of the recorded log, which is the scale information for the data. Note that the gamma ray (GR) units are listed as *GAPI* for gamma–American Petroleum Institute standard units. These units, API for short, are related to the gamma radiation recorded from API calibration test pits.

Figure 4–12 shows the information listed at the end of the recorded log. The scales are repeated. Also note that the logging engineer has added the information that the sonic log has been calibrated in *sandstone* units. This refers to the selection of sonic log parameters listed on the heading. These are commonly used parameters for converting interval travel time to porosity using the time-average equation. However, recall that both ΔTMA and the compaction correction may be different from one sandstone to another.

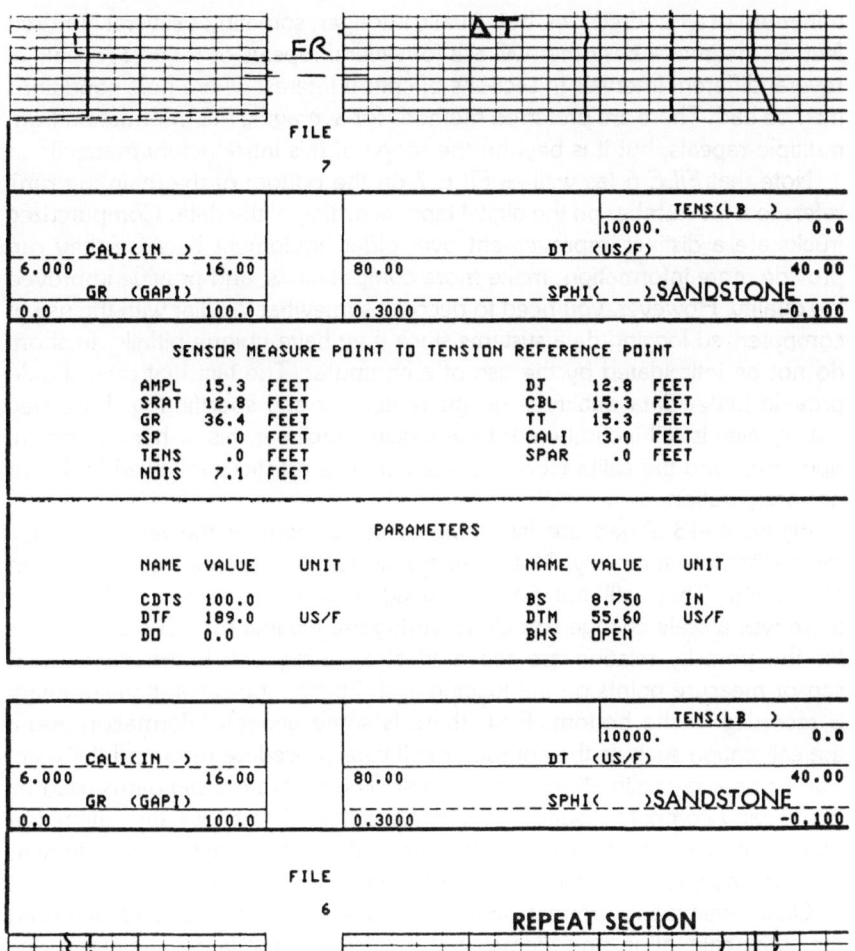

Figure 4–12.   End of well log information.

Information about the location of the various sensors on the tool relative to the tension reference point is recorded next. Below is some parameter information that is a repeat of the sonic log porosity–travel time parameters and the bit size: 8.75 in. Note that the compaction correction is listed as a *CDTS* of 100.0. This is the ΔT value in shale used to determine the lack of compaction correction (divide 100 by ΔTSHALE).

The bottom part of Figure 4–12 is the top of the repeat section of the log. Again, it must start with the scale information at the top. The repeat section is normally logged first, at the bottom of the hole. It may be relatively short, a minimum of a hundred feet, but usually is longer, sometimes several hundred feet. Sometimes a customer will ask for several repeats or request repeats of several different intervals in sections of critical interest where hole conditions may be bad. There are preferred methods for averaging this information from multiple repeats, but it is beyond the scope of this introductory material.

Note that *FILE* 6 (as well as *FILE* 7 on the bottom of the main log run) refers to a file number on the digital tape recording of the data. Computerized trucks are a distinct improvement over older equipment because they can provide more information, make more computations, and provide improved log quality. However, you need to become somewhat familiar with the use of computerized logging data systems since they have unique pitfalls. In short, do not be intimidated by the use of a computer. The fact that they should provide better data quality is no guarantee for any specific log. Improved data quality is still incumbent on the logging engineer, his or her equipment operators, and the calibration and maintenance practices followed by his or her organization.

Figure 4–13 shows the last thing at the bottom of the recorded logs: the calibration summary. Note that the scales are repeated at the bottom of the log. They will not necessarily agree with the scales at the top if there was a scale change sometime during the logging run. The parameters for the porosity relation are repeated also, along with information about sensor measure points on the logging tool. Finally, the calibration summary is recorded at the bottom. First, there is some general information about the calibration such as the computer software procedure used and the date. The gamma ray calibration is shown first. The calibration standards used by the logging company usually have to be obtained to check the calibration *tails* on the log. Frequently, a before and after survey calibration is shown, and the changes in the calibration points can be compared.

On the calibration summary in Figure 4–13, note that the caliper measured the small calibration ring diameter at 8.1 in. and the large calibration ring at 13.0 in. The rings are of known size and can be placed around the caliper end of the tool where the caliper can be extended to get an actual measurement of the known ring sizes. The calibration rings are actually 8.0 in. and 12.0 in. in diameter. This calibration summary implies that the caliper measurements are off by as much as 1 in. at the higher end of the measurements. If this caliper were used without correction to make volume measurements, this error could be significant since volume will be proportional to the *square* of the measured hole diameter. In this case, it would be possible to easily correct the measurement error with the computerized measurement system as long as a *linear* error relationship with the measured caliper is valid. Under the calibrated listing in Figure 4–13, we see that the measured calibration points have been adjusted to the actual calibration ring sizes.

Whether or not the error is linear is open to question. There are only two measurement points, and despite the use of sophisticated electronic

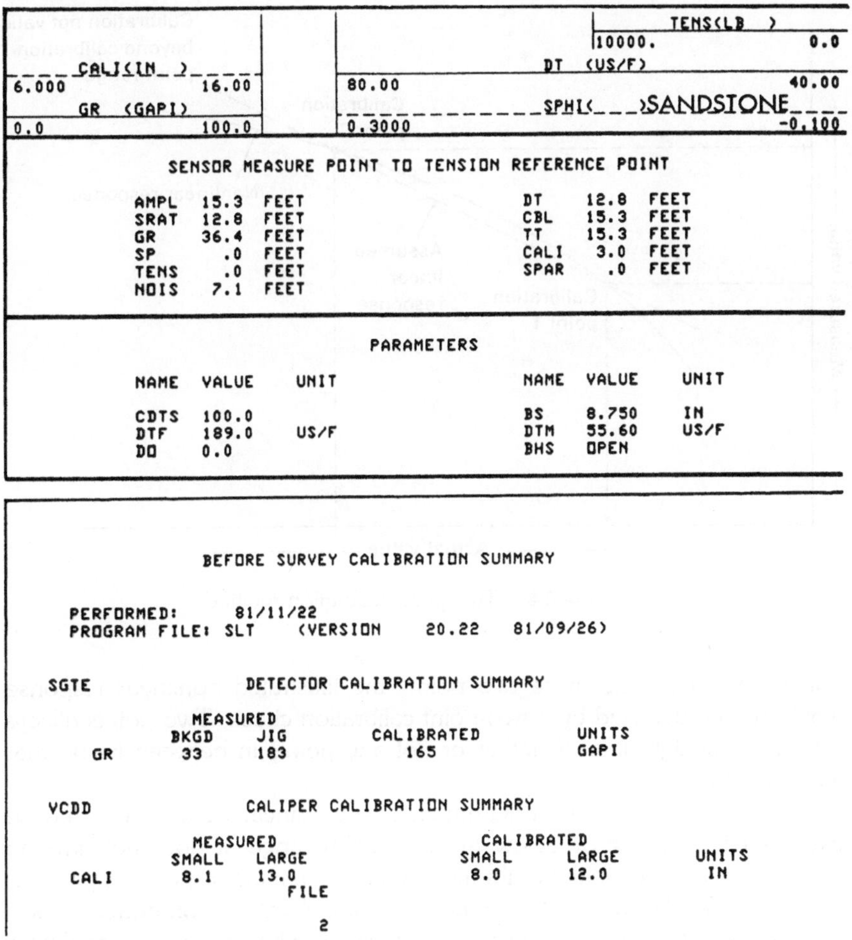

```
                                              |         TENS(LB  )
                                              |10000.              0.0
___CALJ(IN_)___                          DT  (US/F)
6.000             16.00       80.00                           40.00
   GR  (GAPI)                 _____SPHI(___)SANDSTONE____
0.0              100.0        0.3000                          -0.100
```

SENSOR MEASURE POINT TO TENSION REFERENCE POINT

```
        AMPL   15.3  FEET              DT    12.8  FEET
        SRAT   12.8  FEET              CBL   15.3  FEET
        GR     36.4  FEET              TT    15.3  FEET
        SP      .0   FEET              CALI   3.0  FEET
        TENS    .0   FEET              SPAR   .0   FEET
        NOIS   7.1   FEET
```

PARAMETERS

NAME	VALUE	UNIT		NAME	VALUE	UNIT
CDTS	100.0			BS	8.750	IN
DTF	189.0	US/F		DTM	55.60	US/F
DO	0.0			BHS	OPEN	

BEFORE SURVEY CALIBRATION SUMMARY

PERFORMED: 81/11/22
PROGRAM FILE: SLT   (VERSION   20.22   81/09/26)

SGTE   DETECTOR CALIBRATION SUMMARY

```
                MEASURED
                BKGD    JIG       CALIBRATED         UNITS
        GR       33     183          165            GAPI
```

VCDD   CALIPER CALIBRATION SUMMARY

```
                MEASURED              CALIBRATED
                SMALL   LARGE        SMALL   LARGE    UNITS
        CALI     8.1    13.0          8.0    12.0     IN
                        FILE
                         2
```

Figure 4–13.   Calibration summary on well log.

equipment, more measurement points are needed, say, both *between* the extremes of 8 in. and 12 in. as well as above 12 in. Two additional measurements could greatly increase confidence in the calibration. In this particular case, the two points are not too far apart, but that would still leave open to question any measurements *outside* the range of calibration: below 8 in. or above 12 in.

Figure 4–14 illustrates the typical *two-point* calibration method commonly used. This two-point calibration method may not always be adequate. Note that the horizontal axis in Figure 4–14 shows the actual values we would like to measure, whereas the vertical axis shows the measurements that we make. The response of the system is assumed linear between the two calibration

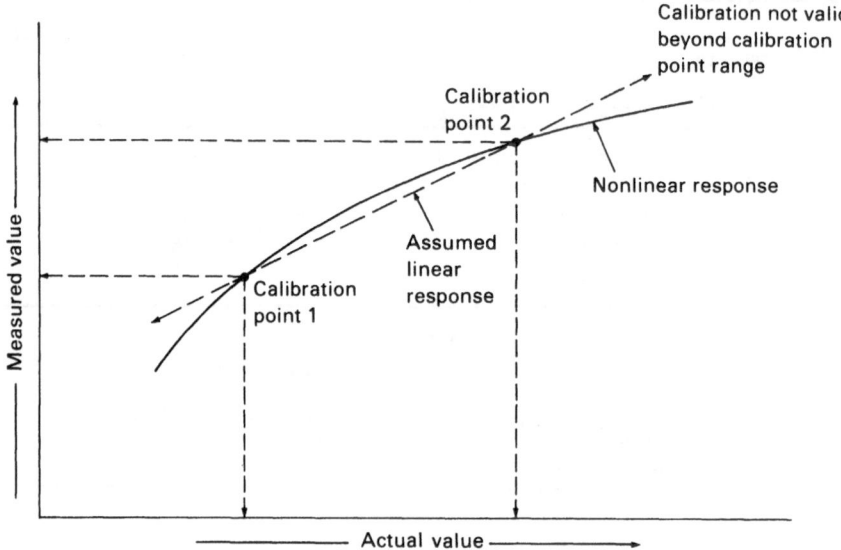

Figure 4–14.   Two-point calibration method.

points, but with only these two points the illustrated nonlinear response would not be detected by a two-point calibration check. Two points *always* establish a straight line. Whether or not any points in between lie on that line is another question.

Another problem is the assumption that the linear calibration response extends either above or below the calibration range. It does not have to extend above or below the calibration range, even if the response *is* perfectly linear between the two calibration points. Many electronic components operate over essentially linear ranges, but they can also be operated on nonlinear portions of their operating characteristics due to either catastrophic malfunctions, design deficiencies, component aging, or improper operation. Some improvement may be introduced to this rather antiquated two-point calibration procedure. At least one service company, Welex, now uses a three-point calibration with their neutron device.

Until improved calibration procedures are adopted, the weaknesses of the two-point calibration method have to be accepted, and allowances have to be made for the possibility of some miscalibration, which is an important reason for using interpretive techniques that minimize the effects of errors. Two such methods are noted in the discussion of answers to the problems in Chapter 3: using ratios to eliminate some unknowns and calculating a possible range in results corresponding to the known uncertainties in data or parameters. Other methods such as statistical techniques and cross-plotting methods can identify possible errors and can compensate for some of them.

PRACTICAL CONSIDERATIONS
AND SONIC LOG ACCURACY

Each sonic logging tool is designed for operation at an optimum speed that is as fast as is commensurate with obtaining reliable data. This is usually around 60 ft/min maximum or 3,600 ft/hr. Recall that logging speed is easily checked by noting the distance between the 1 min timing interruptions on the left side of track 1 of the log. Sonic logs may be run at somewhat slower speeds, for example, when run in combination with another tool. Note that a gamma ray recorded on a sonic log run will not have the resolution of a gamma ray log recorded with one of the radioactive tools. These latter tools are pulled up the borehole at considerably slower speeds.

The best calibration for a sonic log is to observe its response in the surface pipe. This speed should be 57 μsec/ft. Another good check is to observe the response in formations having known matrix travel times. For example, anhydrite has a matrix travel time of 50 μsec/ft and halite has a travel time of 67 μsec/ft. Another check is to compare sonic logs from comparable shales in different wells. Some of these shales will show consistent travel times over a given area. However, the shales may not show exact agreement due to gradational or small mineralogical changes.

Modern sonic logs are borehole compensated and are not adversely affected by enlargements, such as washouts. This means that in many so-called bad-hole conditions some estimate of porosity may be derived from the sonic log that is not possible with any other log.

The resolution of the sonic log is governed by the spacing between the two receivers. For this reason, sonic logs have excellent thin-bed responses since the common receiver spacing in today's tools is 2 ft. Another advantage of sonic logs as porosity devices is that the tool response is little affected by material in the surrounding rock above or below the receiver spacing. Only a severe loss of amplitude between the transmitter and the receivers could lead to uncertainty in the travel time estimation. *Other porosity devices are all affected by material outside the detector spacings.*

We covered earlier the necessity of assigning proper values to the relational constants for sonic porosity devices. In some areas with wildcat wells, this can present some difficulty. Advanced cross-plotting methods can help in some cases. A possible difficulty might be that the relational constants will be sensitive to lithology and grain contact area variations. This disadvantage is somewhat offset by the fact that linear relations usually exist between porosity and ΔT over a range of porosities of commercial interest.

Cycle skipping refers to the inability of the threshold detection device at the receiver to properly sense the correct arrival of the acoustic signal. For example, in a formation where there is a substantial loss of acoustic signal, it is possible that the first compressional wave arrival will be detected correctly at the receiver nearest the transmitting transducer, but not correctly detected on a lower amplitude signal at the far receiver. The first arrival may be picked up one or more cycles later on the far receiver than the corresponding first arrival at the near receiver. Likewise, a low amplitude on *both* receivers may give a similar invalid result. Noise spikes can also result in false triggering of the threshold detection devices. This effect can usually be identified on a sonic log by the rapidly alternating response from a high ΔT to a low ΔT. Figure 4–15 shows an example of an acoustic log where cycle skips probably produced the responses between 5,576 ft and 5,591 ft.

According to Willis and Toksoz, the sonic log response may be off by as much as 10 μsec/ft or approximately a 21% error in a velocity of 18,000

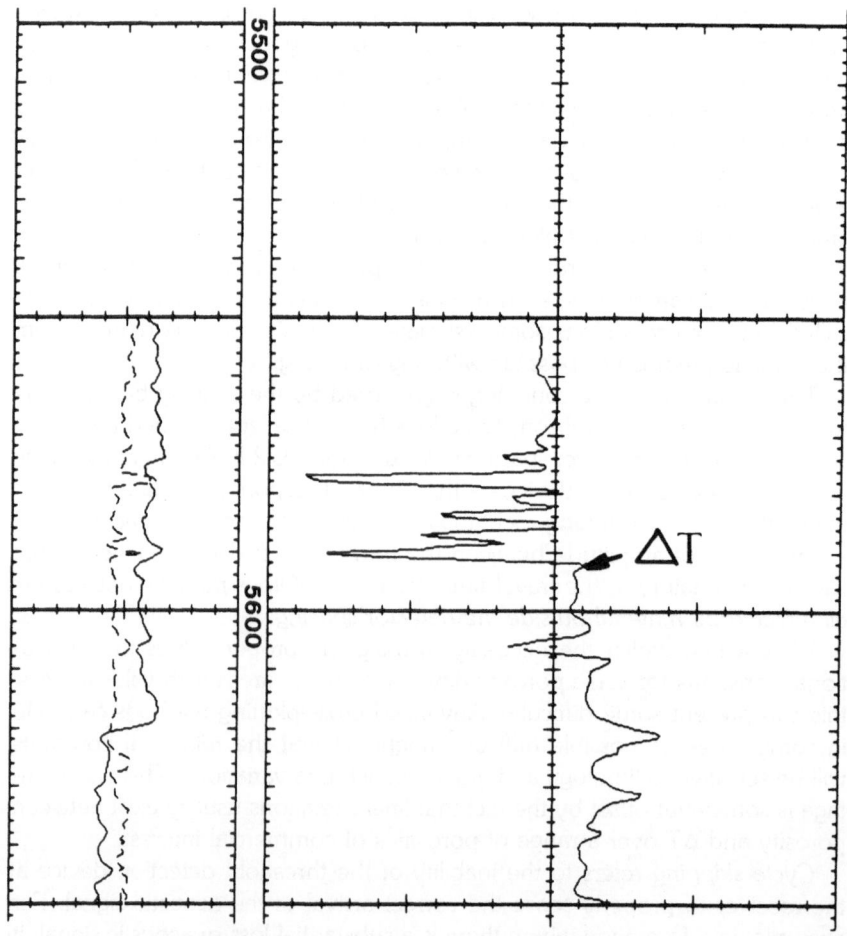

Figure 4–15. Cycle skipping on acoustic log.

ft/sec.[25] This would be an uncertainty of about one-fifth of a cycle (or almost half of either the positive peak or negative peak of the cycle) for a 20 kHz signal. However, the sonic log is usually considered to have an accuracy of 1 or 2 μsec/ft in terms of its ability to predict porosity. Larger errors approaching 10 μsec/ft probably occur only where signal amplitudes are greatly reduced.

Sonic logs have excellent thin-bed resolution. That is, they are able to provide accurate data for very thin beds. The spacing between the two receiving transducers determines the thin-bed resolution of the sonic log. The primary problem is, that for beds thinner than the tool spacing, *the response may not even follow the trend* of the acoustic wave velocities. In fact, the sonic log response will show a *reversed* response in thin beds. That is, when ΔT is increasing, the thin-bed response will show a decreasing ΔT! Figure 4–16 illustrates the response of both a 1 ft spacing tool and a 3 ft spacing tool where idealized 1 ft alternating thin beds with ΔT values of 60 μsec/ft and 50 μsec/ft are shown.

The 1 ft spacing log follows a trend of alternately increasing and then decreasing ΔT values, although it only just attains the true value for the 1 ft thin beds at the *peaks and valleys* of its response. The 3 ft spacing device, on the other hand, shows a characteristic reversal in its response, moving to lower ΔT values opposite higher ΔT beds and moving to higher responses opposite lower ΔT beds. Nowhere within the alternating thin beds does it reflect a correct ΔT value! Consider what this response would mean in terms of a resistivity tool that had a response such as that of the 3 ft sonic, and its resistivity readings with the 1 ft spacing sonic log were being used as a porosity tool to calculate water saturations. There would be no correspondence at all between the porosity readings and the resistivity readings. Of course, most resistivity tools do not exhibit this response reversal. A notable exception is the so-called normal resistivity device.

Figure 4–17 illustrates more aspects of sonic log response characteristics, using a 2 ft spacing log. Note that the thin-bed response for this more realistic bed sequence is not as severe as for the alternating thin beds of Figure 4–16. However, the tool still fails to give correct readings except for beds at least 2 ft thick. Also note the 2 ft *transition* at abrupt changes from low to high ΔT values. Tool readings during this 2 ft transition at these bed boundaries cannot be routinely used for quantitative data. This effect is common to most logging tool responses. Tool readings during these transitions are not reliable for calculation of rock properties. The transition interval thickness is governed by the tool response characteristics, particularly its thin bed response. Note that the tool is correctly *averaging* between the two zones, but the data are still meaningless between the two beds. The worst effect (except for the interval 8,020 ft to 8,023 ft where a thin bed of 6 in. is essentially ignored) in the thin beds for this example is that the tool response *never reaches the full true value* for other thin intervals.

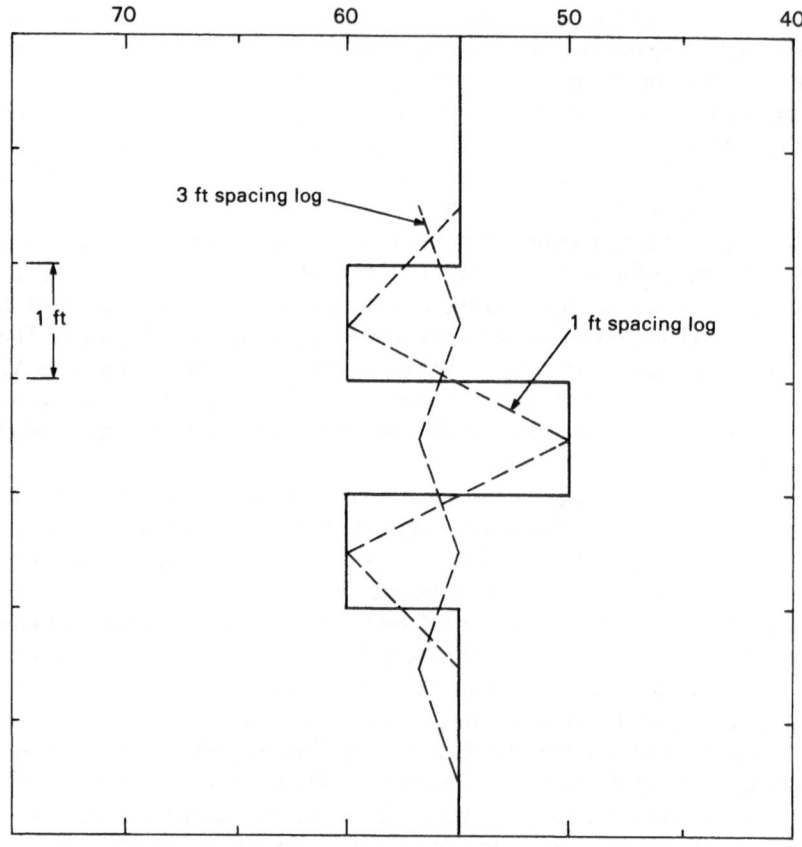

Figure 4–16. Reversed response of sonic log in thin bed.

This generally observed behavior of log responses has led to a practical scheme for picking values suitable for log calculations. In bedding sequences such as those illustrated in Figure 4–17, the peak or valley values are used. If the bed is thick enough, an average (or average of the central portion of the bed) may be used, particularly for porosity devices. No values are used from the transition portion of the response curve where the response is changing from one extreme to another.

COMPENSATION FOR BOREHOLE EFFECTS

Primarily two compensations are used in sonic logging devices: compensation for changing borehole size and the use of long spacings from the transmitter to the receivers to compensate for the affects of altered shales

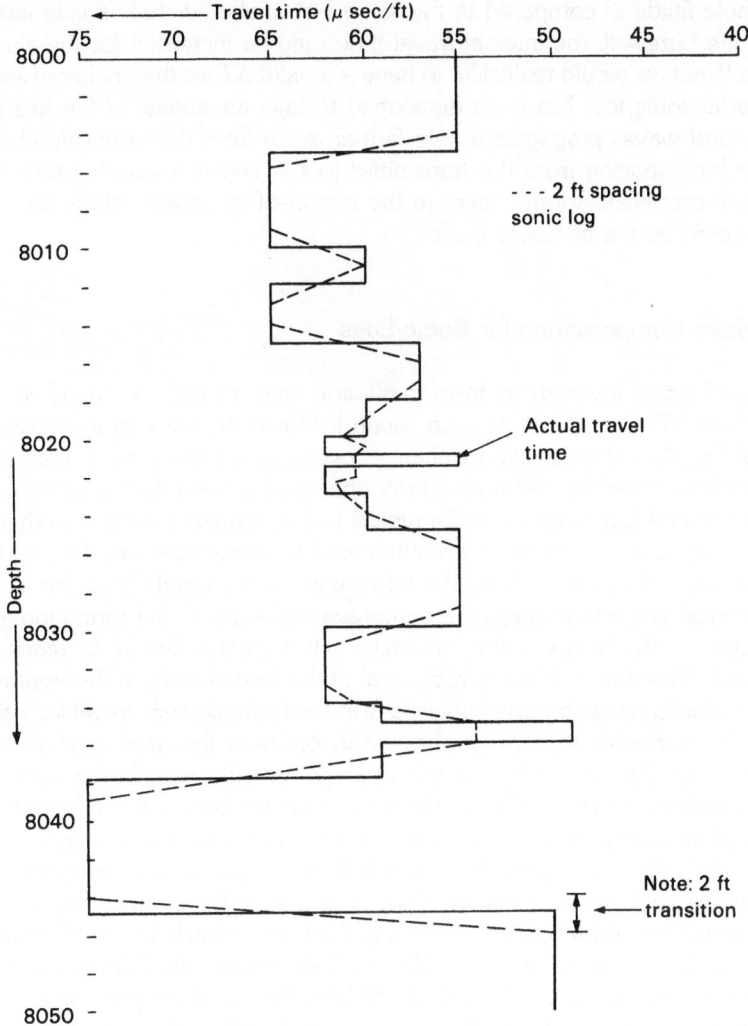

Figure 4–17. Sonic log thin-bed response.

on acoustic velocity. Compensation for changes in the size of the bore-hole is necessary because the fluid travel time at one receiver could be different from that at the other receiver. In such a case, the fluid delay times would be unequal and would not cancel out. The measured interval travel time would be affected by the increased (or decreased) fluid travel time.

In some sensitive shales, the shale may be altered near the borewall where the sound waves travel. If these shales contain additional water from the borehole fluids as compared to the same but unaltered shale a little farther from the borewall, the interval travel time may be increased for the altered shale. What we would really like to have is a valid ΔT for the unaltered shale. A special sonic tool has been developed to take advantage of the fact that the sound waves propagate a little farther away from the borewall when a rather long spacing from the transmitter to the two receivers is used. This tool will propagate sound waves in the region of the shales where they are not altered by the borehole fluids.

Borehole Compensation for Sonic Logs

A typical sonic log with its instrumentation may process as many as five complete ΔT measurements each second. However, none of these will be useful if significant borehole size changes occurring between the two receivers are not accounted for. What about size changes between the transmitter and two receivers? Let us examine the typical timing sequence for a hypothetical sonic logging tool with one transmitter and two receivers (see Fig. 4–18).

The signal that travels from the transmitter to the borehole at the necessary critical angle to propagate a compressional wave in the formation after refraction at the borewall–fluid interface will require a time t_{f_1} to reach the borewall. This time will be the reciprocal of the fluid velocity in the borehole. After refraction, the compressional wave in the formation will travel for a time t_1 in the formation to a point above and opposite the *near* receiver. This point is also determined by the critical angle for refraction. In this case, the compressional wave traveling in the rock along the borehole is continuously propagating compressional fluid waves back into the borehole. Figure 4–18 shows only those ray paths for the two fluid compressional waves that will be detected at the two receivers. These result from the compressional wave traveling in the formation and refracting fluid waves back into the formation opposite the receivers. Only two of these fluid waves out of the many propagating back into the borehole fluid will be detected at the two receivers.

The critical angle for refraction (α in Fig. 4–18) will be determined in accordance with Snell's law as the inverse trigonometric sine function for the ratio of the travel time of sound in the formation to the travel time of sound in the fluid. After reaching the refraction point opposite the near receiver, the formation compressional wave travels on down to the refraction point opposite the *far* receiver. It travels for a time ΔT. This is the interval travel time we will measure. It is the time required for the compressional wave to travel through the formation for a distance corresponding to the receiver spacing, as long as the formation compressional velocity does not change over this distance. If it does change, the angle of refraction at the receivers will be different. Hence, the fluid travel times t_{f_2} and t_{f_3} will be different.

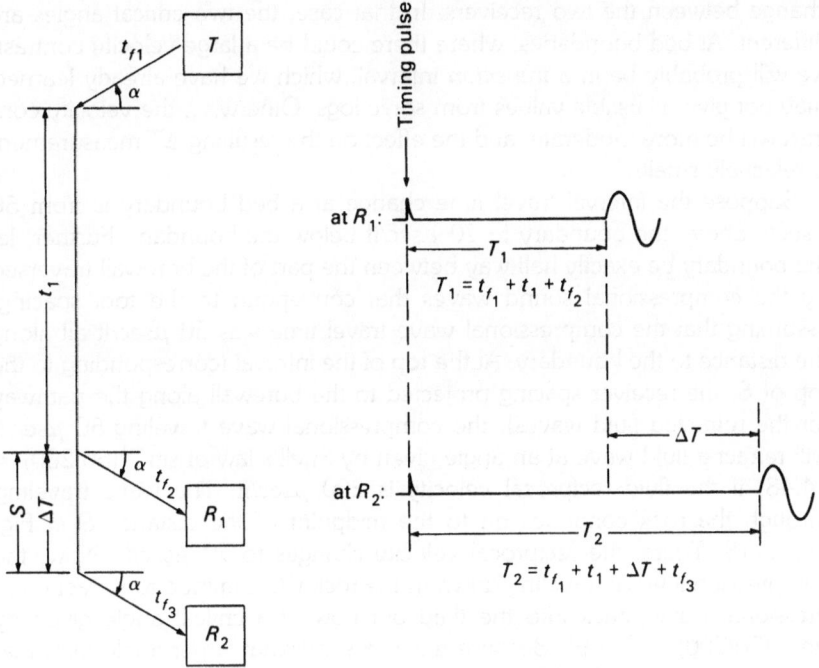

Figure 4–18. Sonic log timing.

Total travel time from transmitter to the near receiver is

$$T_1 = t_{f_1} + t_1 + t_{f_2} \tag{4–7}$$

The total travel time to the far receiver is

$$T_2 = t_{f_1} + t_1 + \Delta T + t_{f_3} \tag{4–8}$$

ΔT will be the interval travel time for the receiver spacing. The difference in the time of arrival of the same signal at the two receivers is then

$$T_2 - T_1 = \Delta T + (t_{f_3} - t_{f_2}) \tag{4–9}$$

If the fluid travel times at the two receivers are identical, the measured travel time will equal the correct formation interval travel time for the receiver spacing distance. Note that the fluid travel time at the transmitter cancels out. However, there are two cases where the fluid travel times at the receivers might not cancel out. First, the formation compressional wave velocity may change between the two receivers. In that case, the two critical angles are different. At bed boundaries, where there could be a large velocity contrast, we will probably be in a *transition* interval, which we have already learned may not give us usable values from sonic logs. Otherwise, the velocity contrast will be more moderate, and the effect on the resulting ΔT measurement is relatively small.

Suppose the interval travel time change at a bed boundary is from 50 μsec/ft above the boundary to 70 μsec/ft below the boundary. Further, let the boundary be exactly half way between the part of the borewall traversed by the compressional sound waves that correspond to the tool spacing, assuming that the compressional wave travel time was 50 μsec/ft all along the distance to the boundary. At the top of the interval (corresponding to the top of S, the receiver spacing projected to the borewall along the pathway for the refracted fluid waves), the compressional wave traveling 50 μsec/ft will refract a fluid wave at an angle given by Snell's law of $\sin^{-1}(50/200) = 14.48°$ if the fluid reciprocal velocity is 200 μsec/ft. The wave traveling through the rock continues on to the midpoint of the distance S in Figure 4–18. There, the reciprocal velocity changes to 70 μsec/ft. Now, the compressional wave traveling down in the rock will continue to refract compressional waves back into the fluid but now at a critical angle given by $\sin^{-1}(70/200) = 20.49°$. Because this is a somewhat larger angle than that for refraction of the fluid waves back to the near receiver, the wave in the rock will not have traveled a complete additional foot of distance below the midpoint of S.

Let the borewall be a rather extreme 3 in. from the two receivers, which we will also assume to be centralized in the hole. For a 3 3/8 in. diameter tool, this would correspond to a 9 3/8 in. hole. Let the separation (3 in. for this example) be denoted by R. The time for the wave to travel from the borewall to the two receivers will be given by $R \cdot \sec(\alpha)/5000$ fps. The angle α is different for the two receivers, so the difference, $t_{f_3} - t_{f_2}$ in Eq. 4–8, will not be zero. For this example, the difference will be 1.7 μsec.

I mentioned earlier that the distance traveled opposite the two receivers will be 1 ft plus something less than 1 ft because of the different angle of refraction for the faster velocity opposite the second receiver. This reduced distance is also found from the geometry of the problem (see Fig. 4–19). The difference between the actual vertical distance (measured from the receiver to the point opposite the location of the refraction of the fluid wave back to the receiver) traveled in the formation from the separation S of the two receivers is given by $X_2 - X_1$ where the vertical distance X_i is determined by $R \cdot \tan(\alpha)$ (where R is the distance separating the tool from the borewall). For this problem the difference between the spacing of the receivers and the actual distance traveled along the borewall between the points of refraction of fluid waves to the two receivers is .0289 ft.

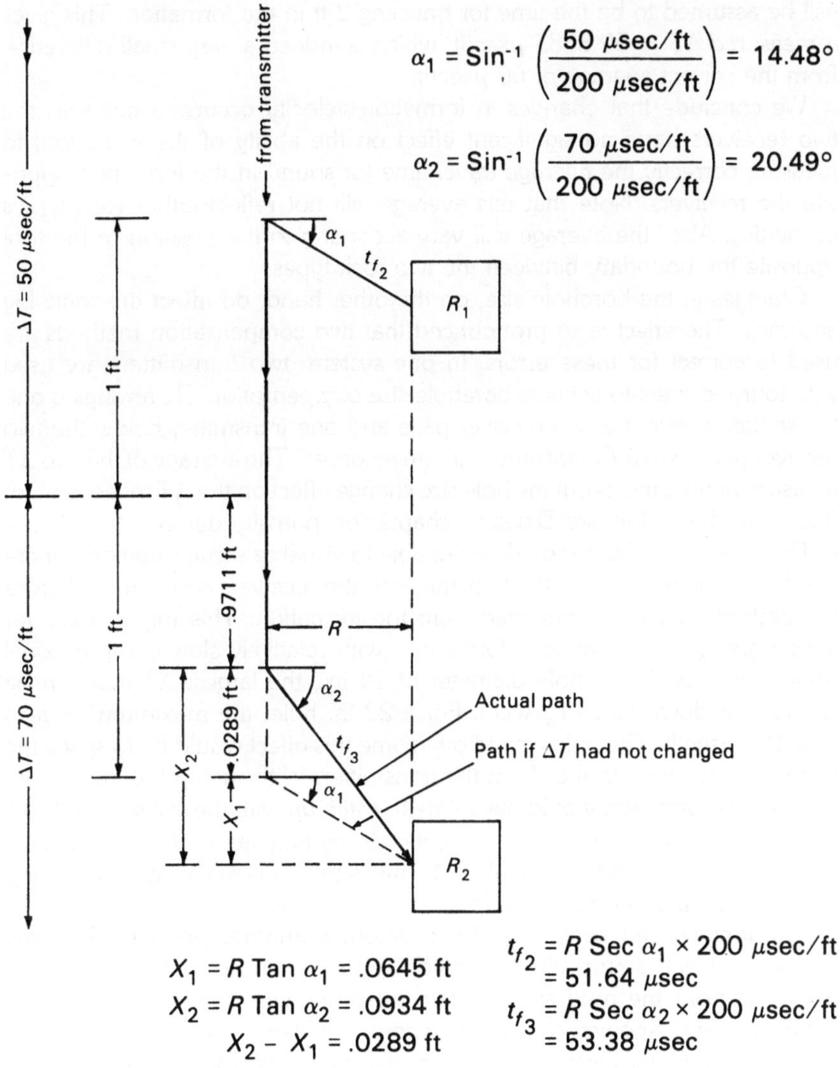

$$\alpha_1 = \text{Sin}^{-1}\left(\frac{50\ \mu\text{sec/ft}}{200\ \mu\text{sec/ft}}\right) = 14.48°$$

$$\alpha_2 = \text{Sin}^{-1}\left(\frac{70\ \mu\text{sec/ft}}{200\ \mu\text{sec/ft}}\right) = 20.49°$$

from transmitter

$\Delta T = 50\ \mu\text{sec/ft}$

$\Delta T = 70\ \mu\text{sec/ft}$

1 ft

1 ft

.9711 ft

.0289 ft

α_1

t_{f_2}

R_1

R

α_2

t_{f_3}

α_1

X_2

X_1

Actual path

Path if ΔT had not changed

R_2

$X_1 = R\ \text{Tan}\ \alpha_1 = .0645$ ft
$X_2 = R\ \text{Tan}\ \alpha_2 = .0934$ ft
$X_2 - X_1 = .0289$ ft

$t_{f_2} = R\ \text{Sec}\ \alpha_1 \times 200\ \mu\text{sec/ft}$
$\quad = 51.64\ \mu\text{sec}$
$t_{f_3} = R\ \text{Sec}\ \alpha_2 \times 200\ \mu\text{sec/ft}$
$\quad = 53.38\ \mu\text{sec}$

Figure 4–19. Geometry for travel time changes.

Thus, the compressional wave will travel 1 ft at 50 μsec/ft and .9711 feet at 70 μsec/ft. This is a total travel time of 50 μsec + 67.98 μsec = 117.98 μsec. Our sonic log system will, however, measure the travel time as 117.98 μsec *plus* the additional fluid travel time required to reach the far receiver, that is, 1.7366 μsec. This total measured time of 119.7166 μsec will be assumed to be the time for traveling 2 ft in the formation. This gives a *measured* ΔT = 59.8583 μsec/ft, which is indeed a very small difference from the correct average of 60 μsec/ft.

We conclude that changes in formation velocity occurring between the two receivers have no significant effect on the ability of the sonic tool to measure correctly the *average* travel time for sound in the formation opposite the receivers. Note that this average will not reflect either rock type's properties. Also, the average will vary according to the position of the tool opposite the boundary between the two rock types.

Changes in the borehole size, on the other hand, do affect the sonic log accuracy. The effect is so pronounced that two compensation methods are used to correct for these errors. In one system, two transmitters are used with four receivers to achieve borehole size compensation. By arranging one transmitter above the two receiver pairs and one transmitter below the two receiver pairs, two ΔT measurements are recorded. The average of the two ΔT measurements cancels out the hole size change effect on the ΔT measurement (for more discussion see Dewan's chapter on porosity devices[26]).

From Figure 4–18 it should be possible to visualize some situations where the direct fluid wave from the transmitter to the receivers could arrive *before* the desired fluid wave refracted from the formation. This might occur for large borehole sizes opposite formations with relatively slow compressional wave velocities. For a hole diameter of 14 in., the largest ΔT that can be recorded is down to 150 μsec/ft. For a 22 in. hole, the maximum value is only 100 μsec/ft. The only way to overcome this effect would be to space the two receivers much farther from the transmitter. With such a tool the slower wave in the formation would have time to *catch up* with the still slower direct fluid waves. If they do not catch up, the strong amplitude of the direct fluid wave will make them undetectable. ΔT measurements made with these *long spacing* tools are referred to as *long-spaced sonic logs*.

The long-spaced sonic log also overcomes another problem. In some shales (in which sound already travels relatively slowly in the formation), the fluids from the mud system alter the properties of the shale near the borewall. Hicks showed that plastic stress conditions exist in clays around the well bore.[27] This results in a lower effective stress closer to the well bore accompanied by a slower wave velocity closer to the well bore. The wave velocity increases with increasing transmitter to receiver spacing under these conditions. Consequently, a ΔT more closely related to the undisturbed shale is made by using a sonic tool with longer transmitter to receiver spacings. This is important where sonic data is used to obtain synthetic seismograms. In Schlumberger's long-spaced sonic device, both a 10 ft and 12 ft spacing log (spacing measured from transmitter to near receiver) are recorded. This is in contrast to the 3 ft spacing commonly used with Schlumberger's *borehole-compensated* device. In both long and shorter spaced tools, the receivers are still spaced only 2 ft apart. Compensation for borehole size changes is made by having a computer memory retain a travel time measurement for one transmitter–receiver pair that can be compared to another measurement a short distance up the hole.[28]

The long-spaced tool is also used when full acoustic waveforms are recorded. The longer spacing allows the various events recorded on the log to be spread out further to facilitate identification. The long-spaced tool does have some shortcomings. There will be a smaller amplitude of received signal that makes triggering by electrical noise spikes in the system more likely. In formations with naturally occurring lower amplitude waves, cycle skipping may also be a problem.

CEMENT BOND EVALUATION AND OTHER SONIC LOG APPLICATIONS

Many other sonic log applications are based on measured characteristics other than compressional wave travel time: travel times for other events (shear wave, etc.), amplitudes for the various events, and frequency and attenuation measurements. These measurements can be used to identify fractures, assess permeability, evaluate cement bond quality, and determine mechanical rock properties such as Young's modulus or Poisson's ratio.

Cement Bond Logs

Acoustic wave amplitude can be used to evaluate the quality of the cement bond of casing to formation. If the bond is poor, the casing will tend to *ring* in response to the acoustic pulse and vibrate with a large amplitude. If the bond is good to both pipe and formation, the casing amplitude will be negligible, and the signals from the formation will dominate in the received wave train. The effect is similar to that of knocking on a wall and listening to the sound to locate the studs behind the wall. If no stud is present where the wall is knocked (no cement bond), a hollow, drumlike sound is heard, whereas the sound will be sharper opposite any studs behind the wall. The log was sometimes referred to as a *clank-clunk* log.

Unfortunately, cement bond logs (CBLs) are not perfect. Their proper use is very dependent on operator experience and selection of various detection thresholds. The apparent cement bond quality will depend on thresholds such as which portion of the wave is used to measure amplitude. In a study described by Pickett, several logging companies ran CBLs over the same interval in the same well using several possible combinations of detection parameters. Each individual run resulted in a completely different interpretation of cement bond quality![29] Those interested in pursuing cement bond interpretation would do well to study Pickett's discussion, which clearly illus-

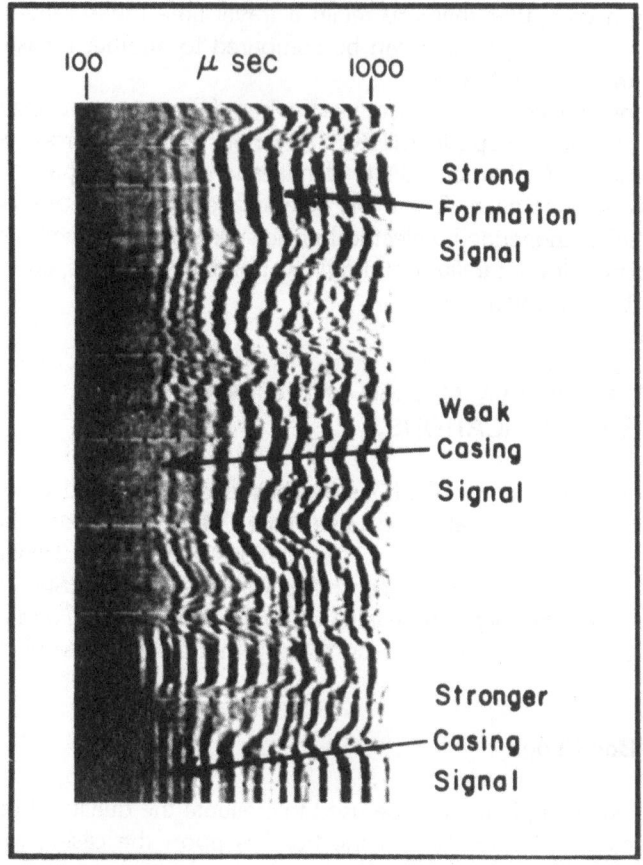

Figure 4–20.　Cement bond log. *(Courtesy of Schlumberger Ltd.)*

trates how the different interpretations relate to the received waveforms and the detection parameters.

A typical CBL is illustrated in Figure 4–20. A gamma ray log is usually recorded in track 1 for correlating the bond log to the other logs. Note the received wavetrain is recorded in a variable density log (VDL) format. In this presentation, the positive peaks of the waveform are shown as dark bands, whereas the negative peaks are shown as white bands. Larger peaks go to the extremes of black (positive) or white (negative), whereas smaller peaks are various shades of gray. This presentation is also known as an acoustic microseismogram or acoustic character log. The variable density display is useful for visual identification of the various events on the log. However, for computer work the actual waveform is preferred because the actual waveform amplitudes may be easily digitized. The VDL display has a serious limitation. If relatively small peaks and valleys of weaker signals are to be illustrated, the larger peaks and valleys of stronger signals will be *clipped*, and these stronger waves will look no different than the weaker ones. This can hinder proper identification of various events on the waveform in some cases.

It is also customary to record the relative amplitude of the received acoustic signal on the CBL. Note that this amplitude will probably come from some predetermined portion of the received waveform. The idea is to include primarily the received pipe signal that occurs at 57 μsec/ft. Strong amplitudes here will only occur if the pipe is not adequately bonded to the formation. The only problem with this approach is that in some formations (such as carbonates), we frequently have travel times nearly the same as that for the pipe. We might have a good bond and still see a large apparent pipe amplitude that actually results from strong acoustic signals received directly from the formation. For this reason, these high amplitude points must be checked (usually shown on a relative amplitude scale as *poor bond*) to see if there is a significant formation signal. The presence of the formation signal implies a good bond.

Schlumberger recently developed a tool that uses eight ultrasonic transducers spaced circumferentially around the tool (Cement Evaluation Tool or CET). With this system, it appears that a much improved cement evaluation may be made. The tool also provides useful information about cement and casing strength as well as squeeze job problems and is a useful acoustic caliper that can show casing collapse. This tool may become the best choice for cement bond evaluation. It is currently run in conjunction with the conventional CBL. Other logging companies are presently developing and introducing their own version of this tool.

Borehole Televiewer

Another sonic log application is the borehole televiewer, a specialty tool used by some companies. It provides an acoustic *picture* of the borewall. An acoustic transducer rotates as the tool is pulled up the hole, and continuous amplitude signals are recorded. The strength of reflection of the signal from the borewall varies according to the acoustic properties of the rock at the borewall. The recorded log is shown as if the inside of the borewall was like a tin can that was cut vertically and then unfolded on a flat surface. The horizontal axis represents the azimuth from 0° on the left to 360° on the right. The tool has proven useful in fracture identification.

Fracture Identification and Permeability Assessment

Acoustic amplitudes are often greatly reduced opposite fractured (the fractures may significantly enhance permeability) zones. This behavior often is responsible for the characteristic *cycle skipping* already discussed. However, we have also noted that other causes, such as reflections and diffractions from lateral changes in rock properties, can lead to destructive interference and resultant loss of amplitudes in the received signal.

Mechanical Rock Properties

If we know the acoustic compressional and shear wave velocities as well as the formation density, we can calculate the mechanical properties of the rock such as Young's modulus or Poisson's ratio. This information can then be used in various applications such as hydraulic fracture system design and lithology identification. The ratio of the compressional to shear wave velocity is also related to the rock type (quartz, limestone, or dolomite) and is sensitive to the presence of gas in the rock's pore system.[30] The presence of gas in the pore system has a pronounced effect on compressional wave velocity (slows the wave), whereas the shear wave velocity is not affected. Thus the velocity ratio changes in gas-bearing zones.

CORRELATING SONIC LOG DATA
TO CORE DATA

Correlating log data to core data can be a problem. We should not always expect a good correlation. Neither should we assume that core data represents the ultimate in data reliability. Depending on the rock type, the opposite may be true. Logs and cores sample different representative volumes of rock. To illustrate the possible severity of the core sampling problem, Pickett completed a carefully designed experiment in which he selected seven small depth intervals, ranging from 0.4 ft to 1.0 ft thick, from a carbonate tidal flat section and took as many core plugs as practical from each piece to measure core porosity.[31] Thus, the average porosity from all the samples of an interval would have been a close approximation to the *true average* porosity for the interval. In these heterogeneous carbonates, the standard deviations in each interval were large, varying from 11.4% to 35.9% of the average. The average of the seven standard deviations was 22%. If the data are assumed to follow a *normal* distribution, this means that 67% of the time we can expect a single plug to vary as much as 22% of the actual porosity. For example, if the true porosity is 20%, we can expect single plug core measurements of this heterogeneous carbonate to vary from 15.6% to 24.4%, 67% of the time. We can expect the single plug core porosity to vary 90% of the time from 12.8% to 27.2%! The odds of getting the true porosity are quite small. Since actual core porosity measurements are repeatable with a precision of 0.1%, we are not seeing an error in core porosity measurement, but rather a sampling error in assuming that a 0.4 ft to 1.0 ft interval of heterogeneous carbonate rock can be faithfully represented by a single thumb-size plug. There can be other practical problems with core data. Often pieces of core are missing for some intervals, and the cored interval may not exactly match the logged interval. For example, the log data may appear to be stretched compared to the core data. Careful depth comparison and shifting, if necessary, may give an improved correlation of core data to log data. Another problem can arise if a piece of core is inadvertently (or carelessly) placed upside down relative to the rest of the core pieces; maybe it is even placed out of proper sequence as it is layed out on the ground! Assuming the core has none of these problems, we must still be aware that logs and cores sample different relative volumes of rock.

One property of sonic logs that must be considered to successfully correlate them to core data is that a sonic log gives a running average of the porosity. Figure 4–21 illustrates the principle. For a 3 ft spacing sonic log

For 3 ft spacing sonic:

$$\text{core average} = \frac{1}{3} \left[\phi(9000') + \phi(9001') + \phi(9002') \right]$$

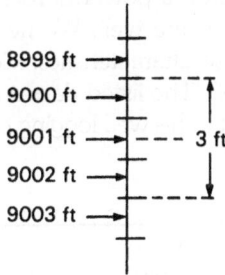

For 2 ft spacing sonic:

$$\text{core average} = \frac{1}{2} \times \left\{ \frac{1}{2} \left[\phi(9000') + \phi(9002') \right] + \phi(9001') \right\}$$

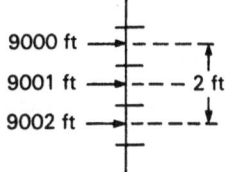

Figure 4–21. Correlating sonic logs to core data. (Note: Each sample is assumed to occur at the middle of a sample interval.)

(commonly used in many of the older sonic tools) measuring the interval travel time at 9,001 ft depth (the *measure point*), we see that the actual depth interval over which travel time is evaluated extends from 8,999.5 ft depth to 9,002.5 ft depth. I use the convention of representing the data point at *X* feet by an interval extending a distance plus or minus one-half of a sample increment from the point. For most core data the sample increment is 1 ft. If we want to compare core porosity data corresponding to the 3 ft spacing sonic log data of Figure 4–21, we would compute a 3 ft running average of the core data, assuming that the core data are recorded for each foot of depth. Thus, the core porosity 3 ft average at 9,001 ft depth is the sum of the core porosities at 9,000 ft, 9,001 ft, and 9,002 ft, divided by 3. For the 2 ft spacing sonic log data, the approach is similar: the core data running average is made up of the sum of one-half of the core porosity at 9,000 ft plus the core porosity at 9,001 ft plus one-half of the core porosity at 9,002 ft, with the sum then divided by 2.

ROCK TYPING WITH THE SONIC LOG

Although the sonic log has a disadvantage in that it is sometimes difficult to relate travel time response to porosity, it can still be a powerful rock typing tool even when it is the only porosity tool run in the well. We need only capitalize on a knowledge of its unique response characteristics. Consider the example sonic log and core data of Table 4–2. The listed data are taken from a similar problem used by George R. Pickett in his well logging courses.

Table 4–2. Acoustic Log Rock Typing Problem

Zone	Lithology	Core Porosity (%)	ΔT (μsec/ft)	R_t (ohm-meters)
1	Dolomite, anhydritic, vuggy	2	44.5	250
2	Dolomite, chalky	10.5	63	9.1
3	Dolomite, chalky	11	63	8.3
4	Dolomite, anhydritic	3	45	110
5	Dolomite, chalky	10	62	10
6	Dolomite, chalky, vuggy	2.5	44.2	160
7	(No core data)	—	56	12

Note: $M = N = 2$; formation water salinity = 25,000 ppm NaCl; formation temperature = 187°F.

Problem: Would you recommend completing this well assuming adequate thickness for zone 7 and a cutoff water saturation of less than or equal to 50%?

The data might be representative of what we would encounter in a typical evaporite sequence. The zones listed would be intervals we had selected from the logs using data from the *peaks and valleys* (avoiding transition values) of the recorded waveform. Perhaps the core data were averaged (it is hoped in the proper fashion as discussed earlier) to improve the correlation of the core data to the sonic log responses. Assume that the resistivity data were recorded using a suitable logging tool. We are also given the information that both the porosity and saturation exponents (*m* and *n*) are equal to 2. The formation water salinity is 25,000 ppm NaCl, and the formation temperature is 187°F. Should we recommend completing this well, assuming that the critical water saturation is 50% and that zone 7 is thick enough to provide adequate storage for commercial quantities of hydrocarbons? Solving this type of problem is exemplary of both practical formation evaluation and using thought processes somewhat more sophisticated than a *cookbook* or *turn the crank* solution. This is more like the real world of log analysis. I illustrate the solution to this problem using the porosity relationship in the slope-intercept form rather than the more complicated time average relation, which would require unrealistic values for either fluid velocity or compaction factor in this evaporite–carbonate sequence. (For those who wish, it is certainly possible to make the appropriate conversions to the time-average parameters and work the problem that way.)

From the salinity, we should satisfy ourselves that the water resistivity would be 0.10 ohm-meters at 187°F. Then, the first step would be to plot the data for zones 1 through 6 on a graph such as that in Figure 4–22. If we fit a rough visual line through the data, it appears to have an intercept of 39 μsec/ft and a slope of 2.26 μsec/ft/% porosity or 226 μsec/ft/fractional porosity. For future reference, we will call this relation A. If we examine Table 4–1, we find that the intercept of 39 μsec/ft represents too fast a travel time for real rocks and that the slope of 226 is much too high for carbonates. With the two well-defined groups of points: 1, 4, and 6; and 2, 3, and 5; we should also consider that there are two distinct rock types, one of which could be vuggy dolomite and the other chalky dolomite. Points 2, 3, and 5 are identified as chalky on the core description included in Table 4–2. Although zone 6 is described as chalky, it is also listed as having vugs. Zone 1 is described as having vugs and anhydrite, whereas zone 4 is described as having anhydrite. However, these descriptions may not be totally accurate. The listed description is one person's opinion based on a cursory visual examination. The well-defined grouping of the two sets of points with the relative impossibility of them being the same rock type supports our conclusion of two distinct rock types. For those who may be wondering, if you did try the time-average formula, you would get a compaction factor of 1.5 for relation A along with the same matrix travel time of 39 μsec/ft. The travel time, as we already know, is too fast for

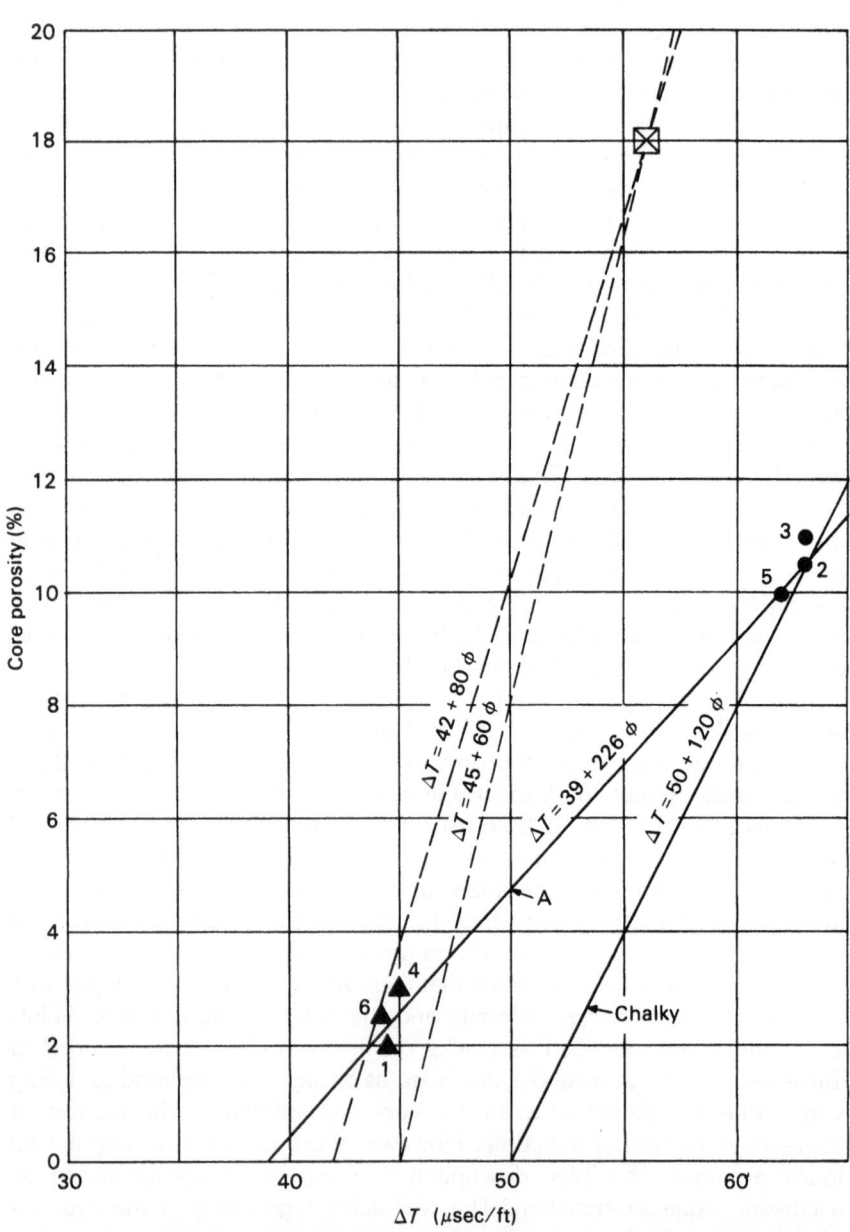

Figure 4–22. Graph of data for sonic log problem.

real rocks and a compaction factor (or *lack of compaction factor*) of 1.5 is too high for most carbonates. For many dolomites, it is necessary to use a compaction factor of less than 1 (or else use unrealistic fluid travel times). Let us draw upon experience (as summarized by Table 4–1) with sonic log porosity calibration constants. For carbonates, we expect matrix travel times of 42 μsec/ft to 50 μsec/ft and relational slopes of 60 μsec/ft to 120 μsec/ft/fractional porosity. The largest slopes will occur for chalky rocks with small pore spaces. It is also likely that these chalky rocks will have higher matrix travel times. Using this knowledge, let us construct a line from a coordinate at zero porosity and 50 μsec/ft travel time and extend it through the group of points 2, 3, and 5. This will take it roughly through the coordinate at 10% porosity and 62 μsec/ft travel time. We will call this the *chalky* relation. It has an intercept of 50 μsec/ft by construction, and we see that the resultant slope is 120 μsec/ft/fractional porosity. In order to pass a line through the group of points (points 2, 3, and 5) to a travel time of less than 50 μsec/ft for zero porosity, we will end up with a line having a slope of greater than 120 μsec/ft/fractional porosity. This would be too large for a carbonate.

Let us see if zone 7 might fit the *chalky* relation. For its travel time of 56 μsec/ft, the chalky relation gives us 5% porosity. For $m = 2$, this gives us a formation factor: $F = 400$. With $R_w = 0.10$ ohm-meters, this gives us $R_o = FR_w = 40$ ohm-meters. For zone 7's resistivity of 12 ohm-meters, we would calculate an S_w several times greater than 100%. Thus, the rock of zone 7 cannot be a chalky dolomite. It does not belong with the same rock type as zones 2, 3, and 5. Note also its anomalous occurrence relative to the relation hypothesized for zones 2, 3, and 5.

At this point, we can now entertain the possibility that zone 7 is a different rock type, perhaps the same as zones 1, 4, and 6. To find out where it might fall on the graph of Figure 4–22, let us find out what porosity is required to calculate an S_w less than or equal to 50% for zone 7. This assumes that 50% is a good cutoff value for water saturation. For an $n = 2$, this implies the resistivity index I is greater than or equal to 4, or since $I = R_t/R_o$ and $R_t = 12$ ohm-meters, we would have $R_o \leq 3$ ohm-meters. For $R_w = 0.10$ ohm-meters, we have $R_o = FR_w$ and thus $F \leq 30$. For $m = 2$, this means porosity must be greater than or equal to 18%. Is this possible for the known ranges of matrix travel time for carbonates? Let us start with the minimum value for slope B in the porosity relationship: 60 μsec/ft fractional porosity for carbonates. If we extend a line from the coordinate of 18% porosity at 56 μsec/ft to the zero porosity line with a slope of 60 μsec/ft/fractional porosity, we find the intercept is 45 μsec/ft at zero porosity. To get this value, solve the relation $\Delta T = A + B\phi$ for A where $B = 60$, $\phi = .18$, and $\Delta T = 56$ μsec/ft. The *exact* value is 45.2 μsec/ft, but the rounded value of 45 μsec/ft is adequate for our purposes. The constructed line, $\Delta T = 45$

μsec/ft + 60 μsec/ft/fractional porosity is drawn on the graph of Figure 4–22.

Now let us try a second alternative for zone 7: extend a line from the minimum ΔT for carbonates of 42 μsec/ft at zero porosity to the point 18% at 56 μsec/ft for zone 7. We draw this line on the graph of Figure 4–22 and find the slope B is 80 μsec/ft/fractional porosity. For this alternative trend, the equation is $\Delta T = 42 + 80\,\phi$. Either this equation or the previous one could be valid for vuggy dolomites. We could also draw a line through both the group of points 1, 4, and 6 and zone 7 at 18% porosity and 56 μsec/ft. The equation for this line would be $\Delta T = 43 + 65\,\phi$. Perhaps zone 7 is the same rock type as zones 1, 4, and 6, but with less pore filling anhydrite (accounting for its rather large porosity in comparison with zones 1, 4, and 6). Since at least 18% porosity would give us an S_w of no more than 50%, we should probably complete zone 7. Unfortunately, we must also consider that there is another possibility; zone 7 is a rock type we have not encountered previously in the cored interval. Maybe it is a granular carbonate. For granular carbonates, B is about 100 μsec/ft/fractional porosity. If we use this B value with the minimum possible ΔTMA for carbonates (42 μsec/ft), we can calculate the maximum possible porosity for this rock type (a granular carbonate). Of course it is somewhat unrealistic to expect a granular carbonate to have a travel time as fast as 42 μsec/ft, but it does allow us to estimate the best possible porosity for zone 7 if it should be a granular carbonate. For $B = 100$ and $\Delta TMA = 42$ μsec/ft, we get 14% porosity for zone 7. This gives us $F = 51$ for $m = 2$, and $R_o = 5.1$ ohm-meters if $R_w = 0.10$ ohm-meters. $I = R_t/R_o = 12$ ohm-meters / 5.1 ohm-meters = 2.35, or $S_w = 65\%$ for $n = 2$. This is too high an S_w for completion even if sufficient reserves are calculated with $\phi = 14\%$ and $S_w = 65\%$. The rock will likely flow only water.

Thus, we have the following conclusions: (1) zone 7 cannot be a chalky dolomite; (2) if zone 7 is a granular carbonate, we need to plug and abandon the well since completion is unjustified; (3) if zone 7 is a vuggy carbonate, we should complete the well. Everything considered, zone 7 would be a good bet for completion. In practice, we would also look for evidence of vuggy carbonate in the rock cuttings from zone 7. We could also run a drill stem test. It might also be possible to run another porosity tool over zone 7 that would be affected differently by vugs. However, we have assumed in this example problem that another porosity tool was not available or else impractical for some other reason.

SUMMARY

Sonic logs require fluid in the borehole to work. The accuracy of sonic logs is no worse than 1–2 μsec/ft in terms of their ability to predict porosity. The chief disadvantage of sonic logs is that it is sometimes difficult to determine the appropriate calibration constants to relate porosity to the measured travel time. A unique advantage of sonic logs is that they are not influenced by material outside the tool spacing. However, in beds thinner than the tool spacing, sonic logs can show a *reversal*: travel times decreasing with increasing porosity. Ordinarily, travel time increases with porosity. In some formations, the acoustic signal amplitudes may be low enough that *cycle skipping* or larger errors in travel time occur.

Porosity (ϕ) may be computed from the sonic log by

$$\phi = (\Delta T - A)/B \tag{4-5}$$

where

$\phi = $ fractional porosity
$\Delta T = $ travel time measured by the tool in microseconds per foot
$A = $ travel time for the rock matrix in microseconds per foot
$B = $ a lithology/stress-related constant expressed in units of microseconds per foot per fractional porosity

Typical values for A and B are given in Table 4–1. A is also commonly designated with the symbol ΔTMA or ΔTM in the literature. The constant B increases with decreasing contact areas and decreasing effective stress. Thus, it is largest for finer-grained rocks and smallest for vuggy rocks.

The time-average formula may also be used to compute porosity, but it requires that three calibration parameters, ΔTMA, ΔTF, and a lack-of-compaction factor, be specified to establish the linear relation of Eq. 4–3 that is equivalent to Eq. 4–5. Despite its success, this formula with its three relational parameters is derived from the unrealistic sandwich model of alternating layers of porosity and rock matrix. Because of this, it is necessary to use unrealistic values for ΔTF or the lack-of-compaction factor to establish a porosity relation for some rock types, particularly carbonates.

The sonic log transform of Raymer et al. is a more modern and empirical approach to relating porosity to sonic log responses over a wide range of porosities.[32] Although it is somewhat premature to judge the utility of this transform, it appears to be based on fitting curves to a wide range of porosity data from different rocks. It may prove useful for ballpark estimates where no other information is available. Still, it would seem that Eq. 4–5 has served well not only for ballpark estimates, but for estimates where more information about the porosity relation was known. Implementation of the transform is somewhat tedious and may be better done by computer, although Dewan's approximation to the transform is simple enough.[33] However, the approximation may only cause more error in some situations.

Sonic logs normally are not run faster than 50 or 60 ft/min in the borehole. There is no specific calibration check for the sonic log other than checking the electronics (voltage and current measurements within the tool). A good check on the tool is to compare the recorded travel time in the casing to the ΔT for steel casing: 57 μsec/ft. Other comparisons may be made opposite beds of known lithology such as anhydrite: 50 μsec/ft. Another method of data quality verification would be to compare sonic data from several different wells over a common zone with relatively constant acoustic log behavior. Pickett has also suggested the use of a full waveform log as a quality check opposite zones with low amplitudes where larger ΔT errors are more likely.[34]

Unfortunately, most logging tool calibrations are based on a two-point calibration that *assumes* a linear relation must exist between the two points. There is no assurance of nonlinearity with the two-point method. We must remember that any data falling outside the calibration interval may also be suspect, even if the relation is linear between the calibration points. For this reason, modern interpretation techniques must be based on methods designed to reduce the influence of these potential errors. These methods include various data cross-plots, statistical techniques, using ratios to eliminate some unknowns, and accounting for uncertainties by calculating a range of results rather than a single value.

The behavior of the sonic log in thin beds leads naturally to a preferred method for selecting values from the log suitable for calculation: pick values from the *peaks and valleys* in thinner beds and use an average (or central average) for thicker zones. No values are to be used from the transition portion of the log where the response is changing from one extreme to another. Although correctly *averaged* over these transitions, the values have no correspondence to the rock properties on *either* side of the transition, and each logging tool type behaves differently over the same transition zones.

Proper correlation of log data and core data requires careful depth matching and some type of averaging or filtering of the core data to make them more closely resemble the response characteristics of the log they are to be compared to. Even after all precautions have been taken, it is frequently difficult to obtain a good correlation in heterogeneous rock types where a single core plug every foot is not adequate to represent the entire foot of core. Also, there may be instances where the log is simply not reflecting accurate values such as the acoustic log with weak signal strength illustrated by Pickett.[35]

In some instances we can take advantage of the different calibration constants necessary to relate acoustic log data to porosity for different rock types and use the tool for rock typing, even when it is the only porosity tool run in the hole. Much of the success of this approach is dependent on taking advantage of our knowledge of the acoustic log response and the basic petrophysical relationships for formation factor, porosity, and water saturation.

The most widely used application of acoustic logs is to measure porosity. However, they are also frequently used for cement bond evaluation and as an aid to fracture identification. Modern, full-wave acoustic logging tools are now more often used.

PROBLEMS

4–1(a). Convert the following interval travel times to interval travel times in units of microseconds per meter:

$$60 \ \mu \ \text{sec/ft}$$
$$110 \ \mu \ \text{sec/ft}$$

4–1(b). Convert 195 μsec/m to the corresponding English units.

4–1(c). A shear wave velocity is 7,500 ft/sec ($\Delta T_s = 133.33$ μsec/ft). What is the velocity in meters per second?

4–2. You are comparing an acoustic log to a rock core of the same interval and note that the acoustic-log-derived porosities do not agree at all with the core porosities. In fact, where porosity is increasing, the acoustic log travel time is decreasing, and vice versa. There is no evidence of cycle skipping on the acoustic log. Considering you have already carefully matched the depths and used an appropriate running average of the core data to make it resemble a sonic log running average response characteristic, what is the likely cause of the mismatch? What would you recommend be done to correct this problem for future acoustic logs in the same area?

4–3. For the same porosity, which rock type would you expect would have the fastest velocity (lowest travel time), a chalk or a vuggy carbonate?

4–4. A commonly used set of parameters in the time-average formula for sandstones is $\Delta TMA = 51.3$ μsec/ft, $\Delta TF = 189$ μsec/ft, and $c_p = 1$. Convert these three parameters describing the straight-line porosity relation to the slope and intercept parameters A and B used in Pickett's equation for porosity. After conversion, do the values for A and B seem reasonable for sandstone? Which gives more accurate results: A and B or the three time-average parameters?

4–5. You have a set of acoustic log data and the corresponding core data. You find the data are easily represented by a straight line with an intercept of 42 μsec/ft at zero porosity and a slope of 60 μsec/ft fractional porosity.

What calibration constants would you use with the time-average formula to calculate porosity for this acoustic log?

4–6.   On the acoustic log of Figure 4–9, calculate the porosity for the interval from 8,415 ft to 8,418 ft if $A = 50$ μsec/ft and $B = 120$ μsec/ft/fractional porosity. Also calculate the porosity for each foot in the same interval. Why would you want to use the peak value at 8,416 ft to represent the whole interval instead of using the foot-by-foot calculations?

REFERENCES

1. S. N. Domenico, "Elastic Properties of Unconsolidated Porous Sand Reservoirs," *Geophysics* **42**, no. 7 (1977): 1339–1368.
2. J. E. White, *Seismic Waves, Radiation, Transmission, and Attenuation* (McGraw-Hill, New York, 1965), p. 37.
3. Warren G. Hicks and James E. Berry, "Application of Continuous Velocity Logs to Determination of Fluid Saturation of Reservoir Rocks," *Geophysics* **21**, no. 3 (1956): 739.
4. Schlumberger, *Log Interpretation, Volume 1, Principles* (Schlumberger Ltd, New York, 1972), p. 48.
5. Hicks and Berry, "Application of Continuous Velocity Logs," p. 739.
6. Ibid.
7. M. A. Biot, "Theory of Propagation of Elastic Waves in a Fluid Saturated Porous Solid, Part I," *Journal of the Acoustical Society of America* **28** (1956): 168–178.
8. J. Geertsma and D. C. Smit, "Some Aspects of Elastic Wave Propagation in Fluid Saturated Porous Solids," *Geophysics* **26**, no. 2 (1961): 169–181.
9. M. R. J. Wyllie, A. R. Gregory, and L. W. Gardner, "Elastic Wave Velocities in Heterogeneous and Porous Media," *Geophysics* **21**, no. 1 (1956): 41.
10. Schlumberger, *Log Interpretation*, p. 39.
11. M. P. Tixier and R. P. Alger, "Log Evaluation of Nonmetallic Mineral Deposits," paper presented at the 8th Annual Logging Symposium of the Society of Professional Well Log Analysts, 1967.
12. Geertsma and Smit, "Some Aspects of Elastic Wave Propagation," pp. 169–181.
13. Sylvain J. Pirson, *Handbook of Well Log Analysis for Oil and Gas Formation Evaluation* (Prentice-Hall, Englewood Cliffs, N.J.), 1963, p. 32.
14. John T. Dewan, *Essentials of Modern Open-Hole Log Interpretation* (Penwell Publishing, Tulsa, Okla., 1983), p. 151.
15. Author's class notes from course in petrophysics taught by George R. Pickett at the Colorado School of Mines, Golden, Colorado, 1977.
16. J. Geertsma, "Velocity-Log Interpretation: The Effect of Rock Bulk Compressibility," *Society of Petroleum Engineers Journal* (December 1961): 235–246.
17. Biot, "Theory of Propagation of Elastic Waves," pp. 168–178.
18. Author's class notes, 1977.
19. Ibid.
20. Douglas W. Hilchie, *Applied Openhole Log Interpretation (for Geologists and Engineers)*, Rev. (Douglas W. Hilchie, Inc., Golden, Colo., 1982), pp. 6–13.
21. L. L. Raymer, E. R. Hunt, and J. S. Gardner, "An Improved Sonic Time-to-Porosity Transform," paper presented at the 21st Annual Logging Symposium of the Society of Professional Well Log Analysts, Lafayette, July 1980.

22. Dewan, *Essentials of Modern Open-Hole Log Interpretation*, p. 153.
23. Raymer et al., "An Improved Sonic Time-to-Porosity Transform."
24. Ibid.
25. Mark E. Willis and M. Nafi Toksoz, "Automatic P and S Velocity Determination from Full Waveform Digital Acoustic Logs," *Geophysics* **48**, no. 12 (1983): 1631–1644.
26. Dewan, *Essentials of Modern Open-Hole Log Interpretation*, p. 153.
27. W. G. Hicks, "Lateral Velocity Variations Near Bore Holes," *Geophysics* **24**, no. 3 (1959): 451.
28. Dewan, *Essentials of Modern Open-Hole Log Interpretation*, pp. 158–161.
29. George R. Pickett, "Acoustic Character Logs and Their Applications in Formation Evaluation," *Journal of Petroleum Technology* (June 1963): 659–667.
30. Ibid., pp. 158–161.
31. Author's class notes, 1977.
32. Raymer et al., "Improved Sonic Time-to-Porosity Transform."
33. Dewan, *Essentials of Modern Open-Hole Log Interpretation*, pp. 141–142.
34. Pickett, "Acoustic Character Logs."
35. Ibid.

CHAPTER 5

Electrical Properties (Induction) Logging

Resistivity, an electrical property, was the first rock property logged in a well, and it is still one of the most important properties logged. Recall from Chapter 3 that most water saturation calculations are based on establishing a ratio of resistivities. If we compare a rock's resistivity to the resistivity it would have at 100% water saturation, we can calculate the water saturation from Archie's equation. Then we can infer that the balance of the pore spaces are occupied by hydrocarbon. However, we must resist the temptation to draw the apparently logical conclusion that the resistivity curves recorded on a log are *saturation* curves. A low resistivity does not by itself mean there is a high water saturation. We must still compare the resistivity to the value it would have if saturated with water. This requires knowledge of the rock's porosity as well as its resistivity.

Many logging books cover resistivity and electrical properties logging first because of historical development as well as their importance to saturation estimates. I elected to discuss sonic or acoustic properties first. I felt it was a simpler starting point that could more easily demonstrate the idea of a logging tool response function. Although resistivity is measured directly, what we really want to know is saturation. We cannot measure saturation directly, so we have to convert resistivity measurements to water saturations.

In Chapter 7, I discuss the other commonly logged rock properties: nuclear or radioactivity. Other properties are also logged, such as gravitational anomalies, but these are not dealt with in this beginning text.

RESISTIVITY

We have already examined the concepts of resistivity and resistance to electrical current flow, so now let us examine how we can measure resistivity in a rock below the surface, using a tool in a borehole. We will use an approach based on that of Hubert Guyod.[1] Consider a cylindrical-shaped piece of material. This material can be a rock as well as any other material. The cylinder might have an appreciable cross-sectional area as in Figure 5–1a, or it might be a long wire with a small cross-sectional area as in Figure 5–1b. We could also have a very short cylinder with a large cross-sectional area such as the disk shown in Figure 5–1c. All are cylinders and we could calculate the resistance any of them offers to the flow of electrical current from the same equation (see Eq. 3–1 of Chapter 3) if we know the cylinder dimensions and the resistivity of the material. Now, let us warp the cylindrical disk of Figure 5–1c into the shell of Figure 5–1d. If the dimension l is small, the areas A and A' on either side of the shell will be nearly equal. The equation

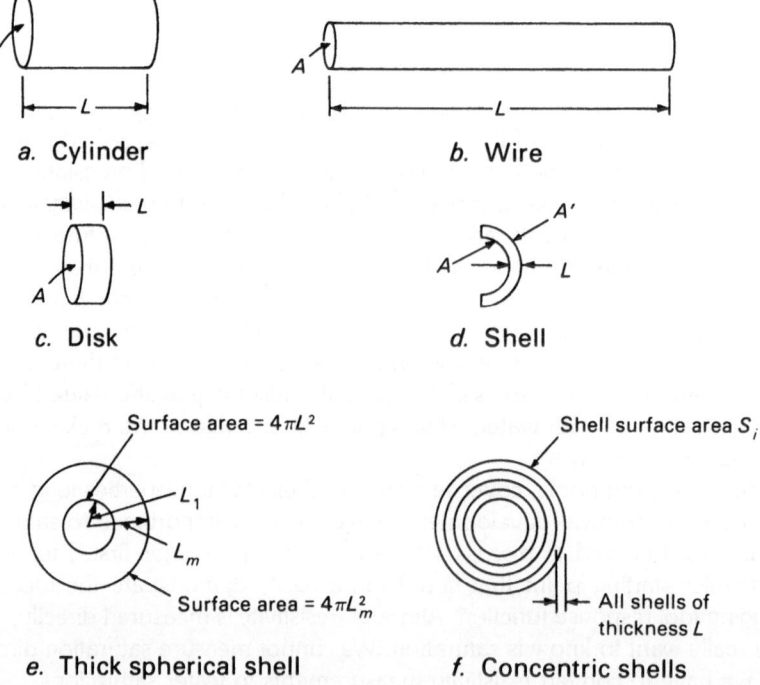

a. Cylinder

b. Wire

c. Disk

d. Shell

Surface area = $4\pi L^2$

L_1

L_m

Surface area = $4\pi L_m^2$

e. Thick spherical shell

Shell surface area S_i

All shells of thickness L

f. Concentric shells

Figure 5–1. Cylinders warped into a shell. *(After Hubert Guyod, Guyod's Electrical Well Logging, ©1944 by Welex, a division of Halliburton Company, Houston, Tex., 1944.)*

will give the resistivity for this shell the same as it does for the cylinders. Next, consider the spherical shell shown in cross section in Figure 5–1e. If the inner radius of the shell is L_1 and the outer radius is L_m, the area of either surface would be given by

$$\text{Area} = 4\pi L^2 \tag{5–1}$$

with L in meters if we are going to use units of ohm-meters for resistivity. We also need to substitute the appropriate radius for L in the equation. Let us divide the thick spherical shell into many concentric thin shells (see Fig. 5–1f) of surface areas S_1, S_2, ... and S_n and equal thicknesses L. Then, we can use Eq. 3–1 of Chapter 3 along with the area from Eq. 5–1 to find the resistance of each thin shell.

Since the total resistance in a series path is the sum of the individual resistances in the path, we can find the resistance of the complete spherical shell of Figure 5–1e from

$$\text{total resistance} = \sum_{k=1}^{n} R_t \times \Delta L_k/\text{Area}_k \tag{5–2}$$

where area is given from Eq. 5–1 in square meters, resistivity is in ohm-meters, and L_k is in meters. The resulting resistance will be in ohms with these units. We can use this equation to calculate the resistance between two equipotential spherical surfaces in a homogeneous and isotropic medium. The homogeneous and isotropic conditions assure us that the medium has the same resistivity throughout and that we can accurately measure it regardless of the direction of measurement or current flow. *Equipotential surface* refers to a surface whose potential (measured relative to some reference point) is the same everywhere on the surface. Here, the implication is that the reference point is the center of two concentric spherical surfaces whose common center could be the location for a current source.

For homogeneous and isotropic material, the equipotential surfaces would be spheres with radii L_1 for the inner or smaller spherical surface, and radius L_m for the larger or outer spherical surface. In Figure 5–1e, then, the two equipotential surfaces are represented by the inner and outer surfaces of the thick spherical shell. If we take Eq. 5–1 for the area of the spherical shells and let the number of shells approach an infinitely large number while the thicknesses ΔL approach zero, and take the sum to those limits, we will express the resistance as an integral

$$\lim_{n \to \infty} \sum_{k=1}^{n} (R_t \Delta L_k/\text{area}_k)$$

$$= \int_{L_1}^{L_m} (R_t/4\pi L_k^2)dL$$

$$= (R_t/4\pi) \times (1/L_1 - 1/L_m) \tag{5-3}$$

with units as for Eq. 5-2.

From Ohm's law we know that for a given current I_C, the voltage drop or potential change across this resistance will be

$$V = I_C \times \text{resistance}$$

$$= (I_C R_t/4\pi)(1/L_1 - 1/L_m) \tag{5-4}$$

With units as before, the potential is expressed in volts.

This equation can also be developed from Laplace's equation. It is also possible to develop equations for the *electrode* potential for various size electrodes and expressions for the resistivity provided by older electrical logs that were referred to as electrical survey or *ES devices* for short. These included the *normal* and *lateral* devices. They were the early resistivity devices, along with the single-point resistance tool and Spontaneous Potential (SP) log that we examine later. For reasons that will soon be clear, these older ES tools were discontinued in favor of improved resistivity devices. Nevertheless, some familiarization with them and the basic principles of electrical resistivity measurement is desirable since these older devices may still be encountered in data from wells drilled before 1960 (and sometimes afterward). In many cases these older ES devices or the SP log were the only logs run in a well.

Although we are limited in terms of the calculations that can be made from a single resistivity measurement without porosity data, recall from the problems for Chapter 3 (see Problem 3-8 and its solution) that saturation can sometimes be successfully calculated anyway. Those interested in further study and more theoretical development on a simpler level can use a reference such as Guyod.[2] Hilchie also provides a discussion on how to use the older ES logs in practice, including correction charts for various bed geometries and invasion effects (called *departure curves*).[3]

The single-point resistance measurement mentioned above was recorded from a single electrode lowered into a borehole. It essentially responded only to the resistance of the fluid in the borehole, as altered somewhat by the rock beds being logged. For this reason, it has proven useless as a resistivity tool, but it did have a distinct advantage of showing thin-bed boundaries and rock type changes clearly, which made it a good correlation tool.

The *lateral* device was a long spacing tool. Typically, the current electrode was placed 16 ft to 18 ft from the potential measuring electrodes. Thus, it did not work well in thin beds and rarely provided accurate measurements of resistivity except opposite very thick beds. It did, however, distinctly identify the bottom of a bed.

Only a *normal* ES device is described in detail. Figure 5–2 illustrates a typical *normal* resistivity device in a borehole. A current source, shown as a battery, is connected to the ground by electrode A_2, which we could call the

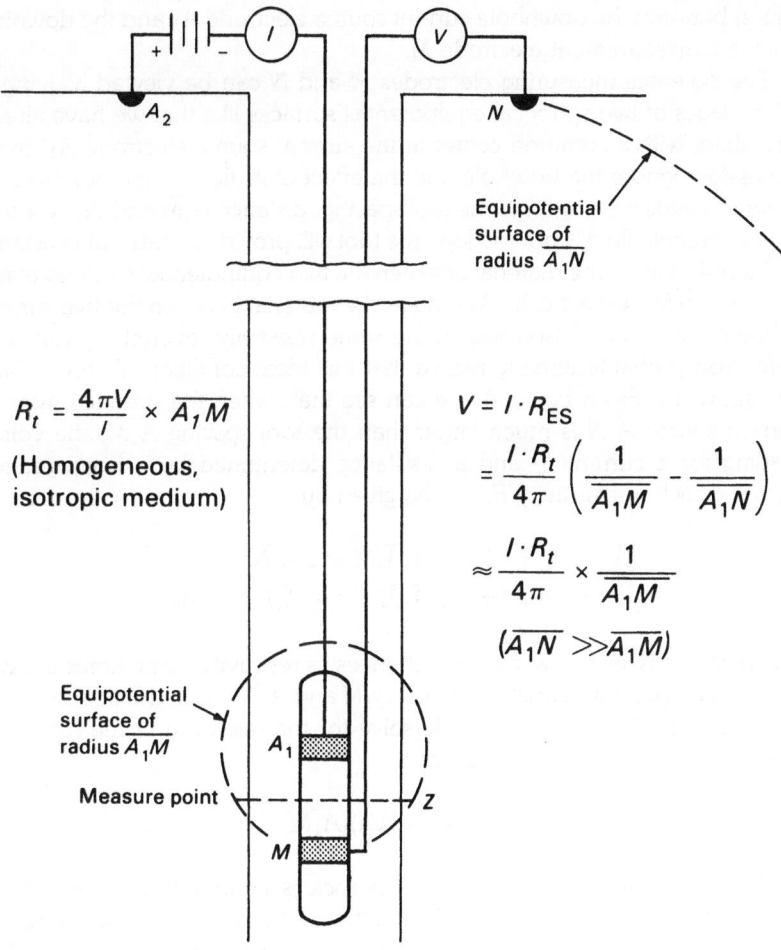

$$R_t = \frac{4\pi V}{I} \times \overline{A_1 M}$$

(Homogeneous, isotropic medium)

$$V = I \cdot R_{ES}$$

$$= \frac{I \cdot R_t}{4\pi} \left(\frac{1}{\overline{A_1 M}} - \frac{1}{\overline{A_1 N}} \right)$$

$$\approx \frac{I \cdot R_t}{4\pi} \times \frac{1}{\overline{A_1 M}}$$

$$(\overline{A_1 N} \gg \overline{A_1 M})$$

Equipotential surface of radius $\overline{A_1 N}$

Equipotential surface of radius $\overline{A_1 M}$

Measure point

A_1 = Current source
A_2 = Current return
M and N = Potential measurement electrodes

Figure 5–2. Normal resistivity device.

current *return* or *drain*. The other end of the battery is connected through meter I (used to read current) to a source electrode on the logging tool at A_1. The connection to the electrode on the tool is made through a wire that is part of the logging cable from the surface to the tool downhole. A voltmeter V is connected at the surface to electrode N and to the logging tool (via another wire in the downhole cable) at another electrode M. It has been found from experience that such a tool records a measurement related to the resistivity of the formation opposite the midpoint (point Z in the figure) between the downhole current source electrode A_1 and the downhole potential measurement electrode M.

The potential measuring electrodes M and N can be viewed as lying on the surfaces of two spherical equipotential surfaces like that we have already described, with a common center at the current source electrode A_1. In this discussion, ignore the borehole and the effect of its fluid. If the borehole size is not too large compared to the tool spacing, distance $\overline{A_1M}$ and the resistivity of the borehole fluid is not too low, the tool will provide a reasonable estimate of the resistivity of the material between the two equipotential surfaces of radii $\overline{A_1N}$ and $\overline{A_1M}$, respectively. As long as the material between the two surfaces is homogeneous and isotropic of the same resistivity, everything will be all right. You probably already realize that this ideal condition is very unlikely for real rocks. From Eq. 5–4, we can see that, when the distance from the current source $\overline{A_1N}$ is much larger than the tool spacing $\overline{A_1M}$, the voltage reading for a current I_C and a resistance determined by a homogeneous isotropic rock of resistivity R_t will be given by

$$V = (I_C R_t)/4\pi \times (1/\overline{A_1M} - 1/\overline{A_1N})$$
$$= I_C R_t/4\pi \times 1/\overline{A_1M} \qquad \overline{A_1N} >> \overline{A_1M}$$

The problem is likely that the rock changes its resistivity many times between the two equipotential surfaces of radii $\overline{A_1M}$ and $\overline{A_1N}$. From the above expression for large $\overline{A_1N}$, we can readily solve for the resistivity of the rock given the voltage V and current readings I.

$$R_t = (4\pi V/I)(\overline{A_1M})$$

We can determine what volume of rock is contributing to the voltage reading by finding what value of $\overline{A_1N}$ in the (as a multiple of $\overline{A_1M}$) expression $(1/\overline{A_1M} - 1/\overline{A_1N})$ corresponds to a particular fraction of the value of $1/\overline{A_1M}$. This latter value is the value of the expression when $\overline{A_1N}$ is infinitely large. We can solve for X from the equality $(1/\overline{A_1M} - 1/X) = $ fraction $\times (1/\overline{A_1M})$. For example, if we want to know what distance X from the measure point will give a voltage reading that is 90% (fraction = 0.9) of the maximum possible voltage reading, we will find $X = 10$ times the spacing $\overline{A_1M}$. For a voltage reading that is one-half of the maximum, we will find that X is twice the tool spacing. Thus, 90% of the tool reading is coming from rock that is within 10 times the tool spacing from the measure point, while only half of the tool response is due to rock within two tool spacings of the measure point on the tool. Figure 5–3 illustrates this *geometrical factor* concept for the *normal* electrode configuration resistivity logging tool. The *normal* configuration refers to the single current source and single potential electrode downhole on the tool with the other single current electrode and single potential electrode at the surface. However, the adjective *normal* is not intended to imply anything special about this particular tool configuration compared to other electrode arrangements.

In terms of well log responses, the normal electrode configuration provides an *apparent* resistivity measurement. The geometric factor developed for the normal configuration is predicated on the assumption that all the material between the two equipotential surfaces is isotropic and homogeneous, having the same resistivity throughout the enclosed space. If the resistivity does change within the enclosed space, the two equipotential surfaces will not be

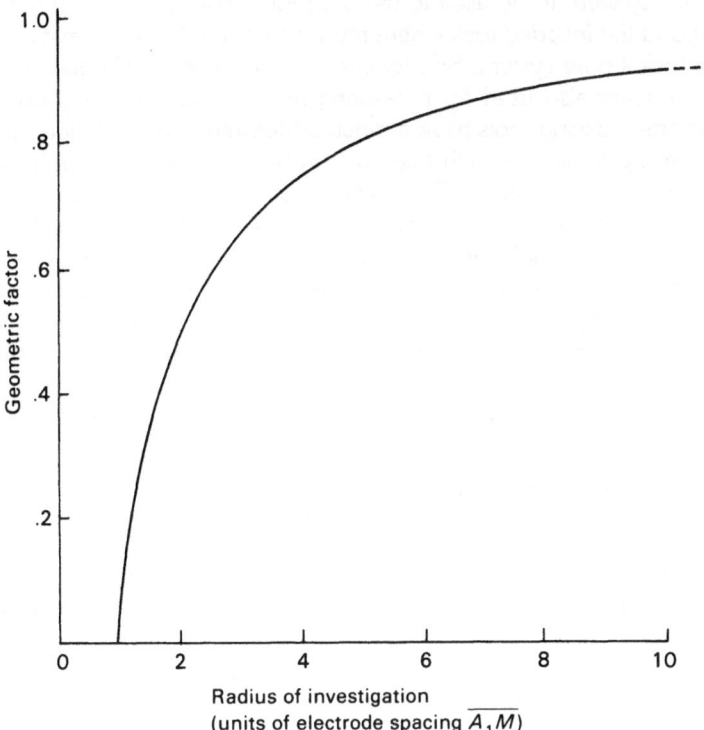

Figure 5–3. Geometric factor for ES device.

spherical. Fortunately, modern resistivity devices, such as the focused current and induction logs, reduce the volume of material affecting the resistivity measurement. The ideal is to reduce the effect of adjacent beds above and below the bed of interest and effects from the borehole. In practice, you must regard resistivity measurements as apparent resistivities, even with modern tools that are more accurate.

The *lateral* configuration of electrodes is actually quite similar to the *normal* tool. With the lateral device, one more electrode (it actually does not matter which: current or potential because of a reciprocity principle in electrical measurements) is placed downhole. This slight difference in configuration, however, gives a quite different response pattern. The lateral devices were made with spacings of the order of 20 ft so that they responded from rock considerably farther from the borehole, beyond the invaded portion of rock where borehole fluids might be found in the pore system. However, these lateral devices were useless except for very thick beds (thicker than the tool spacing). For thinner beds they had *blind* zones below bed boundaries where the rock opposite the tool did not even affect the response of the tool.

The normal device tools were usually made with $\overline{A_1M}$ spacings of 64 in. when they were to be used to measure the resistivity of the undisturbed rock beyond the *invaded* rock where mud system fluids could have migrated into the rock's pore system. Shorter spacing tools with $\overline{A_1M}$ equal to 16 in. and 18 in. were also used for measuring resistivity closer to the borehole. These shorter spacing tools have a much better thin-bed resolution than the longer spacing tools, although they will probably not read a *true* resistivity of the uninvaded formation. Even with more modern resistivity tools, we can generally count on longer spacing tools responding from rock farther away from the borehole, whereas shorter spacing tools provide a better bed resolution. For resistive beds that are thinner than the tool spacing, the normal devices displayed a so-called crater pattern where they did not respond to the formation resistivity. In fact, the recorded resistivity showed a reversal in resistive thin beds. However, these *craters* could be used to identify thin-bed boundaries. The thickness of the crater was equal to the tool spacing $\overline{A_1M}$ plus the bed thickness. Figure 5–4 illustrates these concepts.

Although the normal devices did not perform well when low resistivity, salt mud systems were used in the drilling program, these tools were commonly used in most older logging suites. A short-spacing version of the normal device was still used in combination with more modern induction logging devices well into the 1970s. Despite the limitations of the normal device with the short spacing (not necessarily recording the flushed zone resistivity or the undisturbed zone resistivity), elaborate interpretation schemes were developed to combine the *short normal*, as it was called, with the induction log readings. These two devices, the 16 in. normal and induction log, were recorded simultaneously in the borehole and were sometimes known as an

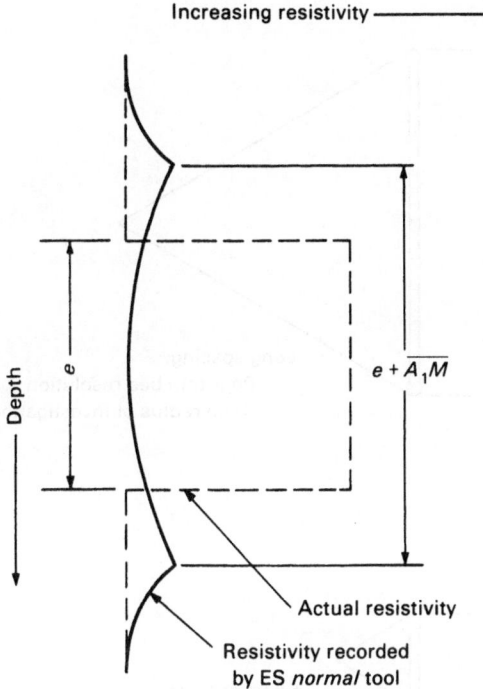

Figure 5–4. Cratering on normal resistivity device.

Induction-Electric Log survey (IES). The interpretation schemes were advertised to provide not only both undisturbed zone resistivity and its ratio to the flushed zone resistivity but also relative saturation information. No doubt these schemes worked in some cases. Fortunately, cross-plotting methods were later developed that, in retrospect, provide a better interpretation that is less subject to the necessary assumptions of the more elaborate schemes and is more resistant to errors in tool readings.[4]

Figure 5–5 illustrates the concepts of tool electrode spacing versus radius of investigation. This is common to modern electrical devices as well: the shorter the electrode spacings, the better the thin-bed resolution, but the more likely the resistivity reading is being affected by fluids migrating into the formation from the borehole fluid if there is sufficient porosity in the rock. Unlike acoustic logging tools, electrical devices are affected by rock both above and below the electrodes or *sensors*. If you need to obtain a measurement of resistivity from one of the old ES surveys, the 64 in. normal or *long normal* curve is the best for most situations because the lateral device only responded adequately for very large bed thicknesses. You can assume the resistivity is equal to the apparent resistivity recorded by the normal

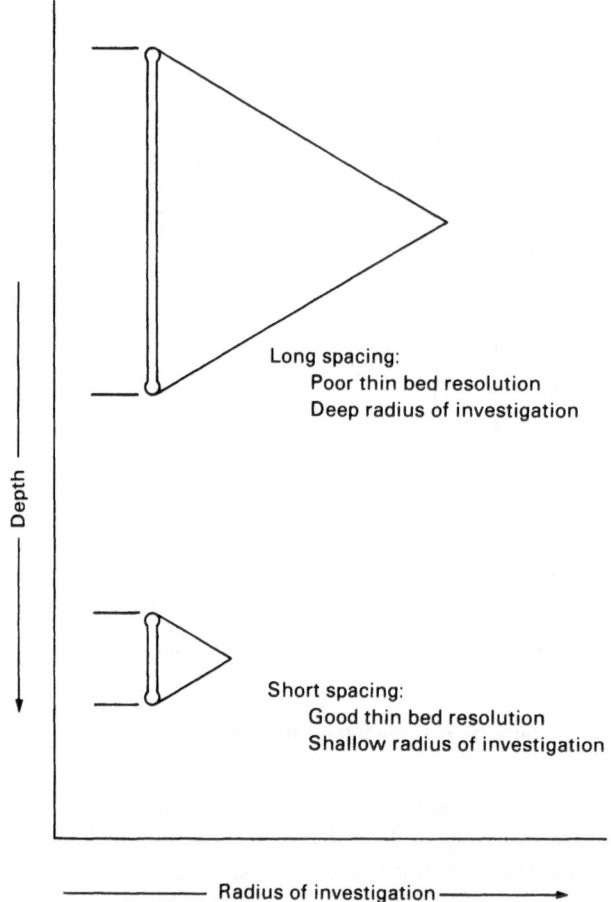

Figure 5–5. Tool spacing versus radius of investigation.

device, or you can use the appropriate service company chart books if you anticipate bed-thickness or invasion problems.

Figure 5–6 shows the typical borehole environment. The borehole is shown with a mud system whose resistivity is denoted R_m. If there is any porosity and permeability in the rock, a mudcake usually forms on the borewall. Its resistivity is denoted by the symbol R_{mc}. If the rock is permeable and porous, the water in the drilling mud, known as mud filtrate, will move through the mudcake on the borewall into the pore system of the rock formation. Its resistivity is denoted R_{mf}. This mud filtrate will displace formation water, and often a large part of any contained hydrocarbons, from the pore system. As a result, the flushed zone near the borewall contains

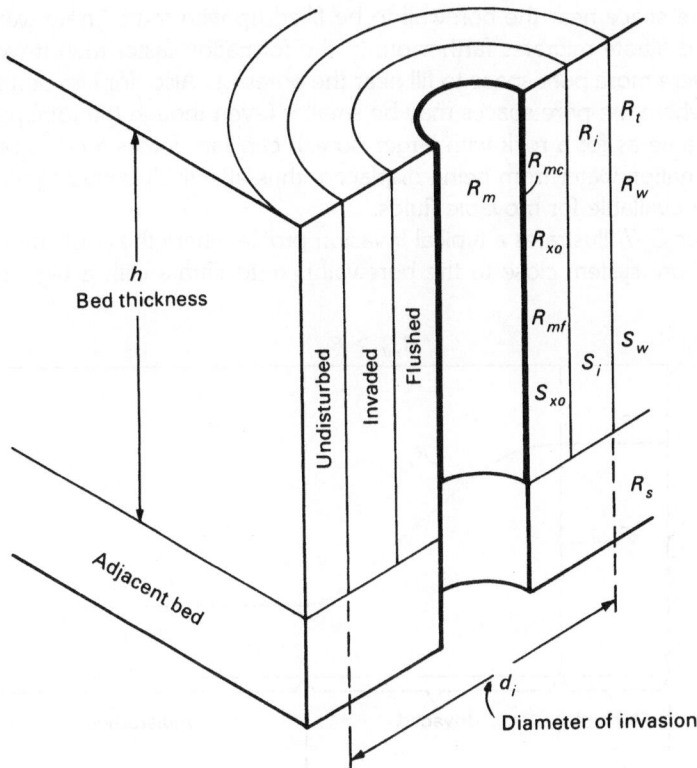

Figure 5–6. Borehole environment.

this mud filtrate instead of formation water. This causes the flushed zone to have a different resistivity (if $R_{mf} \neq R_w$) than the undisturbed rock farther away from the borewall. The flushed zone resistivity is denoted by R_{xo} and the saturation of the flushed zone by S_{xo}. Farther out, not all the formation water has been displaced by mud filtrate. This is known as the *invaded* zone whose resistivity is denoted by R_i and whose saturation is denoted by S_i. In this zone, the pore system will contain both mud filtrate and formation water. Finally, we have a zone that is not contaminated by mud filtrate from the mud system. Only formation water (plus any hydrocarbon that may have been present) is contained in the pores of the rock. Here, we denote the formation water resistivity by R_w and the undisturbed zone saturation by S_w. At the point where the invasion process terminates, we identify the *diameter of invasion* d_i. Figure 5–6 also shows the typical symbol used for resistivity of the adjacent or surrounding beds next to the zone of interest: R_s.

It is of interest to note that the diameter of invasion is considered to be smaller for the more permeable and porous rocks. This is because there is

less pore space near the borewall to be filled up with mud filtrate water, so
the mud filtrate migrates farther out in the formation faster than it would if
there were more pore space to fill near the borewall. Also, for less permeable
rocks where the pore spaces may be smaller (even though the total porosity
is the same as for a rock with larger pores), capillary forces tend to prevent
the formation water from being displaced, thus effectively reducing the pore
volume available for movable fluids.

Figure 5–7 illustrates a typical invasion profile where the contained water
in the pore system close to the borewall is mud filtrate with a higher resis-

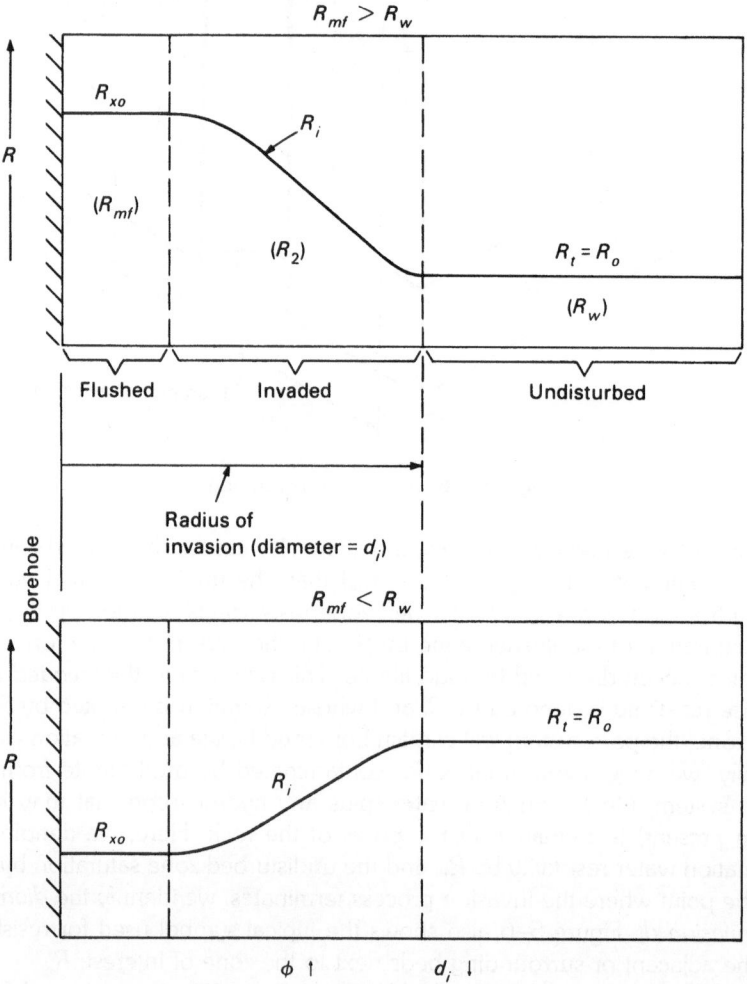

Figure 5–7. Invasion profile. (Note: Invasion in water zones requires permeability
and porosity.)

tivity than the formation water farther out from the borehole. The resistivity of the rock will reflect this difference in the resistivity of the contained fluids. For this reason, we need to find the undisturbed zone resistivity in order to calculate the water saturation or hydrocarbon saturation. Even if the mud filtrate resistivity fortuitously is the same as the formation water resistivity, we still need a resistivity measurement in the undisturbed zone where fluid saturations have not been altered by fluid invasion from the mud system. In Figure 5–7, if the mud filtrate resistivity is less than the formation water resistivity, the resistivity profile will be reversed, with the resistivity of the invaded zone being smaller than the undisturbed zone (see the lower part of Fig. 5–7).

Figure 5–8 illustrates typical borehole resistivity measuring tools. The electrode spacing will determine the *radius of investigation* of the tool, that is, whether it responds primarily to R_t, R_i, or R_{xo}. Note that the tools usually used to measure either R_t or R_i are centered in the borehole, whereas tools designed to measure R_{xo}, and usually referred to as R_{xo} devices, are

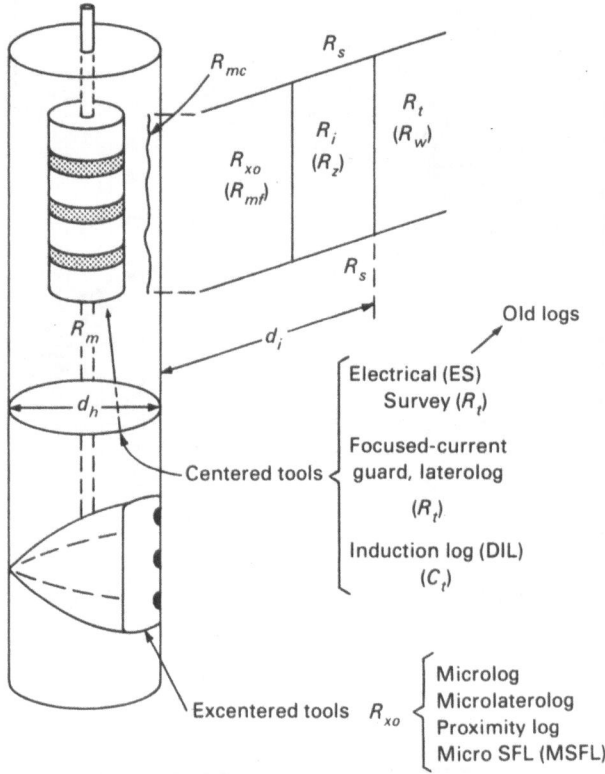

Figure 5–8. Borehole resistivity tools.

excentered or pressed against the borewall. Their readings will be valid only so long as the tool sensors remain against the side of the borewall. If the pad loses contact, the tools will essentially be reading the mud resistivity in the borehole. Note the symbol R_z used to refer to the fluid resistivity in the invaded zone.

The centered tools include the ES devices just discussed as well as the focused-current (laterolog or guard) logs and induction logs that are discussed shortly. Note that there are several R_{xo} devices—tools that have been developed to measure R_{xo}. The microlog was an older pad-mounted device that used a *micronormal* device with an electrode spacing of 1 1/2 in. along with a *microinverse* device with an electrode spacing of 2 in.

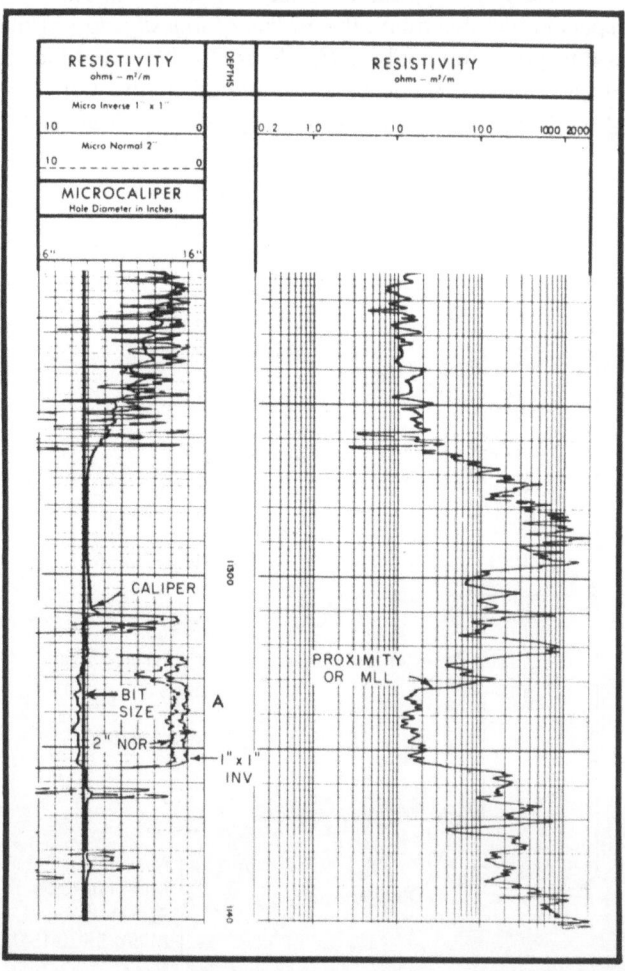

Figure 5–9. Microlog example. *(Courtesy of Schlumberger Ltd.)*

The micronormal device was a miniature normal type electrode configuration tool, whereas the microinverse was a miniature lateral electrode spacing device. Opposite a borewall with mudcake, the devices indicated a positive separation with the microinverse tool reading somewhat higher. These microlog devices were mounted on different types of pads (D pad and H type hydraulic pad) and charts were developed for each that used elaborate interpretation schemes whereby it was supposed that R_{xo} could be calculated from the tool readings. In practice, the tool turned out to be a very useful mudcake and permeability indicator, if not a quantitative resistivity device. Figure 5–9 is an illustration of a typical recorded microlog.

INDUCTION LOGGING

Electrical surveys, using the normal electrode configuration and lateral electrode configuration, require that a conductive mud system be used in a borehole. Nonconductive, oil-base muds were introduced in the Rangely, Colorado, oil fields in 1948. The induction log was developed as a resistivity tool in response to the introduction of nonconductive mud systems.

Another useful application of the induction log today is in air- or gas-drilled holes, which also do not contain the conductive fluid necessary for ES surveys. Because of the capability of focusing the electromagnetic field from the induction log, it also has a better response characteristic than the old ES log, which it has replaced. The induction log is a conductivity tool and responds to conductivity measurements with an accuracy of $\pm$ 2 millimhos per meter. However, at least one logging service company states that its tool can be calibrated downhole to a precision better than this. Being a conductivity tool, it is excellent for accurate measurements of low resistivity rocks where fresh muds (or oil-based, air, or gas) are used. The induction log is the preferred resistivity logging tool for rocks with resistivities less than 100 ohm-meters. It is the only tool that can be used in nonconducting mud systems.

Induction logs were initially interpreted using the geometric factor concept developed by Doll.[5] Figure 5–10 illustrates the concept for induction logs. If we generate an alternating current in a coil, such as the transmitter coil in Figure 5–10, this current will, in turn, generate a time-varying magnetic field about the transmitter coil. If the magnetic lines from this field cut through any other conductive loop, an electric current will be induced in the conductive loop. This induced current will, in turn, propagate a magnetic field that will now induce a current back at the receiver coil in the tool. This conductive loop around the tool will be comprised of a complete conductive path through the rock. There will be many such possible paths. They are often called ground loops.

A current will also be induced in the receiver coil direct from the magnetic field propagated by the transmitter (*mutual induction*), however, steps will

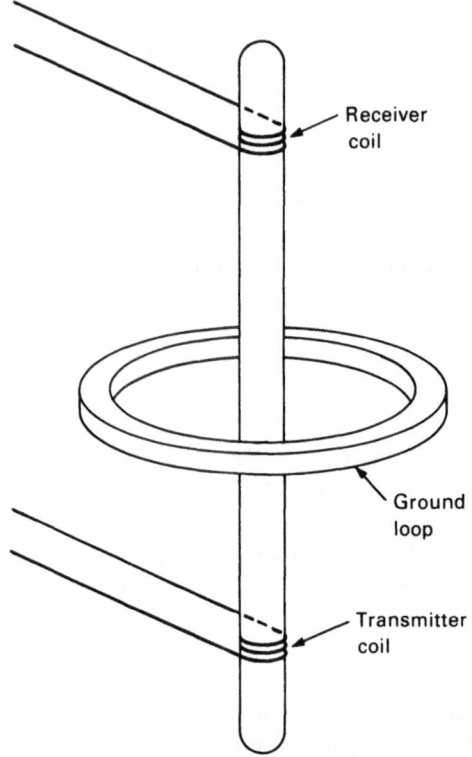

Figure 5–10. Geometric factor concept.

be taken in the instrumentation to eliminate this and any other undesired signals. The desired signal is the signal propagated from the ground loops in the surrounding rock back to the tool receiving coils. With the geometrical factor concept, the apparent conductivity of the rock is simply the sum of the separate conductivities of each ground loop. Unfortunately, this apparent conductivity does not equal the true conductivity of the rock. Each individual ground loop, after having a current induced in it from the transmitter coil, in turn propagates an electromagnetic signal not only back to the receiving coil but to each of the other ground loops as well. The geometrical factor concept cannot account for this additional signal.

According to Doll, the voltage induced in the receiver coil due to the *pth* ground loop is

$$V = (K/L)g_p\sigma_p \qquad\qquad (5\text{–}5)$$

where σ_p is the conductivity of the *pth* ground loop.[6] K is a constant dependent on coil characteristics (dimensions, numbers of turns, frequency of transmitter signal, geometry of the coil, etc.), and L is the main coil spacing (meters). g_p is the geometrical factor that depends only on the geometry of the ground loop and main coil pair. Thus, Eq. 5–5 is of particular interest in that it tells us that the voltage induced at the receiver can be expressed as the product of only three terms: one containing only coil characteristics (K/L), one involving the geometry of the coil pair and ground loop (g_p), and a term equal to the true conductivity of the medium. The problem with this simplified interpretation is that it ignores the *propagation* effect whereby each individual ground loop induces signals not only back at the receiver but into every other ground loop. This is sometimes referred to in the literature as the *skin effect*.

Since most tool designs are based on the geometrical factor concept, provisions are usually made to correct the induction log readings, usually electronically or from departure curves. These corrections are necessarily based on ideal geometries such as an infinite medium, uninvaded thin bed, or infinitely thick invaded bed (see Fig. 5–11). Of course, more sophisticated models may be used for correction with modern, computer-operated tools, but these will likely involve simplifications and assumptions (if they are used on a routine basis) that also may not reflect actual conditions in the surrounding rock.

Let us examine examples of a geometric factor application. For ease of understanding, I have made two geometric factor curves shown as Figure 5–12 and Figure 5–13.[7] We must assume for both of these figures for geometric factor that they pertain to a known borehole diameter. Figure 5–12 is a computed geometric factor curve for a five-coil system (5FF40) in a single thin bed with no invasion. For bed thicknesses greater than 60 in. the geometric factor approaches 1. Note, however, that induction logs may not reflect true resistivity for beds much thinner than 60 in. For thinner beds, the geometric factor is significantly less than 1. A 20 in. thick bed of infinite lateral extent and no invasion has a geometric factor of about 0.3. The thin bed will account for only 30% of the conductivity reading of the tool. The remaining 70% of the response is contributed by the beds above and below the thin bed.

Figure 5–13 shows the geometric factor for an infinitely thick, invaded bed for the same tool (5FF40). This time, the horizontal axis represents the radius of invasion measured from the center of the borehole. For the invaded zone, the geometric factor gives the proportion of the signal coming from the invaded zone. For relatively shallow invasion, we see that very little signal is contributed from the invaded part of the rock. However, for an invasion radius of 60 in., which is very deep invasion, we see that the invaded zone will contribute most of the signal received by the tool. For this case, it is

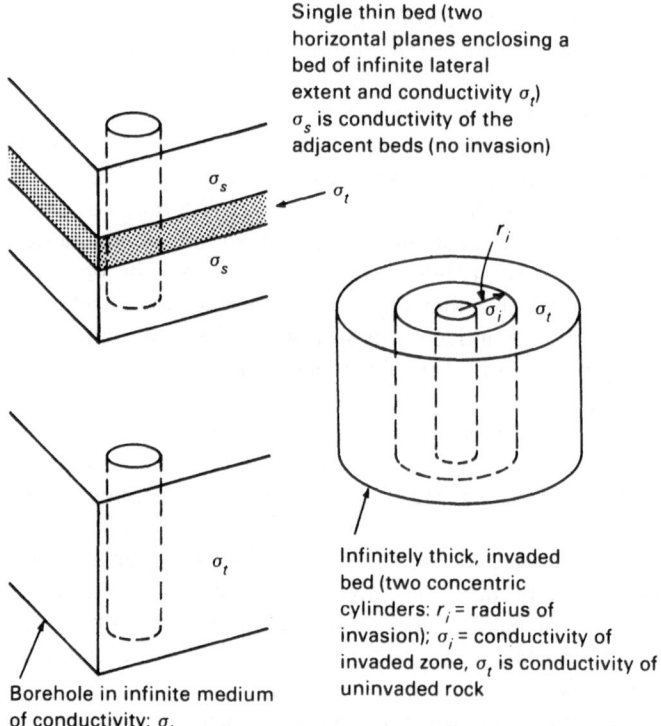

Single thin bed (two horizontal planes enclosing a bed of infinite lateral extent and conductivity σ_t) σ_s is conductivity of the adjacent beds (no invasion)

Infinitely thick, invaded bed (two concentric cylinders: r_i = radius of invasion); σ_i = conductivity of invaded zone, σ_t is conductivity of uninvaded rock

Borehole in infinite medium of conductivity: σ_t

Figure 5–11. Ideal geometries for induction log corrections. (Note: Radius of invasion is measured from borehole center, whereas depth of invasion is measured from borewall.)

unlikely that we will be able to ascertain the true resistivity of the uninvaded portion of the rock. The tool cannot *see* deep enough into the formation.

Commercial induction logging systems used today are mostly six-coil tools, with three transmitters and three receivers. Other designs are also used. The 5FF40 tool is now known as the *medium* induction log (ILM is Schlumberger's abbreviation) in reference to its somewhat less deep radius of investigation in comparison to the *deep* induction log (6FF40 or ILD in Schlumberger's terminology). By recording measurements with these two induction logs simultaneously (known as a *dual induction* tool), some knowledge may be gained of the invasion profile. This allows for correcting the deep-reading tool for invasion effects, if any. This dual induction combination tool is normally run in conjunction with another tool, either a spherically focused log (SFL) or *Laterolog 8* (LL8). Some companies may use a *shallow guard* log. All of these tools read only a few inches into the invaded zone not far from the borehole. These readings, together with the dual induction

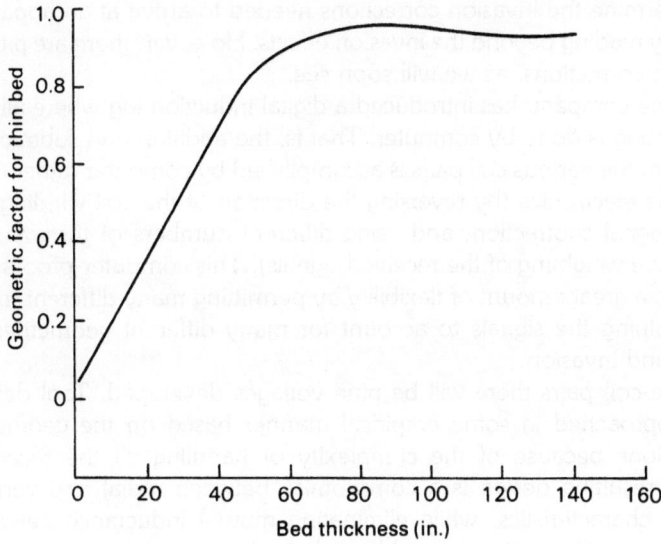

Figure 5–12. Geometric factor for thin bed.

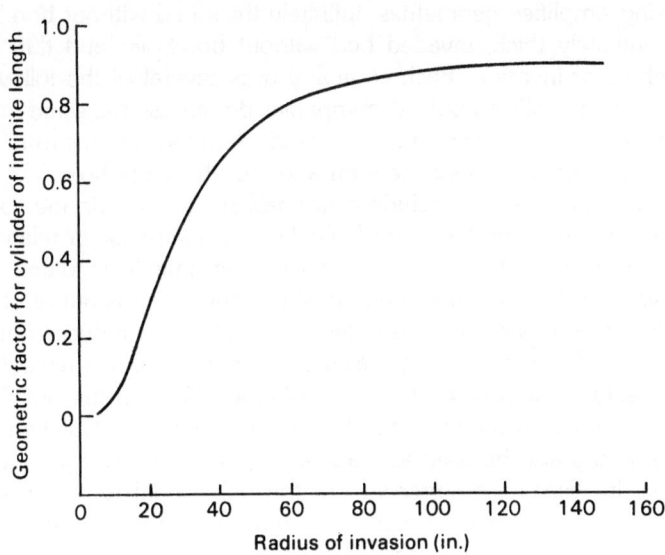

Figure 5–13. Geometric factor for invasion.

reading, determine the invasion corrections needed to arrive at an apparent true resistivity reading beyond the invasion effects. However, there are pitfalls with invasion corrections, as we will soon see.

At least one company has introduced a digital induction log where all the signal processing is done by computer. That is, the addition and subtraction of signals from the various coil pairs is accomplished by computer rather than the downhole electronics (by reversing the direction of the coil windings to accomplish signal subtraction, and using different numbers of turns in the coils to achieve weighting of the received signals). This computer-processing method adds a great amount of flexibility by permitting many different models for combining the signals to account for many different geometries of thicknesses and invasion.

With three-coil pairs there will be nine voltages developed. Tool design is usually approached in some empirical manner based on the geometric factor equations because of the complexity of handling all the received signals. The resulting design is a compromise between radial and vertical investigation characteristics, while eliminating mutual inductance between the main coil pair. Corrections must be made for borehole signal (via charts, downhole electronics, or computer processing for the digital sonic log) from any drilling fluids, and for propagation effects. Because of the complexity and difficulty of finding solutions to Maxwell's equations with appropriate boundary conditions, the corrections for skin effect are usually restricted to the following simplified geometries: infinitely thick bed without borehole or invasion; infinitely thick, invaded bed without borehole; and thin bed without borehole or invasion. Each company uses several of the following manipulations on its tool's output. All companies do not use the same manipulations, so we should not expect different induction log systems to record exactly the same resistivity everywhere on a log for the same hole.

The possible manipulations include a normalization to divide the sonde voltage output by the factor necessary (based on geometric factor relations) to make the result equal to the true resistivity in an infinite medium. The response is simplified by rejecting the part of the signal that is out of phase (phase discrimination) with the transmitter signal. This also reduces mutual inductance. Skin-effect corrections are also possible, but will be truly correct insofar as the actual borehole environment resembles one of the simplified models used. A multiplier based on test tank experiments or calculations for the coil system may also be used to make the apparent resistivity equal to the true resistivity. Finally, a correction for the hole signal (signal derived from the borehole fluids) may be made. It is not unusual for bed thickness corrections and hole signal corrections to be made from charts. We will examine these corrections in more detail shortly. An invasion correction is also available for modern tools that employ two induction tools in one, each with a different coil spacing and correspondingly different radius of investigation. This correction will be discussed shortly.

Figure 5–14 illustrates an induction log that uses a system of coils to propagate an electromagnetic field into the surrounding rock away from the borehole. In addition to the main coil pair shown with a spacing of 40 in. separating them, the induction log uses a system of additional coils to focus the propagated electromagnetic wave so that it penetrates farther from the borehole into the uninvaded portion of rock. These additional coils also eliminate mutual induction between the main transmitter–receiver coil pair. In this figure, electrical rock property measured by the tool as conductivity is labeled σ_t, with appropriate subscripts for the mud column (*m*), invaded zone (*i*), adjacent bed (*s*), and uninvaded zone (*t*). This is to emphasize

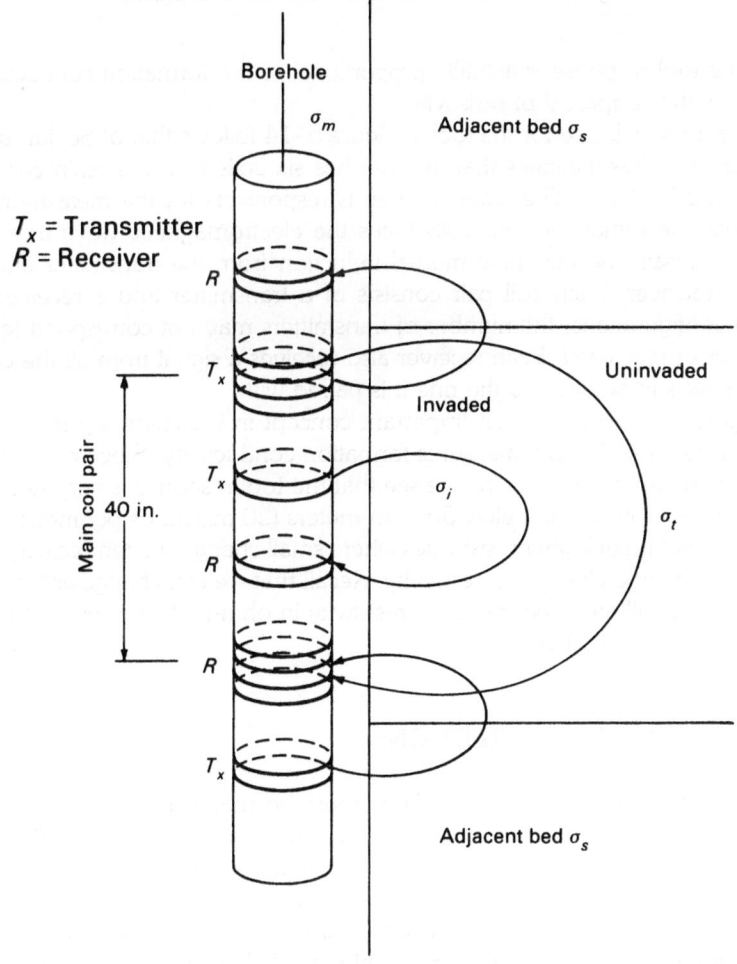

Figure 5–14. Induction log (conceptual) following Schlumberger's 6FF40.

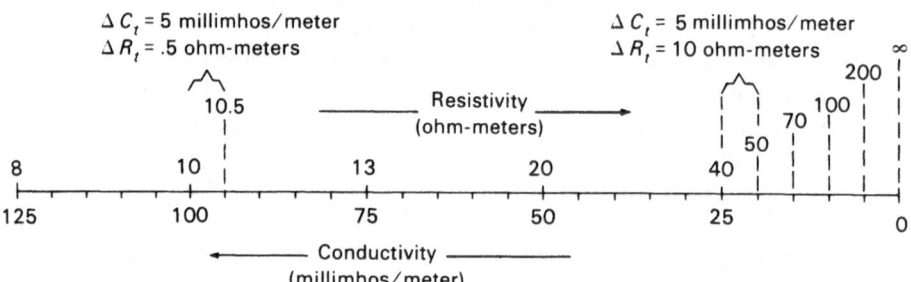

Figure 5–15. Conductivity is reciprocal of resistivity.

that the tool response is actually proportional to the formation conductivity, which is the reciprocal of resistivity.

The nomenclature for the tool in Figure 5–14 follows that of Schlumberger's 6FF40. This indicates that the tool has six coils with the main coil pair separated by 40 in. The main coil pair is responsible for the measurement, whereas the remaining four coils focus the electromagnetic wave from the main coil pair and eliminate mutual induction from the transmitter directly to the receiver. Each coil pair consists of a transmitter and a receiver (in some configurations, the number of transmitters may not correspond to the number of receivers). Each receiver also receives a signal from all the other transmitters in addition to the one it is paired with.

Figure 5–15 illustrates an important concept in induction log interpretation. Recall that the tool measures formation conductivity. Since resistivity is the reciprocal of conductivity, we see that the tool response is very sensitive for low resistivities, say, below 50 ohm-meters (20 millimhos per meter), but lacks resolution at higher resistivities where small changes in conductivity correspond to large changes in resistivity. Recall that we can change either conductivity in millimhos per meter to resistivity in ohm-meters or vice versa by dividing either into 1,000.

INDUCTION LOG CORRECTIONS

Corrections to induction log readings must be made in a specified order according to the particular service company procedures. The corrections for the Schlumberger induction tools will be illustrated. Although most logging service companies follow similar procedures, it is best to consult the particular company manuals or ask technical representatives for their correct procedures. Corrections for the Schlumberger induction tools are applied in the following order.

1. Borehole corrections, if necessary, are applied to compensate for the signal contributed by the fluid in the borehole. This correction is applied according to the borehole size and mud system resistivity.
2. Bed thickness corrections, if necessary, are applied after hole signal corrections are applied. These corrections are applied to thin resistive beds sandwiched between conductive beds or thin conductive beds sandwiched between resistive beds.
3. Invasion corrections, if necessary, are applied only after all other necessary corrections are made. These corrections are required if fluids from the mud system invade deep into the rock away from the borehole where they can influence the deep-reading resistivity tool. The correction is applied according to the mud system resistivity and the apparent depth of invasion as inferred from different resistivity tool readings.

Borehole Signal Corrections

Borehole signal corrections for the Schlumberger induction logs (ILM and ILD) are made from Figure 5–16. This is chart Rcor–4a from Schlumberger's 1986 chart book. The hole conductivity signal is subtracted, if necessary, from the induction log conductivity reading before any other corrections are made. Note that the correction applies to a particular borehole size and mud resistivity. It is very important to note that some induction log signals are already adjusted for the hole size on the recorded curve. Refer to the log heading or consult the logging engineer to be sure. If there are significant changes in borehole size (differing from the nominal hole size used to correct the recorded signal) in a zone of interest on one of these precorrected logs, it may be necessary to consult the engineer or assume that the nominal hole size was close to the recorded bit size and compute and subtract any necessary additional correction.

Figure 5–16 is a graph of the borehole geometrical factor versus hole diameter in both millimeters (at the top) and inches (at the bottom) on the horizontal axis. A nomograph on the right side of the figure can be used to solve the equation

$$\text{hole signal} = \text{borehole geometrical factor}/R_m \qquad \textbf{(5–6)}$$

where the hole signal to be subtracted is in milli-Siemens per meter (mS/m) if expressed in modern international standard units, or the equivalent units, millimhos per meter (mmho/m), and the mud resistivity R_m is in units of ohm-meters.

Note on the graph that each small square represents 10 mm or 0.4 in. The example in the figure shows that in a 14.6 in. borehole a 6FF40 deep

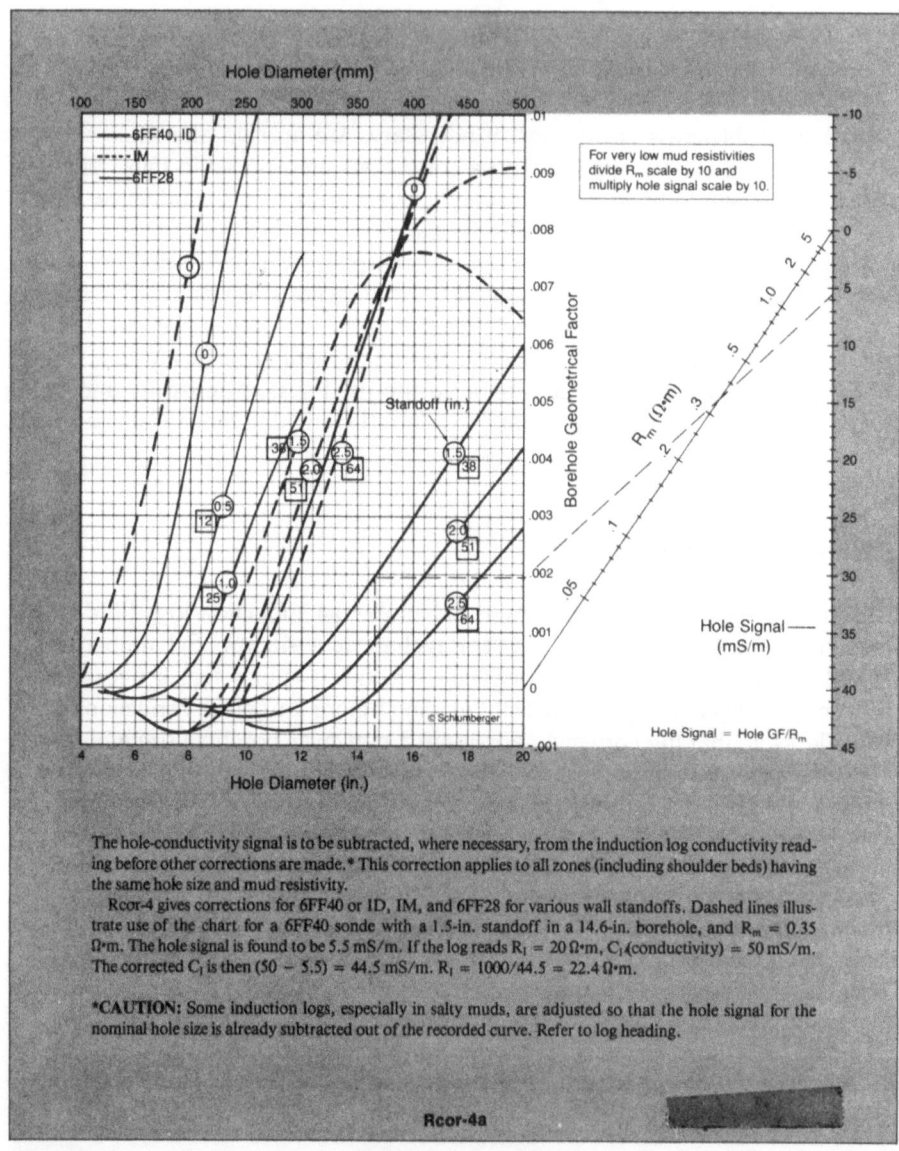

The hole-conductivity signal is to be subtracted, where necessary, from the induction log conductivity reading before other corrections are made.* This correction applies to all zones (including shoulder beds) having the same hole size and mud resistivity.

Rcor-4 gives corrections for 6FF40 or ID, IM, and 6FF28 for various wall standoffs. Dashed lines illustrate use of the chart for a 6FF40 sonde with a 1.5-in. standoff in a 14.6-in. borehole, and $R_m = 0.35$ Ω·m. The hole signal is found to be 5.5 mS/m. If the log reads $R_t = 20\ \Omega$·m, C_t(conductivity) = 50 mS/m. The corrected C_t is then (50 − 5.5) = 44.5 mS/m. $R_t = 1000/44.5 = 22.4\ \Omega$·m.

*CAUTION: Some induction logs, especially in salty muds, are adjusted so that the hole signal for the nominal hole size is already subtracted out of the recorded curve. Refer to log heading.

Rcor-4a

Figure 5–16. Induction log borehole signal corrections. *(Courtesy of Schlumberger Ltd.)*

induction tool with a 1.5 in. standoff and a mud resistivity of 0.35 ohm-meters has a hole signal of 5.5 mS/m. If the log reading is 20 ohm-meters, the conductivity is 50 mS/m (or millimhos per meter) and the corrected reading is (50 − 5.5) = 44.5 mS/m. This is a resistivity of 22.4 ohm-meters. We see that the hole signal effectively reduces the induction log reading below the correct resistivity. Thus, we subtract the hole signal conductivity from the recorded conductivity to make the correction for hole signal.

The *standoff* referred to in the figure is a rubber fin device designed to keep the induction tool from contacting the borewall. Note that for zero standoff, the hole signal correction will be extremely high. Also, without a standoff, it is often difficult to obtain a good repeat log with the induction tool. On the graph, the standoff is indicated in inch units in the circles and millimeters in the squares. The solid curves to the right of the graph pertain to the 6FF40 deep induction tool, whereas the short dashed curves (shown in red ink in Schlumberger's new chart books) to the left are for the 6FF28 deep induction tool. The 6FF28 is a slim-hole tool designed for use in small holes. The longer dashed curves are the curves for the medium induction log hole signal corrections. Note that, for the same mud resistivity and hole size as for the above example for the deep induction tool, the borehole geometric factor is nearly .0074 for a medium induction log (ILM or IM), which means the hole signal correction for the medium induction tool will be over three and a half times that for the deep induction log, which had a borehole geometric factor of slightly less than .002.

Also note from the figure that for most holes, say, less than 12 in. in size, the deep induction tool with a standoff requires no correction. For medium induction logs with hole size less than 9 in., the correction should also be unnecessary. Since the slim-hole 6FF28 will likely be confined to small hole applications, it is likely that no correction will be necessary for it except in washed out zones. Note that the lower the resistivity (higher conductivity) of the mud, the larger will be any necessary corrections to the induction conductivity measurement.

Bed-thickness Corrections for Induction Logs

After any necessary hole signal corrections are made, the induction log readings must be adjusted for any necessary thin-bed corrections. What is a thin bed? Contrary to popular belief, a thin bed for an induction log can be 20 ft thick, as we will soon see. The common rule of thumb that the induction log requires no correction for beds thicker than 5 ft (as implied by geometrical factor theory) can be a dangerous assumption. A further problem arises because bed-thickness corrections must be applied before invasion corrections. The problem is that it is possible to apply routine computer corrections for both invasion and hole signal. However, it is difficult to apply routine bed-thickness corrections. For this reason, many computerized analyses employ routine invasion corrections where bed-thickness corrections have not been applied. Shortly, we will see an example of the dangerous pitfall that can arise from this unfortunately routine practice of using computerized invasion corrections without bed-thickness corrections.

Figure 5–17 shows the correction for thin conductive beds assumed to be bounded by infinitely thick resistive beds above and below of the same resistivity R_s. The figure shows the correction for only the deep induction log (ID) or 6FF40 and the 6FF28 slim-hole tool. No thin conductive bed correction is available in the Schlumberger charts for the medium induction log. To use Figure 5–17 for the 6FF28 tool, multiply the bed thickness by 1.43 before using the graph. Thus, use a graph thickness of 4 ft for an actual 2.8 ft bed with a 6FF28 tool. The graph is only valid for a shoulder bed resistivity (SBR) setting of 1 ohm-meter. This SBR setting can be found on the log heading information (the SBR setting refers to a filtering process designed to improve tool response).

Note that for conductive thin beds, no correction is shown as necessary for beds thicker than 4 ft. The correction increases for thinner beds and smaller ratios of induction reading to adjacent bed resistivity. For example, an induction log reading of 6 ohm-meters in a 3 ft bed sandwiched between beds of resistivity 30 ohm-meters ($R_{ID}/R_s = 0.2$) is corrected to 0.82 × 6 ohm-meters = 4.9 ohm-meters ($R_{IDCOR}/R_{ID} = 0.82$). When using this chart we also have the tacit assumption that the adjacent beds above and below are of both equal resistivity and relatively thick. In practice, it can be difficult to satisfactorily approximate these assumptions. Be sure to watch for beds with rapidly alternating resistivities that are only a few feet thick. In this case, a correction probably should not be made, or at least the full correction should not be used. This is a matter of individual judgment and experience. Corrections using Figure 5–17 should not be made when the corrected induction reading will be less than 1 ohm-meter.

Induction log bed-thickness corrections for resistive thin beds are shown in Figure 5–18 for the deep induction and 6FF28 tools and Figure 5–19 for the medium induction log. Bed thicknesses in both feet and meters are shown on the horizontal scales with the corrected resistivity on the vertical scales. The apparent resistivities R_a are read from the induction log. Note that a separate graph is used for different values of adjacent bed resistivity R_s. Also, these charts are only valid for induction logs run with a shoulder-bed resistivity setting of 1 ohm-meter. These corrections cannot be made routinely by computer. Another problem with these corrections is the

Charts Rcor-5, Rcor-6, and Rcor-7 correct the induction logs (6FF40, ID, 6FF28, and IM) for bed thickness. A skin-effect correction is included in these charts.

To use, select the chart appropriate for the tool type and for the adjacent bed resistivity (R$_s$). For Charts Rcor-5 and Rcor-6, enter the bed thickness and proceed upward to the proper R$_s$ curve. Read the corrected resistivity value (R$_t$) in ordinate.

For Chart Rcor-7, enter the chart with the R$_{ID}$/R$_s$ ratio (apparent ID reading/adjacent-bed resistivity) and go upward to the bed thickness. Read the correction factor (R$_{tRcor}$/R$_{ID}$) in ordinate.

EXAMPLE: R$_{ID}$ = 4.2 Ω·m
R$_{IM}$ = 6.0 Ω·m
R$_s$ = 2.0 Ω·m
Bed Thickness = 3 m

Giving, from the R$_s$ = 2 Ω·m charts
R$_{tRcor}$ = 4.5 Ω·m
R$_{tMcor}$ = 6.2 Ω·m

For the small diameter 6FF28, multiply the bed thickness by 1.43 before entering these correction charts. For example, in a 7-ft bed, the bed thickness used in correcting the 6FF28 reading is 10 ft (7 × 1.43 = 10).

Correction for Thin Conductive Beds
6FF40, ID, 6FF28*

NOTE: These corrections are computed for a shoulder-bed resistivity setting of 1 Ω·m. Refer to log heading.

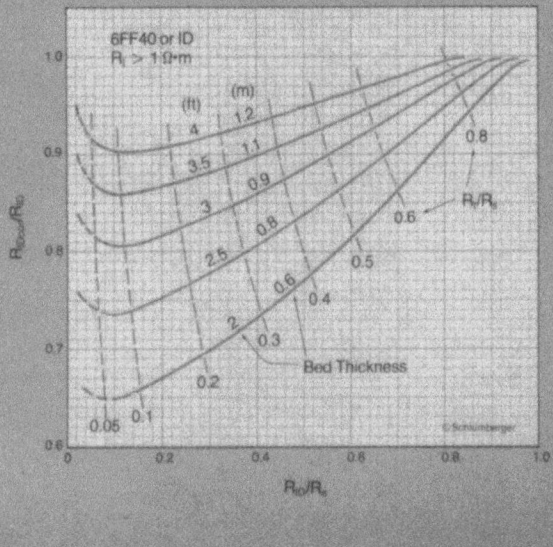

Rcor-7

Figure 5–17. Induction log corrections for thin conductive beds. *(Courtesy of Schlumberger Ltd.)*

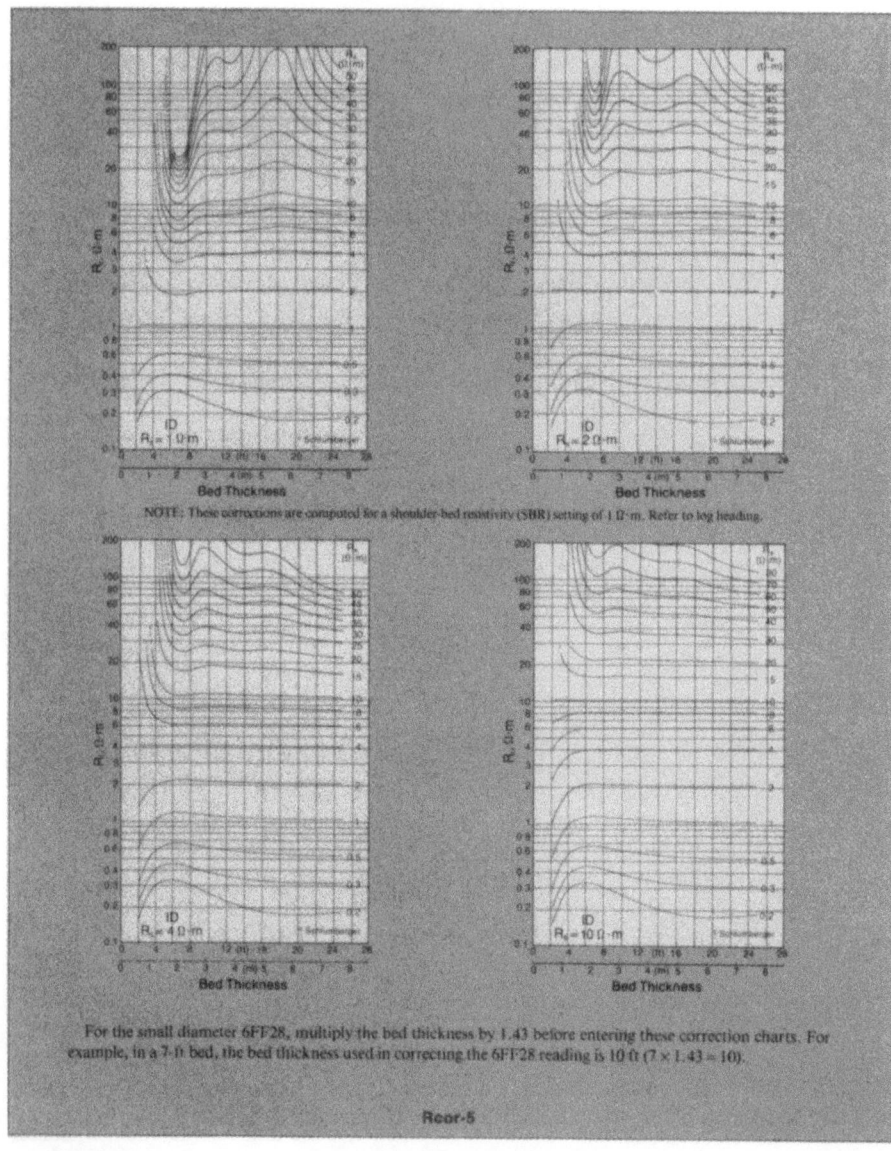

Figure 5–18. Deep induction log bed-thickness corrections. *(Courtesy of Schlumberger Ltd.)*

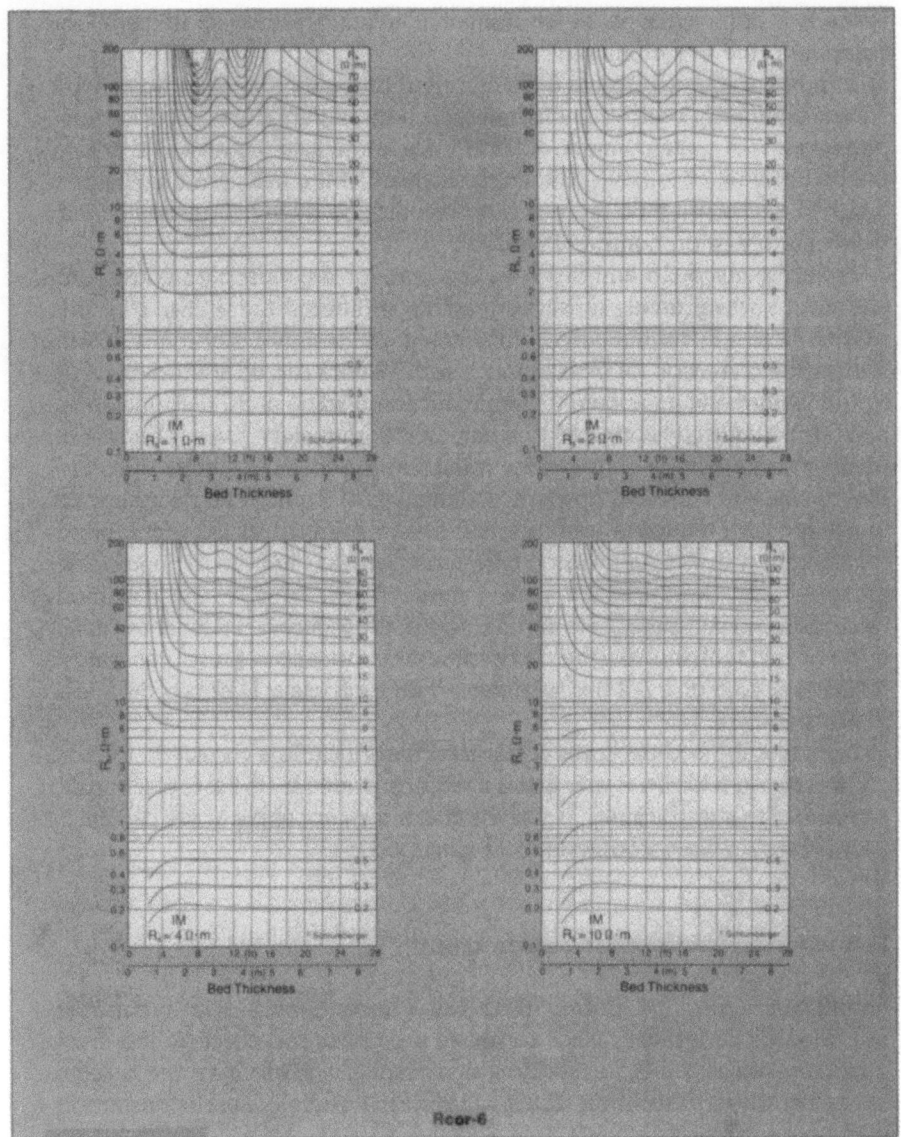

Figure 5–19. Medium induction log bed-thickness corrections. *(Courtesy of Schlumberger Ltd.)*

difficulty in approximating the assumption of infinitely thick beds immediately adjacent to the bed of interest.

In practice, many resistive beds are often bounded above and below by several alternately resistive and conductive beds of only a few feet thickness. Moreover, it is quite common for the beds immediately above and below the bed of interest to have different resistivities. If the beds are dipping, the thin bed effects will be more severe than indicated by the correction charts, which are based on horizontal beds.

From Figures 5–18 and 5–19 we see that, for the most part, corrections are not necessary unless resistivity readings of interest are above 10 ohm-meters. Note that, for the deep induction log with resistivities above 10 ohm-meters, the correction for beds as much as 18 ft thick can be appreciable. For $R_s = 1$ ohm-meter an apparent deep induction reading of 30 ohm-meters in an 18 ft bed must be corrected to nearly 200 ohm-meters. Even for adjacent bed resistivities of 10 ohm-meters, induction log readings above 40 ohm-meters must be corrected upward considerably. Corresponding adjustments to the medium induction tools are less severe but can still be pronounced for higher resistivities sandwiched between low resistivity beds of appreciable thickness. Also note that for beds thinner than 5 or 6 ft, the induction readings may be impossible to correct. Again, use Figure 5–18 for corrections to the 6FF28 tool, multiply the bed thickness by the ratio of the tool spacings, $40/28 = 1.43$. For a 7-ft bed with the 6FF28 tool, use a 10-ft bed thickness in the graph.

Note that the bed-thickness correction charts include a correction for *skin effect* that is in addition to any that is already provided by the tool electronics. Again, we have the difficulty in that the charts may not apply to the particular geometry we have encountered on a given log.

Invasion Corrections for Induction Logs

Mud filtrate from the drilling fluid will invade porous and permeable formations. The invading fluids can have a pronounced effect on the measured resistivities if their resistivity is significantly different from the original formation water. Also, these fluids will displace any movable hydrocarbon in the pore spaces.

The upper part of Figure 5–7 illustrates a typical invasion profile for a fresh mud system in a zone where the porosity is water filled. The term *fresh mud system* is relative since it refers to the circumstance where the resistivity of the mud filtrate R_{mf} is greater than that of the formation water R_w that is displaced. The vertical axis represents the apparent resistivity that would be measured, whereas the horizontal axis represents increasing distance away from the well bore. Close to the well bore, say, within 6 in. to 10 in., the mud filtrate will have totally replaced the original formation water if the zone is porous and permeable. A resistivity tool responding only to this volume close to the well bore will measure the resistivity of this flushed zone R_{xo}.

A little farther from the well bore not all the original formation water will have been displaced, and we will have a mixture of the two fluids in the pores: both original formation water of resistivity R_w and mud filtrate of resistivity R_{mf}. A resistivity tool responding in this *invaded* rather than flushed zone will read a resistivity somewhere between the undisturbed resistivity R_t and R_{xo} of the flushed zone.

Finally, far enough from the borehole we find the undisturbed zone with resistivity R_t where only original formation water of resistivity R_w is present in the pore system. For pore systems containing only water, this resistivity is often denoted R_o. The radial distance from the *borewall* to the beginning of the undisturbed zone is known as the *depth of invasion*, although the term *diameter of invasion* is usually specified. It is easy to see that if the mud filtrate and original formation water have the same resistivity, the flushed zone resistivity, invaded zone resistivity, and undisturbed zone resistivity would all exhibit the same resistivity measurement in *totally water-bearing* zones.

In the invasion profiles of Figure 5–7, complete flushing of formation water near the well bore is assumed. Something less than 100% flushing is always possible in zones with poor permeability, and in *tight* zones with little or no permeability, we may find that there is no invasion profile and the resistivities will be the same any distance from the well bore (assuming rock properties and porosity remain the same within a practical radius of the well bore and the beds are not too thin).

Figure 5–20 illustrates a typical invasion profile in an oil-bearing zone. The upper part of the figure shows the relative saturation profile as we move away from the well bore from left to right. The saturation in the flushed zone is denoted S_{xo} and that of the undisturbed zone S_w. The lower part of the figure shows the corresponding resistivity profile. The resistivity of the invaded zone is denoted R_i. Close to the well bore, all of the original formation water and part of the hydrocarbon have been flushed from the pore system and have been replaced by the invading mud filtrate. Farther away from the bore wall, only part of the original formation water and part of the oil have been flushed. Finally, in the undisturbed zone, no fluids have been displaced.

Closer to the well bore the resistivity R is shown as reading higher where the fresh mud filtrate has flushed all the original formation water and some hydrocarbon. A resistivity tool responding close to the well bore will read R_{xo}. A little farther out, the resistivity will be somewhat lower and still probably somewhat higher than the undisturbed zone, R_t. Here a tool reads R_i. Finally, beyond where the mud system fluids have invaded, a resistivity tool will respond to only the undisturbed R_t. This is a typical invasion profile in a fresh

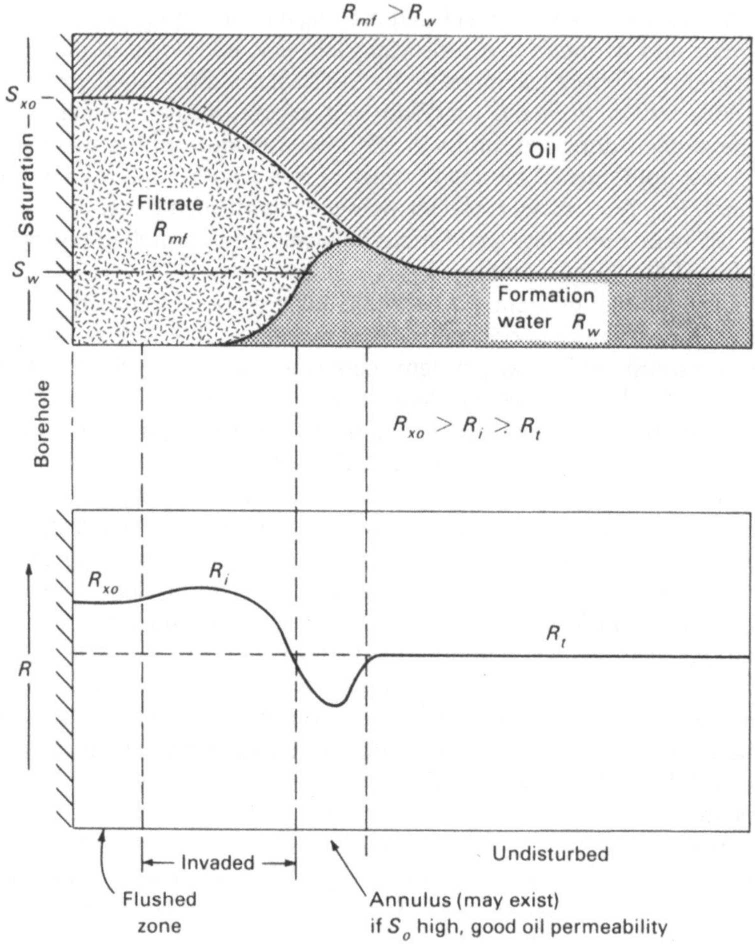

Figure 5–20. Invasion profile in oil zones.

mud system. A logging tool combination used to investigate the invasion profile incorporates a deep induction log (affected least by invasion), a medium induction log that has a somewhat smaller radius of investigation (affected more by invasion), and a short-focus resistivity device such as a spherically focused log or short-spaced laterolog (affected the most by invasion).

Actually, whether or not the resistivity reads higher closer to the borewall in a hydrocarbon-bearing zone when a fresh mud system is used depends not only on flushing and the ratio of R_{mf} to R_w but also on the relative saturations of the flushed and undisturbed zones. From Chapter 3, we know that

$$R_t = IFR_w$$

Similarly,

$$R_{xo} = I_{xo}FR_{mf}$$

where I have substituted the flushed zone parameters as denoted by the subscript xo. Taking the ratio of R_{xo} to R_t

$$R_{xo}/R_t = (I_{xo}/I)(R_{mf}/R_w) = (S_w/S_{xo})^n(R_{mf}/R_w)$$

For many cases, the square ($n = 2$) of the ratio of saturations will be no less than 1/2, so that R_{xo} will be larger than R_t if R_{mf}/R_w is larger than 2, which is true of most mud systems described by the adjective *fresh*. For the extreme example where $S_{xo} = 0.60$ and $S_w = 0.30$, with $n = 2$, the square of the ratio of the saturations is 1/4 and we require that R_{mf}/R_w be greater than 4 if R_{xo} is to measure more than R_t.

In Figure 5–20, an *annulus* of higher water saturation that forms as the formation water is swept outward is illustrated. The annulus is a buildup of an abnormally high concentration of formation water (higher than the initial water saturation S_w). It occurs because the formation water moves faster than the oil in the pore system. The annulus is hypothesized to exist in zones with good oil permeability. The best evidence for the formation of the annulus actually comes from observing well log behavior. Consider a dual induction log system where the deep reading induction tool responds to the undisturbed zone, whereas the medium induction tool responds to the region where the annulus may occur. If no annulus has formed, the resistivity read by the medium induction log in a fresh mud system should be higher than the deep induction reading. However, if the medium induction log is responding to an area close to the radius of the annulus, it may read lower than the deep induction tool in a fresh mud system, whereas the shorter focus tool (say, a spherically focused log or *Laterolog 8*) reads higher than both the induction tools. The fact that the shorter focus tool reads higher than both means that the profile should probably follow that of a fresh mud system with each tool reading progressively a smaller resistivity (deep induction reading less than medium induction) as we move away from the well bore. With the annulus, this usual fresh mud resistivity profile is upset with the result that the medium induction tool responds to the annulus and reads lower than the deep induction tool.

However, the fact that an annulus is not detected is not proof that it does not exist. Detection depends on the annulus forming at about that radial distance where it will affect the medium induction tool reading. Also, an annulus will not necessarily form in all hydrocarbon-bearing permeable rocks. When one does, it usually indicates good oil permeability. However, be careful to account for possible hole signal contributions in larger holes reducing the medium induction log response (see the earlier discussion of borehole signal corrections).

Figure 5–21 illustrates Schlumberger's chart for invasion correction for one particular dual induction–spherically focused log combination that is designed DIS–EA. Note that the chart is applicable only to thick beds with an 8 in. borehole where skin effect corrections have been made and the ratio of R_{xo}/R_m is approximately 100. If these conditions are not satisfied, the true R_t, R_{xo}/R_t, or depth of invasion (d_i) read from the chart will be in error. Since we can usually only correct for borehole size and 8 in. is a common hole size anyway, this will likely be the only one of the conditions we can truly satisfy. For that reason, we may have to regard the correction we make and any data derived from the invasion chart as more qualitative than quantitative.

We enter the horizontal axis of the chart with the ratio of the induction tool resistivities, medium induction divided by deep induction reading, and move up until we intersect a horizontal line extending from the left vertical axis, which is the ratio of the short focus log reading to the deep induction log reading. At this intersection we can read the ratio of the true resistivity to the deep induction reading on one of the dotted lines extending up through the chart (or interpolate between two of the lines if necessary). Also, we can use the dashed lines extending vertically upward through the chart to read an apparent diameter of invasion d_i. Yet another set of solid lines moving across the chart and slightly slanted upward to the right will provide an estimate of the ratio of the resistivity of the flushed zone to the true undisturbed zone resistivity R_{xo}/R_t.

There will usually be a company chart for at least one other ratio of R_{xo}/R_m but you may not find a chart for R_{xo}/R_m close enough to the ratio you find in your practical application. Also, it may not be possible to make the necessary thin-bed corrections to satisfy the thick-bed assumption since the application range of the bed-thickness correction charts is limited.

Invasion charts also exist for other tool combinations such as the dual induction *Laterolog 8*, and appropriate service company charts should be used with them. Be sure to ascertain the assumptions required for the chart you are using.

As an example of using invasion charts, consider a 16 ft bed ($h = 16'$) penetrated by an 8 in. hole with a medium induction tool reading of 36 ohm-meters and a deep induction tool reading of 27 ohm-meters. If the ratio R_{xo}/R_m is approximately 100, and we also have a spherically focused tool (RSFL) reading 2.5 times the deep induction tool reading, we can use the chart of Figure 5–21 for a Schlumberger tool. The ratio of the induction tool readings is $R_{IM}/R_{ID} = 1.33$. Extending a line vertically upward from this

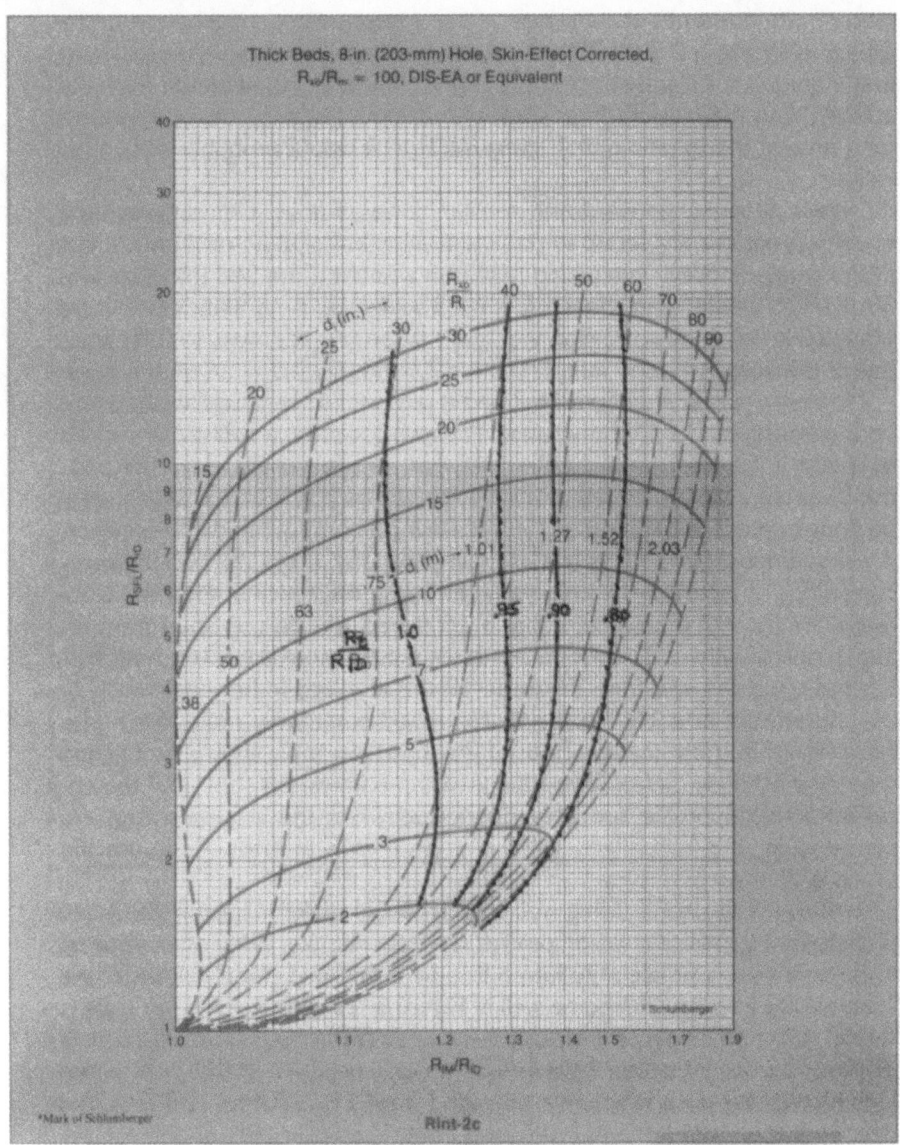

Figure 5–21. Invasion correction example chart. *(Courtesy of Schlumberger Ltd.)*

ratio on the horizontal scale to intersect a horizontal line extended from the ratio of $R_{SFL}/R_{ID} = 2.5$ on the vertical axis, we find the ratio of true resistivity to the deep induction reading R_t/R_{ID} is 0.90, the apparent invasion diameter is 60 in., and the ratio R_{xo}/R_t is somewhere about halfway between the lines for a ratio of 3 or a ratio of 5. Letting $R_{xo}/R_t = 4$ would probably work quite well.

This is all straightforward and okay as far as it goes. Computerized log analysis programs are set up to provide routine invasion corrections just like this example. Indeed, everything might be all right if the bed thickness was 30 ft or so rather than the 16 ft for the example. As it stands, we have provided an invasion correction that reduces the true resistivity to 0.90 times that of the deep induction tool, or 0.90×27 ohm-meters $= 24$ ohm-meters.

However, suppose that our 16 ft bed is sandwiched between two adjacent beds of resistivity $R_s = 2$ ohm-meters. The invasion chart heading has warned us that it is for *thick* beds. Let us see what happens if we apply the bed-thickness correction to these induction log readings first, which is what should be done in practice. Of course, if you are using a computerized interpretation it will be impossible to do this with most programs. From the bed-thickness corrections of Figure 5–18 for deep induction tools, we note from the chart graph for $R_s = 2$ ohm-meters that a 16 ft bed will be corrected from an apparent resistivity of 27 ohm-meters to something more like 50 ohm-meters. Likewise, we can see from Figure 5–19 the medium induction reading of 36 ohm-meters in a 16 ft bed will also reflect something more like a true resistivity of 50 ohm-meters. Both of the induction tools should be reading more like 50 ohm-meters (twice the invasion corrected R_t!) so that the apparent invasion profile suddenly disappears when the induction logs are corrected for thin-bed effect. There was no reason to erroneously use the invasion charts.

In fact, many people using well logs probably make inappropriate interpretations of apparent invasion profiles like this example. What often appears to be an invasion profile may turn out to be a thin-bed effect! Consider the example log from an Oklahoma well in Figure 5–22. The gamma ray scale is 0–150 API units, the spontaneous potential log (SP) scale is 20 millivolts per division, and the induction log is on a commonly used 0.2–2,000 ohm-meter logarithmic four-decade scale. In the sand from 11,033 ft to 11,047 ft, the deep induction tool reads as much as 60 ohm-meters in the top of the zone. A computer-processed analysis of this same log had reduced this reading to 40 ohm-meters via invasion corrections. This was very unfortunate considering the fairly thick conductive bed just below reading only 2 ohm-meters and a thin conductive bed above reading less than 10 ohm-meters. The bed below probably has the more significant effect on the recorded log in the interval of interest. Note how the apparent resistivity decreases in the lower part of the resistive bed (dropping from 60 ohm-meters to just above 40

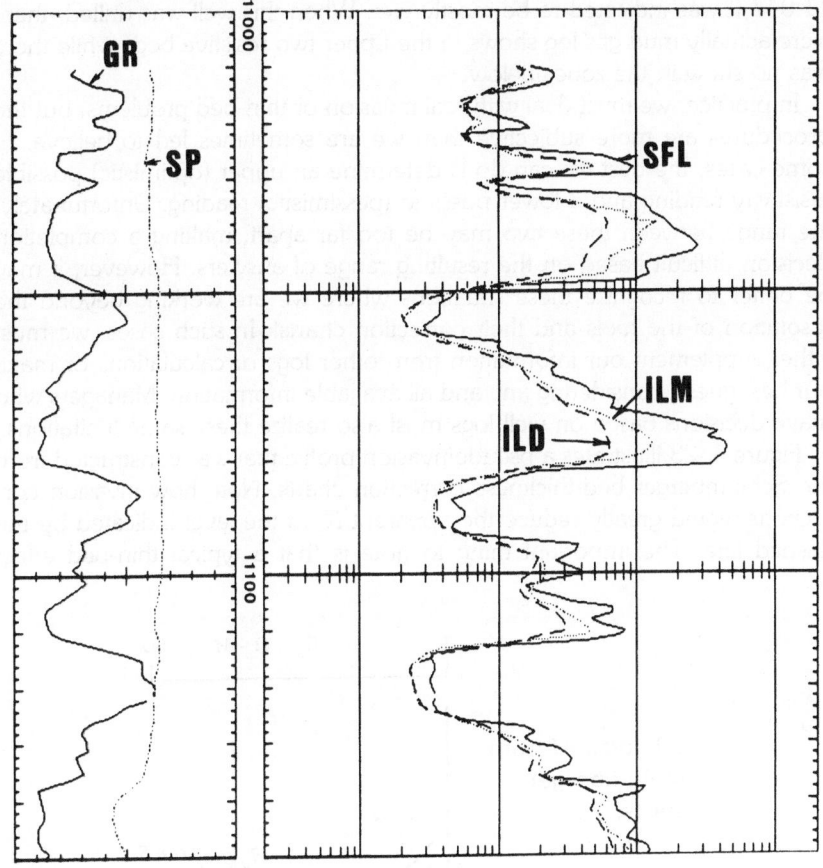

Figure 5–22. Example *invasion* profile in Oklahoma well with invasion corrections.

ohmmeters). This is commonly referred to as *shoulder bed* effect. It is very possible that the entire bed from 11,033 ft to 11,047 ft has a constant resistivity that is even larger than the highest reading of 60 ohm-meters from the deep induction log. The medium induction tool may be giving a more realistic assessment of the actual resistivity in this case: something over 100 ohm-meters. There could still be an invasion effect, but it will be difficult to assess.

The worst part of the computer analysis of this log was that it not only lowered the resistivity of this resistive bed but it also applied less of a correction to the beds below, resulting in an interpretation that showed them to have some hydrocarbon present, whereas the zone from 11,033 ft to

11,047 ft was indicated to be mostly wet. When this well was drilled, there were actually mud gas log shows in the upper two resistive beds while there was no show in the zones below.

In practice, we must deal with real invasion or thin-bed problems, but the procedures are more subjective than we are sometimes led to believe. In some cases, the best we can do is determine an upper (optimistic) possible resistivity reading and a lower possible (pessimistic) reading. Unfortunately, the range between these two may be too far apart, making a completion decision difficult based on the resulting range of answers. However, it may be better to recognize these situations where we are working beyond the resolution of the tools and their correction charts! In such cases, we must either supplement our information from other logs or calculations or make our best guess, considering any and all available information. Managers who make decisions based on well logs must also realize these same limitations.

Figure 5–23 illustrates a pseudoinvasion profile that was constructed from the Schlumberger bed-thickness correction charts. Note how invasion corrections would greatly reduce the apparent R_t to the level indicated by the dashed line. The important thing to note is that a typical thin-bed effect

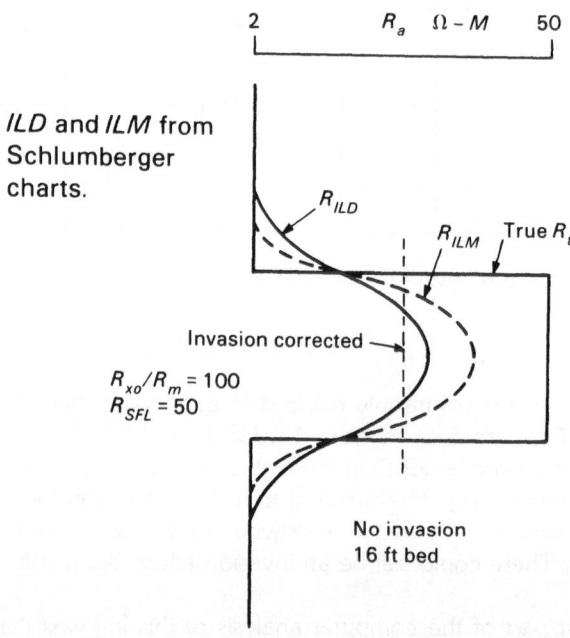

Figure 5–23. Induction log response showing apparent *invasion* profile due to thin-bed effect.

profile looks exactly like an invasion profile. Unless any large resistivity contrasts and the bed thicknesses involved are considered carefully, it is easy to see how a log like that in Figure 5–22 is incorrectly interpreted as an invasion-affected resistivity log. There may be some invasion effect on a set of logs in a sequence of thin beds, but it will be offset by the thin bed effects. Certainly, any reduction of the apparent resistivity will likely be disastrous.

Example of Induction Log Presentation and Correction Problems

Figure 5–24 is a correlation log presentation of the induction log, so-called because it is on a scale of 2 in. per 100 ft, which lends itself more readily to correlation with similar logs from other wells in a given area. Note that only the deep induction (ILD) is shown, although this log was recorded

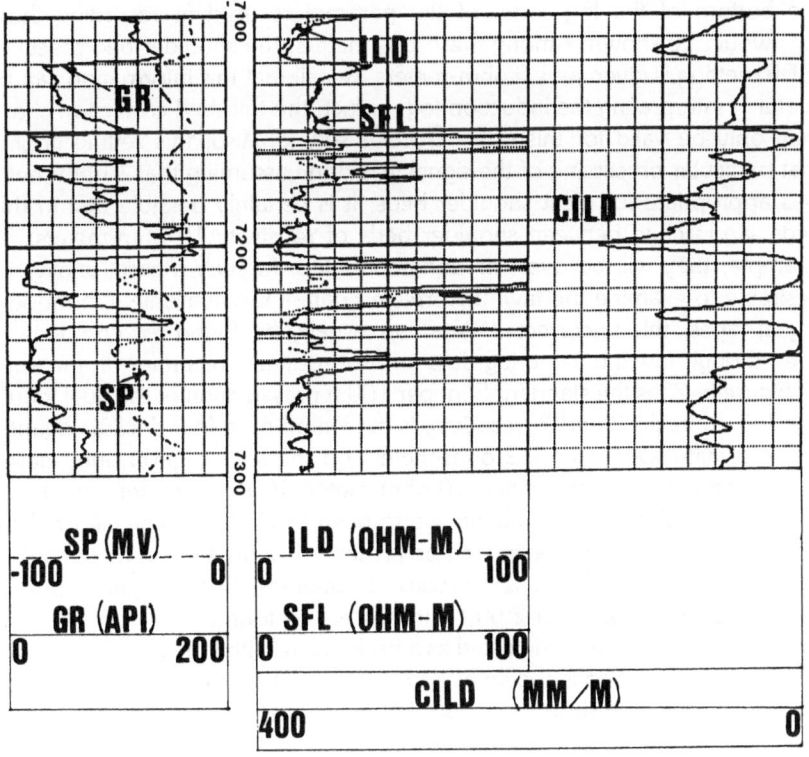

Figure 5–24. Induction log correlation scale presentation.

simultaneously with a medium induction log. The short focus resistivity tool (SFLA) is also recorded on the same scale in track 2 from 0 ohm-meters to 100 ohm-meters. An offscale is also shown, which can be observed, for example, from about 7,235 ft to 7,250 ft. The scale is not shown, but since the trace does not start until just past the first division, which must correspond to the full scale reading of the normal scale (100 ohm-meters), we can infer that the offscale is 100 ohm-meters per division starting at 0 ohm-meters on the left and going to 1,000 ohm-meters on the right of track 2. Where a wide range of resistivity must be shown, following these offscales back and forth can be difficult, especially if working on a log with both upper and lower backup or offscales. For this reason, hybrid resistivity scales and modern style logarithmic scales were developed. The unusual (and now obsolete) hybrid scale will be demonstrated in Chapter 6.

The deep induction log reading is also recorded as conductivity across both tracks 2 and 3, starting at 400 millimhos/meter on the left and going to 0 millimhos/meter on the right. In track 1, the gamma ray and spontaneous potential curves are both recorded on scales as shown in the figure. At the bottom of the log some of the parameters used in recording these logs would be shown. Among these parameters, the shoulder bed resistivity (SBR) setting is given as 1.0 ohm-meters. Recall that this information can be useful in interpreting the induction log. In fact, the thin-bed correction charts may only be valid for this one particular setting. Also, this setting implies that particular processing of the signal was done assuming that 1 ohm-meter was appropriate for most shoulder beds. It is certainly possible that in thin beds sandwiched between shoulder beds of resistivity 10 ohm-meters, for example, they will be over- or undercorrected. In such a case, it may be desirable to examine the raw, uncorrected data if it is available. On modern digital tapes, this uncorrected data can usually be found for comparison, even though raw induction log data are not routinely recorded on the log.

Figure 5–25 is the standard 5 in. per 100 ft presentation of the dual induction log. In tracks 2 and 3, resistivity is presented on a four decade logarithmic resistivity scale from 0.2 ohm-meters on the left to 2,000 ohm-meters on the right. The 1 ohm-meter, 10 ohm-meter, 100 ohm-meter, and 1,000 ohm-meter points on the scale have also been labeled for convenience. This permits most resistivity readings to be presented without resorting to confusing backup scales. Only in very resistive formations, such as tight carbonate rocks, will resistivities over 2,000 ohm-meters be found.

The deep induction is indicated as a broken line with long dashes, whereas the medium induction is shown as a broken line with short dashes, both over tracks 2 and 3. The short focus tool is shown as a solid curve across tracks 2 and 3.

Despite the scales, it is important to remember from previous discussions that induction log readings over 200 ohm-meters may not mean too much.

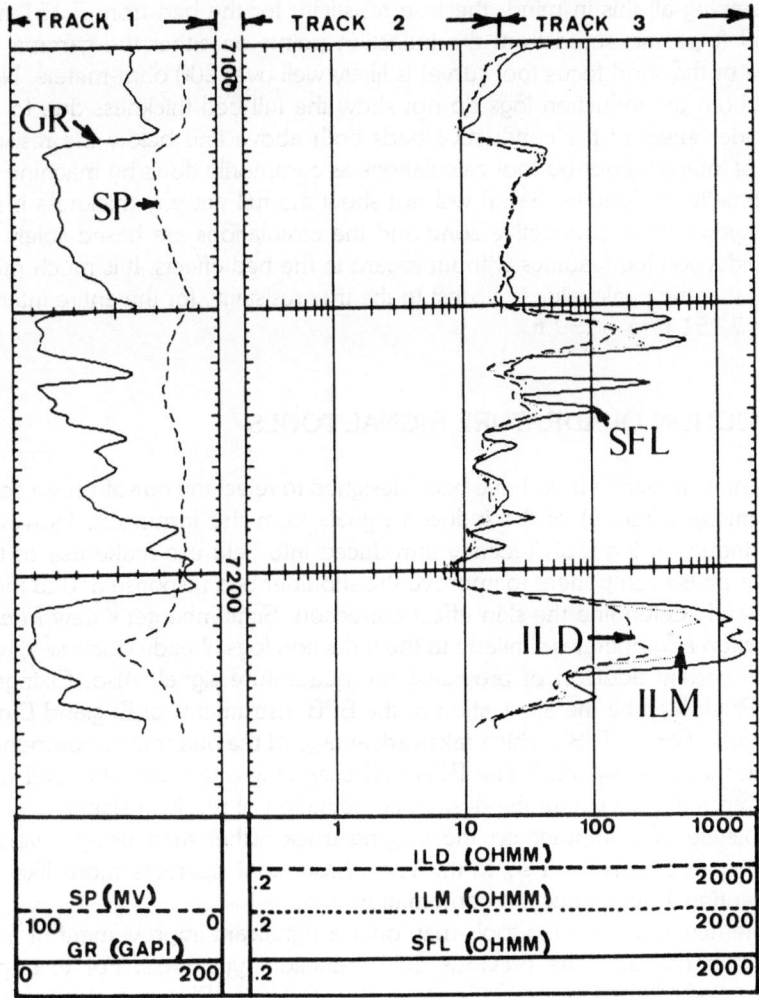

Figure 5–25. Induction log standard logarithmic scale presentation.

Recall that the tool responds to conductivity. Resistivities over 200 ohm-meters correspond to conductivities less than 5 millimhos/meter, which is approaching the *noise* level for the tool. Also, from the bed-thickness charts (see Figs. 5–18 and 5–19), most of the induction log readings over 50 ohm-meters actually represent much larger resistivities that will be hard to accurately determine. Even for adjacent bed resistivities as high as 10 ohmmeters, all deep induction log readings of 50 ohm-meters or more will be corrected (see Fig. 5–18) to true resistivities of over 70 ohm-meters for beds thinner than 16 ft!

Keeping all this in mind, the true resistivity for the bed from 7,151 ft to 7,159 ft (as measured from the inflection points on either the gamma ray curve or the short focus tool curve) is likely well over 200 ohm-meters. Note how both the induction logs do not show the full bed thickness due to the shoulder effect of the conductive beds both above and below the resistive bed of interest. Foot-by-foot calculations as commonly done by machine (or as sometimes done by hand) will not show the full net pay thickness if this is a hydrocarbon productive zone and the calculations are based solely on the induction log readings without regard to the bed effects. It is much more likely that the peak value is closest to the true resistivity for the entire interval from 7,151 ft to 7,159 ft.

INDUCTION QUADRATURE SIGNAL TOOLS

Until now induction tools have been designed to reject the out-of-phase component (*quadrature*) of the induced signals from the formation. However, new induction log tools recently introduced into field use make use of this out-of-phase component to improve the shoulder bed response and to more accurately determine the skin effect correction. Schlumberger's new *Phasor Induction Log* operates similarly to the induction logs already discussed, with the important addition of providing the quadrature signal. Also, Elkington and Patel describe the application of the BPB Instruments of England *Digital Induction Sonde* (DIS), which takes advantage of the quadrature component of the induction signals.[8] The BPB tool uses only one transmitter coil with four receiver coils where the desired combination of received signals is done completely by computer on the logging truck rather than using reversed winding directions, and so forth. The *Phasor* tool operates more like the conventional dual induction combination tool.

The new induction log tools may offer a significant improvement in thin-bed response over the previous dual induction type tools. For example, Figure 5–26 is Schlumberger's chart Rcor-9 for its *Phasor Induction Log* bed-thickness corrections from its 1986 chart book. This chart is set up to be used somewhat differently than the bed-thickness correction charts for Schlumberger's dual induction log (compare Figs. 5–18 and 5–19). Refer back to the example of the 18 ft thick bed with an apparent induction reading of 30 ohm-meters sandwiched between adjacent beds of resistivity 1 ohm-meter. We enter the chart in Figure 5–26 on the vertical scale to the right with the ratio of the *Phasor* induction reading to the adjacent bed resistivity (R_{IDP}/R_s). In this example, the ratio R_{IDP}/R_s is 30/1 = 30. This value actually falls just above the last horizontal line extended from the vertical scale at $R_{IDP}/R_s = 25$. For an 18 ft bed at this latter value, we can see that no thin-bed correction would be necessary. It may be safe to infer that there will be very little correction, if any, necessary for our example with the ratio of 30.

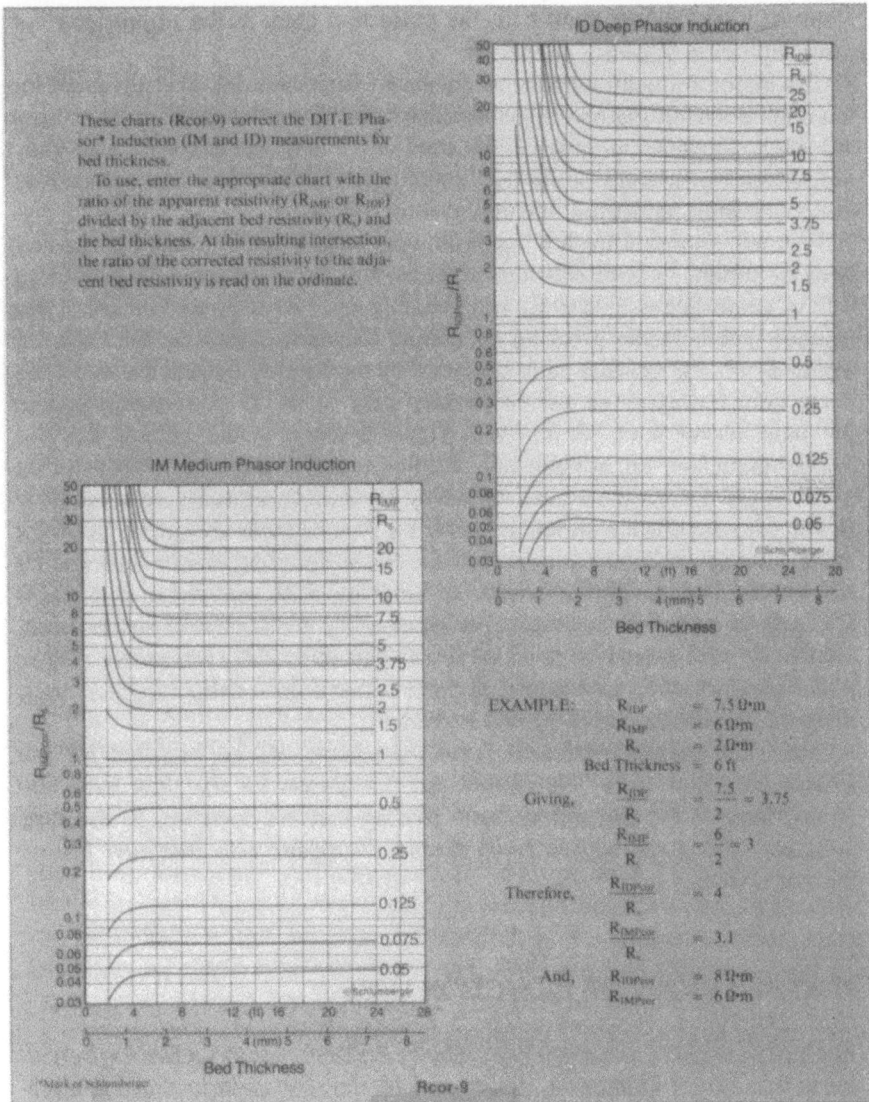

Figure 5–26. Phasor induction log bed-thickness corrections. *(Courtesy of Schlumberger Ltd.)*

Note for beds thinner than 8 ft, the correction starts to be appreciable for resistivity ratios (R_{IDP}/R_s) above 20.

The important point here is the significant improvement over the chart for the dual induction log bed-thickness response. The advantage of the *Phasor* induction charts is that they can be used with many values for R_s since they are set up for the ratio of the tool reading to R_s. Of course, the chart does not cover large ratios (the maximum ratio is 25).

The new *Phasor* induction log promises a large improvement. For resistive beds bounded by very conductive beds, the deep induction log readings will be seriously pessimistic for any readings over 20 ohm-meters unless the resistive bed is thicker than the maximum thickness shown on the charts of Figure 5–18. For resistive beds bounded by moderately conductive beds, the deep induction readings will be useable only up to 50 ohm-meters except for beds thicker than 28 ft. From Figure 5–26 it would appear that the *Phasor* induction log will provide useable measurements of resistivities up to 25 times the shoulder bed resistivity for bed thicknesses as small as 8 ft. For very conductive shoulder beds, the *Phasor* tools should be good for resistivities up to 25 ohm-meters. While this is not much more than the 20 ohm-meter limit for the conventional induction tool, remember that this is for beds of only 8 ft thickness. For moderately conductive shoulder beds, the *Phasor* tool should be good for resistivities up to 250 ohm-meters. While a medium induction tool would perform better than a deep induction tool, it would be affected more by any invasion.

Note that different invasion correction charts will be required for the *Phasor* induction tools. We cannot use the graphs for the dual induction tools. Many of the pseudo invasion profiles that are common to the dual induction-type tools in thin beds should disappear with the new *Phasor* induction tool.

FURTHER STUDY OF INDUCTION LOGS

For further study of induction logs, refer to Anderson for induction log behavior in thin beds.[9] Dumanoir, Tixier, and Martin discuss induction log applications in fresh mud,[10] although as pointed out earlier, Pickett provided improved methods for interpretation via his cross-plotting techniques.[11] Moran and Kunz provide a good discussion of the basic theory of two-coil induction sondes with extensions to multicoil systems.[12] Hardman and Shen discuss problems in dipping beds.[13] Moran perhaps provided the impetus for using the quadrature signal to improve induction log response that we now see with the Schlumberger *Phasor* induction log.[14] Also, Anderson illustrates how induction logs show a reversal in 2 ft thick alternating sand–shale beds.[15]

SUMMARY

With the older electrical survey tools, an apparent resistivity was usually measured that often could not be corrected to a true R_t. The best device to use for these older logging suites was probably the 64 in. normal, or *long normal*, as it was sometimes called. Where practical, long normal responses need to be corrected for invasion, mud resistivity, and bed effects.

Electrode or sensor spacing for logging tools is related to both the radius of investigation and the thin-bed resolution of the tool. Shorter spacings give better thin-bed resolution coupled with shorter radii of investigation. To *read* deeper into the formation, longer spacings must be used and thin-bed resolution is compromised.

The induction log is the only resistivity log that can be used in boreholes with nonconducting mud systems (oil-based muds) or air- and gas-drilled holes. The induction log is best used in formations with resistivities less than 100 ohm-meters (preferably 50 ohm-meters maximum) with fresh mud systems (say, $R_t < 2.5R_{xo}$). In thin, conductive beds, we can usually assume the tool reads the true resistivity if there is no significant invasion. The tool reading is usually not quantitatively reliable in beds less than 5 ft thick under any conditions.

The induction log calibration is good to at least ± 2 millimhos/meter. This means that induction log readings are unreliable above 100 ohm-meters. A common practice is to truncate induction log readings at 200 ohm-meters maximum. This corresponds to a 5 millimhos/meter minimum. Some analysts use higher cutoff limits, but I believe that 200 ohm-meters is the practical limit considering the calibration.

Corrections for hole signal may need to be made if the hole size is larger than 9 in. for a medium induction log and larger than 12 in. for the deep induction log. However, this statement is valid only if a standoff is used on the induction tool. Without the standoff, hole signal corrections will be necessary for much smaller holes as well. Also, if a standoff is not used, the repeatability of induction log readings may be very poor.

Corrections for bed thickness are mandatory in resistive beds where there is a large contrast to the adjacent bed resistivities. These corrections may have to be made for relatively thick beds (refer to Figs. 5–18 and 5–19), especially where the resistivity contrast between the bed of interest and the adjacent beds is larger. Note the SBR setting used on the log to see that it is not unreasonable compared to the low resistivity zones near the zones of interest on the well log.

Nearly all induction log readings above 20 ohm-meters may be affected to some extent. With the *Phasor* induction log, the bed effects are significantly reduced (see Fig. 5–26) except for large resistivity contrasts (R_{IDP}/R_s), where the charts do not apply.

Invasion corrections should be approached much less routinely and more carefully than is the common practice today. Bed-thickness effects yield apparent invasion profiles with the dual induction log combinations. Invasion corrections are to be made only after bed-thickness corrections have been made.

In practice, the problem is that there are not enough bed-thickness charts to cover all possibilities. Moreover, the charts apply only to horizontal beds where the bed of interest is sandwiched between two beds of equal resistivity and very large thickness. The invasion correction graphs apply to specified ratios of R_{xo}/R_m, and a specific application's ratio may not be close to any of the chart-specified ratios. All these restrictions are hard to satisfy in practice. Sometimes the medium induction log may be used as a true resistivity reading for a resistive thin bed if the effects of invasion can be considered negligible.

Because of differences in the approach of tool design, we cannot expect to see the same induction-log-measured resistivity opposite each bed recorded by different service company tools in the same hole. Normalization of different company tools so that all read the same (or appear to have the same calibration) may be necessary, but heed this warning: tool readings from different tools may be different in a resistive bed, yet identical in a conductive bed (and vice versa). If at all possible, normalization should only be done using a bed that has a resistivity similar to the bed(s) of interest.

PROBLEMS

5–1. After any necessary borehole signal corrections are made to the induction log reading, is the corrected resistivity higher or lower than the uncorrected resistivity?

5–2. If you were choosing a resistivity tool to read as close to the borewall as possible, would you want a short or long tool spacing between the electrodes?

5–3. Why cannot resistivity tools other than the induction log be used in a borehole with an oil-based mud system?

5–4. Consider the following porosity data to be used with the resistivity log of Figure 5–25.

Depth (ft)	Porosity (%)
7,151	6
7,152	7
7,153	5.7
7,154	5.5
7,155	6.2
7,156	5
7,157	3

Assume $a = 1$, $m = n = 2$, $R_w = .09$ ohm-meters, and no hole signal, thin-bed effect, or invasion corrections are necessary. Calculate the water saturation for each depth of porosity data. Does the resulting saturation profile seem reasonable? Can you work this problem with another approach other than a foot-by-foot calculation? Could you make a thin-bed correction for the resistivities in the interval 7,151 ft to 7,157 ft? Would a thin-bed correction be necessary?

5–5. You have an induction log reading of 500 ohm-meters in a zone of interest. Can you use this figure as is in your calculations? Explain your answer.

5–6. In a saline mud system with deep invasion, where $R_w > R_{mf}$, would you expect the true resistivity to be higher or lower than the appropriate resistivity tool reading?

REFERENCES

1. Hubert Guyod, *Guyod's Electrical Well Logging*, a reprint of 16 articles that appeared in various issues of the *Oil Weekly* between August 7, 1944 and December 4, 1944 ©1944 by Welex, a division of Halliburton Company, Houston, Texas, 1944.
2. Ibid.
3. Douglas W. Hilchie, *Old (pre-1958) Electrical Log Interpretation* (Douglas W. Hilchie, Inc., Golden, Colo., 1979).
4. Author's class notes from course on well log interpretation presented by George R. Pickett at the Colorado School of Mines, Golden, Colorado, 1975.
5. H. G. Doll, "Introduction to Induction Logging and Application to Logging of Wells Drilled without Oil Base Mud," *Journal of Petroleum Technology* **1** (1949): 148–162.
6. Ibid.
7. Author's class notes, 1975.
8. P. A. S. Elkington and H. K. Patel, "Invasion Profile from the Digital Induction Log," paper presented at the 26th Annual Logging Symposium of the Society of Professional Well Logging Analysts, Dallas, June 1985.
9. Barbara Anderson, "Induction Sonde Response in Stratified Media," *The Log Analyst* **24**, no. 1 (1983): 25–31.
10. J. L. Dumanoir, M. P. Tixier, and Maurice Martin, "Interpretation of the Induction-Electrical Log in Fresh Mud," *Journal of Petroleum Technology* (July 1957).
11. Author's class notes, 1975.
12. J. H. Moran and K. S. Kunz, "Basic Theory of Induction Logging and Application to Two-Coil Sondes," *Geophysics* **27**, no. 6 (1962): 829–858.
13. R. H. Hardman and L. C. Shen, "Theory of Induction Sonde in Dipping Beds," *Geophysics* **51**, no. 3 (1986): 800–809.
14. James H. Moran, "Induction Logging—Geometrical Factors with Skin Effect," *The Log Analyst* **23**, no. 6 (1982): 4–10.
15. Barbara Anderson, "The Analysis of Some Unsolved Induction Interpretation Problems Using Computer Modeling," *The Log Analyst* **27**, no. 5 (1986): 60–73.

CHAPTER 6

Electrical Properties (Focused Current and Spontaneous Potential) Logging

In this chapter I continue the discussion of resistivity logging with focused current resistivity tools and the natural borehole potential or *spontaneous potential* (SP) log.

Modern focused current logs include laterologs, guard electrode logs, and spherically focused logs. They can be designed to measure either the undisturbed resistivity R_t, the flushed zone resistivity R_{xo}, or the invaded zone resistivity R_i. Focused current logs were developed in response to the need for a tool that could function in salt-based mud systems where high resistivities were encountered such as the Williston Basin. The old electrical survey tools were almost shorted out by conductive, salt mud systems. Longer spacings would have improved the response of the old tools, but at the expense of inadequate bed resolution for interpretation purposes.

FOCUSED CURRENT RESISTIVITY TOOLS

Figure 6–1 illustrates how focusing was achieved in one of the earlier focused current tools: the *Laterolog 7*, so-called because seven downhole electrodes were placed on the tool. In addition, a current return electrode and a potential reference were placed further up the tool and at the surface, respectively.

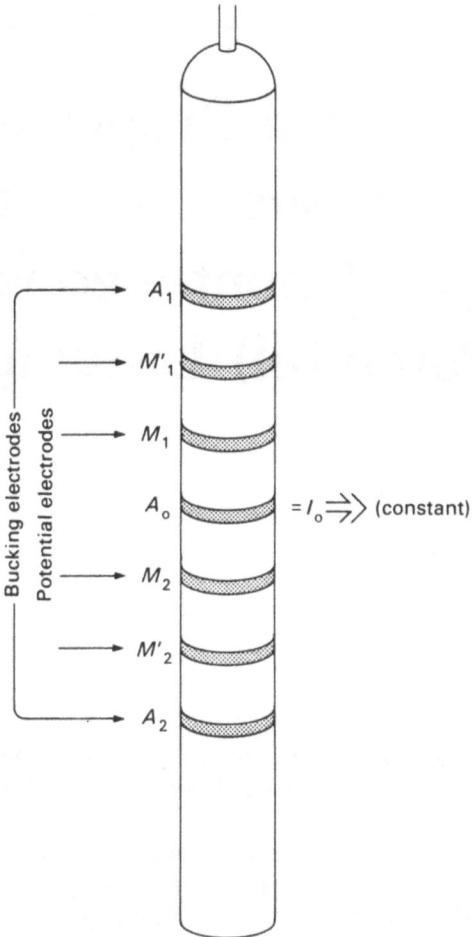

Figure 6–1. Laterolog 7.

A constant current is sent out into the formation from current electrode A_o. A variable current is sent from electrodes A_1 and A_2 (*bucking* electrodes used to focus the current from A_o). This current is adjusted automatically such as to maintain the two pairs of monitoring electrodes, M'_1 and M'_2 and M_1 and M_2 at the same potential. With the current from A_o held constant, the potential measured between any one of the monitoring electrodes and the potential electrode at the surface varies directly as resistivity.

Since no current could flow from A_o between M_1 and M'_1 or between M_2 and M'_2, the current from A_o was forced horizontally away from the tool, deep into the formation. Figure 6–2 illustrates the focusing effect for

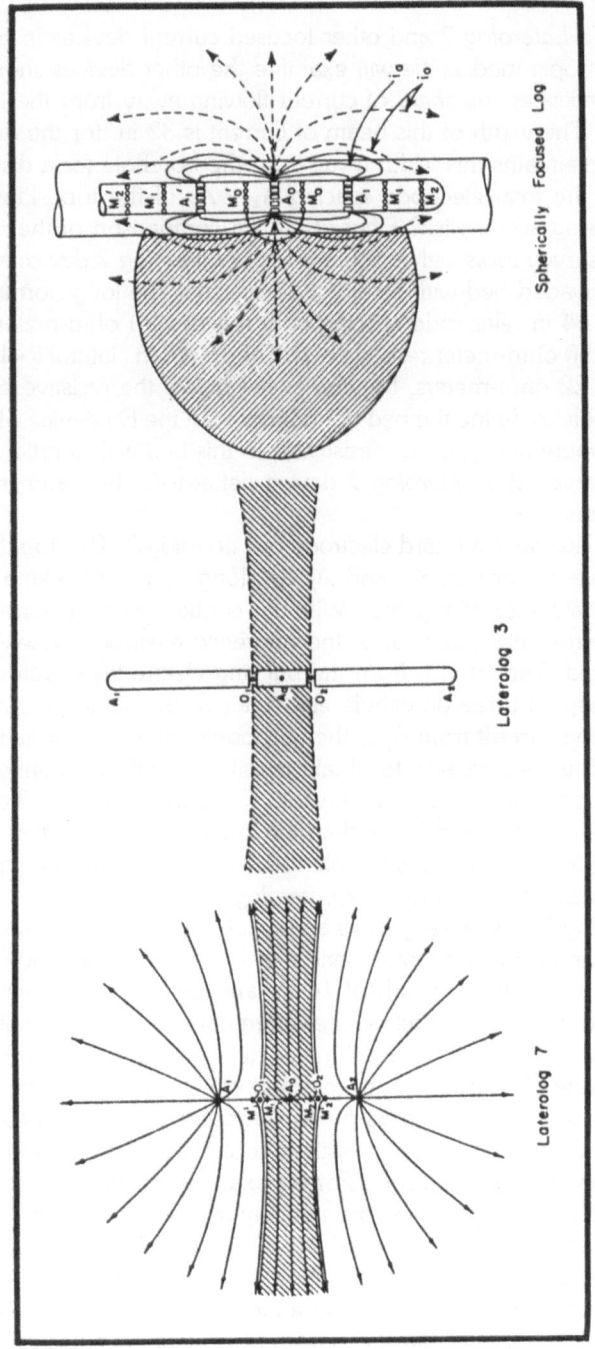

Figure 6–2. Focused current electrode devices. *(Courtesy of Schlumberger Ltd.)*

Schlumberger's *Laterolog 7* and other focused current devices in homogeneous and isotropic media. We will examine the other devices shortly. The shaded area indicates the *sheet* of current flowing away from the tool into the formation. The width of this beam of current is 32 in. for the *Laterolog 7*. The beam maintains this width away from the borehole for a distance at least equal to the total electrode spacing $A_1 - A_2$ of the tool. How much does this focusing accomplish? Figure 6–3 is a comparison of the response of the old ES survey tools with a Schlumberger *Laterolog 7* device in a thin, resistive, noninvaded bed with a salty mud system. The long normal (L.N.) device with a 64 in. electrode spacing only reaches 20 ohm-meters in the center of the 250 ohm-meter resistive bed. The 18 ft 8 in. lateral tool reaches a resistivity of 22 ohm-meters, but somewhat *below* the resistive bed. It is nearly impossible to define the bed boundaries with the ES devices, let alone measure anywhere near the true resistivity! In this bed with a ratio of R_t/R_m of 5,000, however, the *Laterolog 7* device defines the bed and measures 220 ohm-meters.

Figure 6–4 illustrates a guard electrode or *Laterolog 3*. This tool has only three downhole electrodes. A_1 and A_2 are long, current-bucking (guard) electrodes that are shorted together. With this configuration the current from A_o varies, whereas the potential at the reference electrode (elsewhere on the tool) is fixed. The current from the bucking electrodes is automatically adjusted to keep all three downhole electrodes at the same potential. The magnitude of the current from A_o is then proportional to the formation *conductivity* which is reciprocated to obtain resistivity. With this configuration, the current beam is only about 12 in. wide. Thus, the *Laterolog 3* or guard log has a better thin-bed resolution than the *Laterolog 7* and is less affected by invasion of the mud in the borehole. The middle diagram in Figure 6–2 illustrates the current beam from the *Laterolog 3*.

The *Laterolog 8* was developed as an invaded zone resistivity tool to use with the dual induction log combination. It is a shorter electrode spacing laterolog tool with a beam width of 14 in. and performs very well in thin beds. With the two dual induction measurements, it permits a solution of the true formation resistivity where the induction measurements are affected by invasion (at least, it does where thin-bed effects are no problem!). The *Laterolog 8* is sometimes used as an R_t device in thin beds (beyond the resolution of the deeper reading resistivity tools) if it can be assumed that invasion effects are not significant. For an example of this application, see Aguilera's Well No. 1 in his book on naturally fractured reservoirs.[1]

The modern spherically focused (SFL) tool was developed as an improvement to both the old electrical survey short normal tools and the *Laterolog 8*. The old normal resistivity devices were interpreted on the basis of equal, radial current intensity in all directions from the downhole current electrode. However, this assumption is violated by the presence of the fluids in the borehole. The SFL tool uses focusing current electrodes much like the lat-

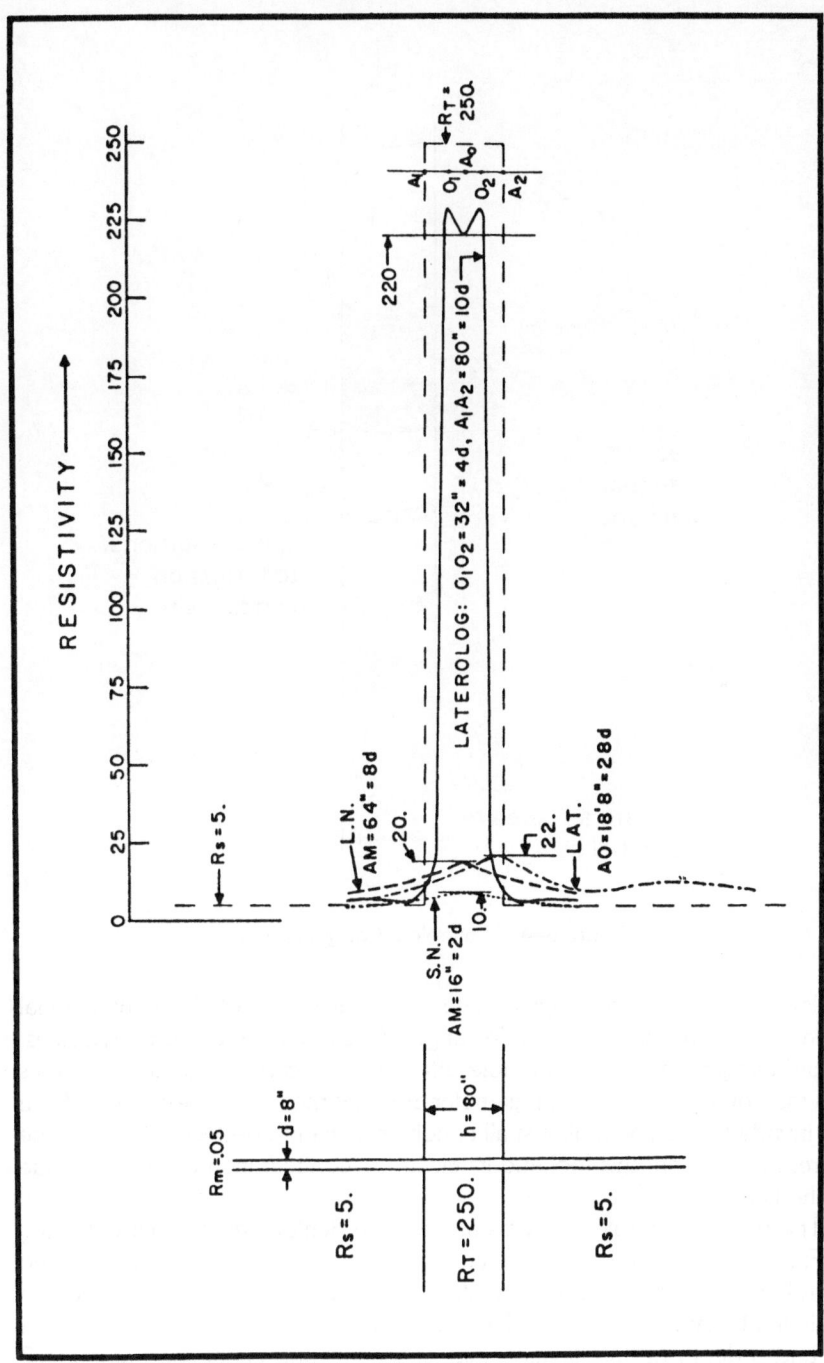

Figure 6–3. Response comparision of focused current and ES devices. *(Courtesy of Schlumberger Ltd.)*

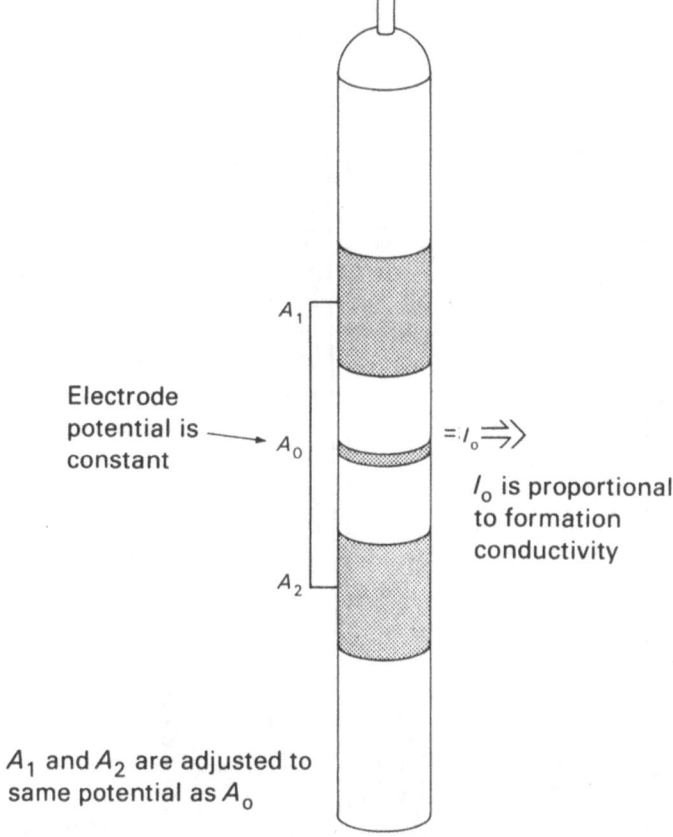

Electrode potential is $\longrightarrow$ A_0 constant

A_1

A_2

$= \cdot I_o \Rightarrow\rangle$

I_o is proportional to formation conductivity

A_1 and A_2 are adjusted to same potential as A_o

Figure 6–4.   Laterolog 3 or guard log.

erologs to maintain an approximate spherical shape of the equipotential surfaces near the tool for a wide range of hole sizes and mud resistivities. Schlumberger claims the borehole effect is practically eliminated for holes less than or equal to 10 in. in diameter and that most of the response is from the invaded zone under almost all conditions.[2] Figure 6–2 also illustrates the current beam from this tool and its electrode configuration on the right side of the figure.

The modern laterolog systems are like the Schlumberger Dual Laterolog of Figure 6–5. Both a shallow focused laterolog (LLs) and a deep focused laterolog (LLd) are recorded simultaneously, using the same electrodes (in a different manner for each). The deep laterolog has a beam width of 24 in. and a depth of investigation that is deeper than either the *Laterolog 7* or *Laterolog 3* tools. One version of Schlumberger's tool records the deep

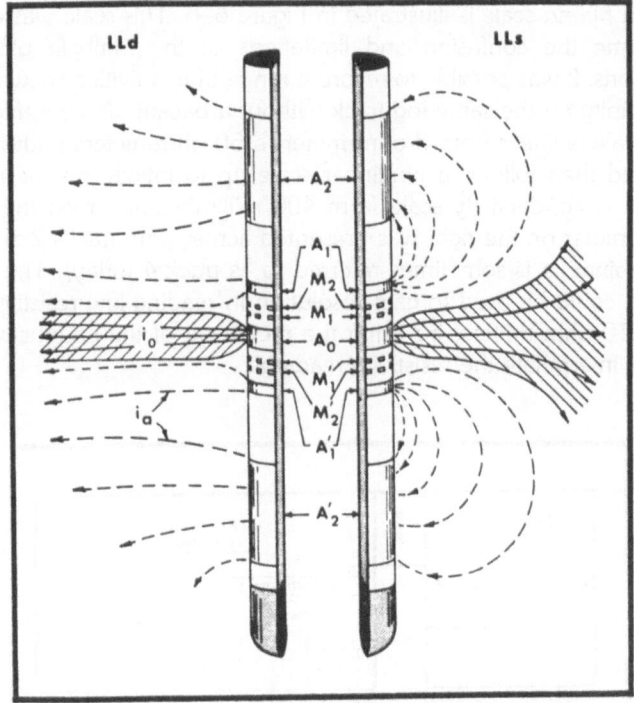

Figure 6–5. Dual laterolog. *(Courtesy of Schlumberger Ltd.)*

and shallow measurements simultaneously, whereas another version records them sequentially. The shallow focused device also achieves a beam width of 24 in. but with a much shallower response characteristic. Its depth of investigation is between that of the *Laterolog 8* and the *Laterolog 7*.

Modern laterologs and guard logs are normally recorded on a four-decade logarithmic scale to account for possible extreme ranges between high and low resistivity contrasts. Older logs were recorded on cumbersome scales that often used multiple backup scales. It was often difficult to keep track of whether you were on the *main* scale or one of the backup scales. Backup scales could be both upper (to cover values higher than the main scale) and lower (to cover values below the main scale) or there might be two upper offscales. Sometimes the backup and the main scale both started at zero on the left side of the track. These were often referred to as *amplified* scales when the backup was a lower scale. For example, the main scale might have covered from 0 ohm-meter to 100 ohm-meters, while the backup, amplified scale was from 0 ohm-meter to 10 ohm-meters. Various combinations and arrangements were used on the older logs. Be sure to note the scales with which you are working.

The old *hybrid* scale is illustrated in Figure 6–6. This scale was designed to overcome the confusion and limitations of the multiple offscale log presentations. It was possible to record a range of resistivities anywhere from zero to infinity on the same log track without a backup. Note in Figure 6–6 that the scale is linear from 0 ohm-meter to 50 ohm-meters midway across track 2 and then follows a nonlinear scale up to infinity on the right side of track 2. A conductivity scale from 400 millimhos/meter on the left to 0 millimhos/meter on the right was presented across both tracks 2 and 3 (this track combination is sometimes referred to as *track 4* today). The conductivity scale could be used to gain resolution in reading low resistivities, say, less than 10 ohm-meters, by using the reciprocal of the conductivity scale reading to interpolate the resistivity readings.

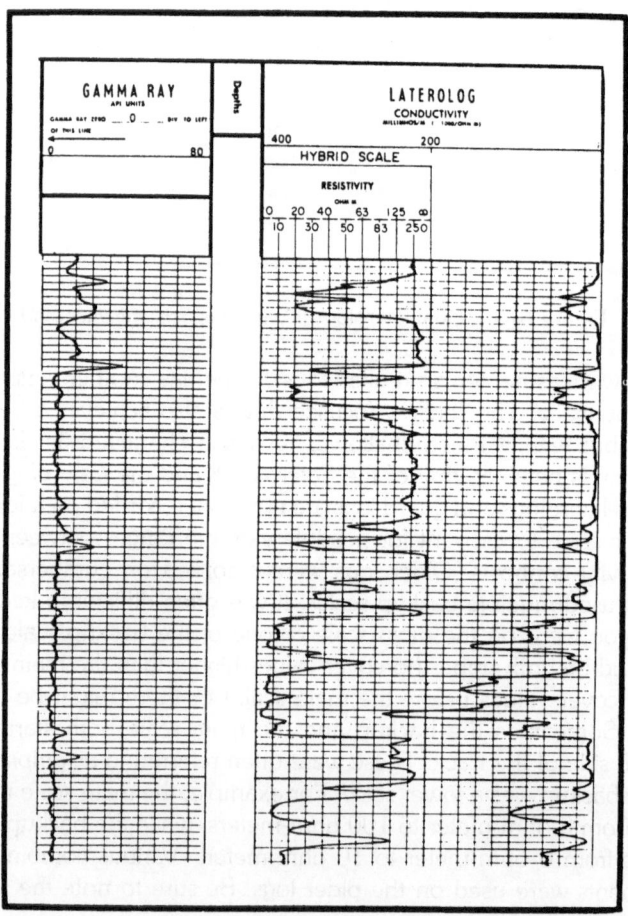

Figure 6–6. Hybrid resistivity scale. *(Courtesy of Schlumberger Ltd.)*

Actually, the nonlinear portion of the hybrid scale is easy to understand if you consider that the right side of track 2 represents the reciprocal of conductivity from 20 millimhos/meter (at 50 ohm-meters) to zero conductivity (infinite ohm-meters) on the right. Then each scale division above 50 ohm-meters represents a change of 4 millimhos/meter. For example, one division from the right side is 4 millimhos/meter. Its reciprocal is 1000/4 = 250 ohm-meters, which is listed on the hybrid scale.

CORRECTIONS FOR FOCUSED CURRENT RESISTIVITY TOOLS

Laterologs and guard logs can be corrected for borehole effects, shoulder bed effects, and invasion effects. Just as for induction logs, the available focused current log correction charts may be limited in their scope of application.

Borehole Corrections

Figure 6–7 shows the corrections for the Schlumberger deep and shallow laterologs for the dual laterolog combination. These charts are for tools centered in the borehole (Schlumberger DLS-B tool), but charts are also available for tools used without standoffs as well as other tool versions. Note the assumption of thick beds. Note also that the deep laterolog tool will measure close to the true apparent resistivity for a wide range of resistivities for the common 8 in. hole size. For very large resistivities (as compared to the mud resistivity), the corrected resistivity will be only about 90% of the tool reading. The shallow laterolog performs about as well considering as a standard that it reads close to what it should in the invaded zone. Corrections from these charts would presumably be applied before the LLd or LLs measurements are used for further calculations or before any other corrections are applied. It would appear for normal (close to 8 in.) hole sizes that no borehole corrections would be necessary for most applications. For very large holes, the tool reading will be somewhat low. The charts for the other tools mentioned also indicate that borehole corrections for the other dual laterolog tools should not be significant in many practical applications.[3]

Figure 6–8 shows the borehole corrections for the *Laterolog 8* and SFL tools. Note that, although the SFL tool is not too adversely affected by large holes, the *Laterolog 8* is severely affected (except at lower resistivity values) for any hole size much over 8 in. In most cases, the chart implies a correction upward for the *Laterolog 8* in enlarged holes. The correction can exceed the chart range of 1.5 times the apparent *Laterolog 8* reading. This chart illustrates the improvement the SFL offers through its enforced spherical focusing.

Deep Laterolog Borehole Correction
DLS-B Tool Centered, Thick Beds

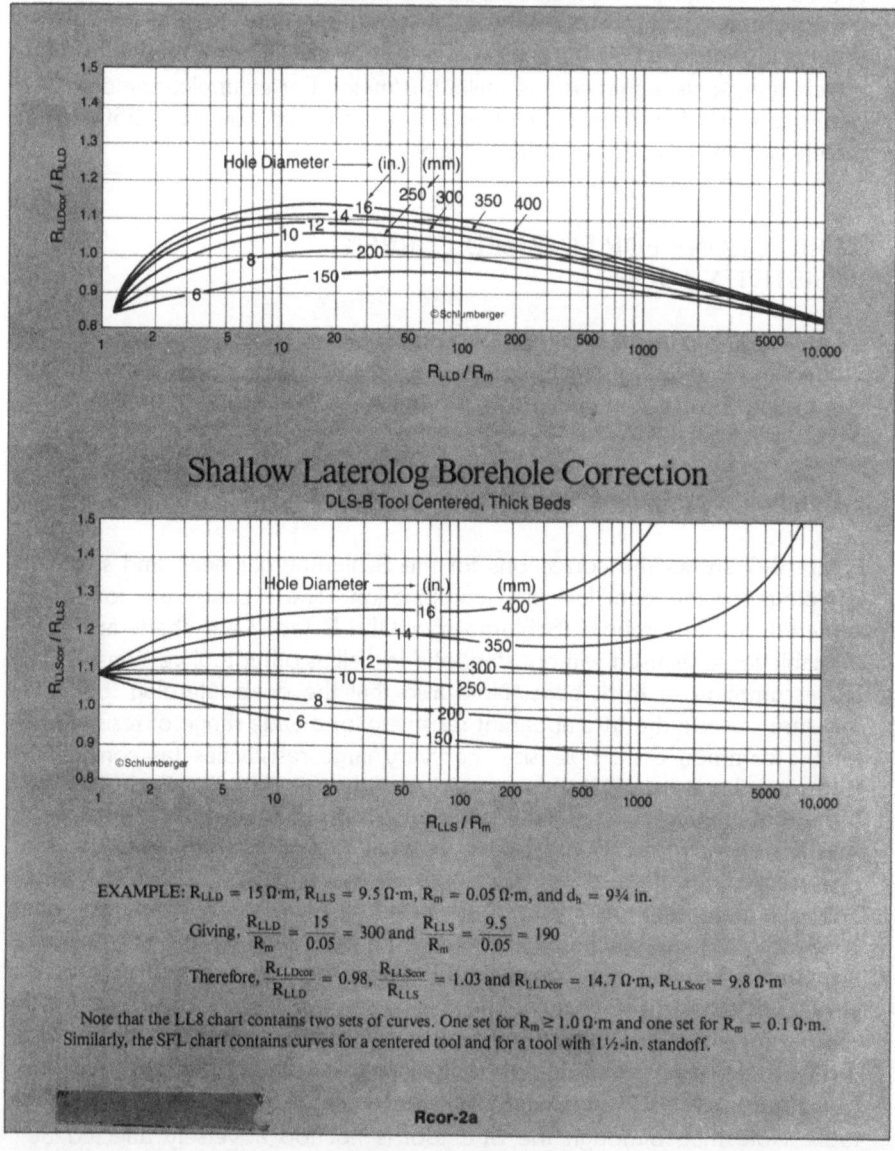

Shallow Laterolog Borehole Correction
DLS-B Tool Centered, Thick Beds

EXAMPLE: R_{LLD} = 15 Ω·m, R_{LLS} = 9.5 Ω·m, R_m = 0.05 Ω·m, and d_h = 9¾ in.

Giving, $\dfrac{R_{LLD}}{R_m} = \dfrac{15}{0.05} = 300$ and $\dfrac{R_{LLS}}{R_m} = \dfrac{9.5}{0.05} = 190$

Therefore, $\dfrac{R_{LLDcor}}{R_{LLD}} = 0.98$, $\dfrac{R_{LLScor}}{R_{LLS}} = 1.03$ and R_{LLDcor} = 14.7 Ω·m, R_{LLScor} = 9.8 Ω·m

Note that the LL8 chart contains two sets of curves. One set for $R_m \geq 1.0$ Ω·m and one set for $R_m = 0.1$ Ω·m. Similarly, the SFL chart contains curves for a centered tool and for a tool with 1½-in. standoff.

Rcor-2a

Figure 6–7. Dual laterolog borehole corrections. *(Courtesy of Schlumberger Ltd.)*

Laterolog-8 Borehole Correction
1½-in. (38-mm) Standoff, Thick Beds

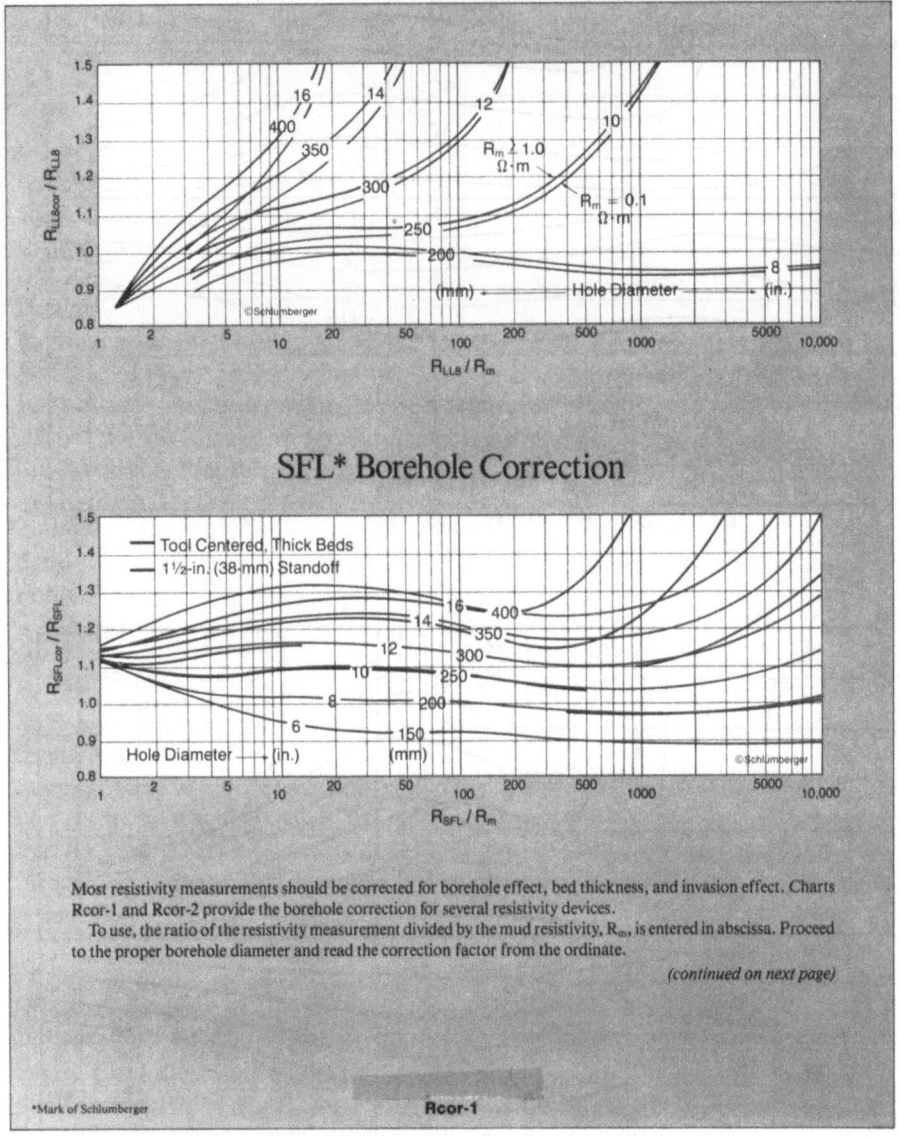

SFL* Borehole Correction

Most resistivity measurements should be corrected for borehole effect, bed thickness, and invasion effect. Charts Rcor-1 and Rcor-2 provide the borehole correction for several resistivity devices.

To use, the ratio of the resistivity measurement divided by the mud resistivity, R_m, is entered in abscissa. Proceed to the proper borehole diameter and read the correction factor from the ordinate.

(continued on next page)

*Mark of Schlumberger

Rcor-1

Figure 6–8. Laterolog 8 and spherically focused log borehole corrections. *(Courtesy of Schlumberger Ltd.)*

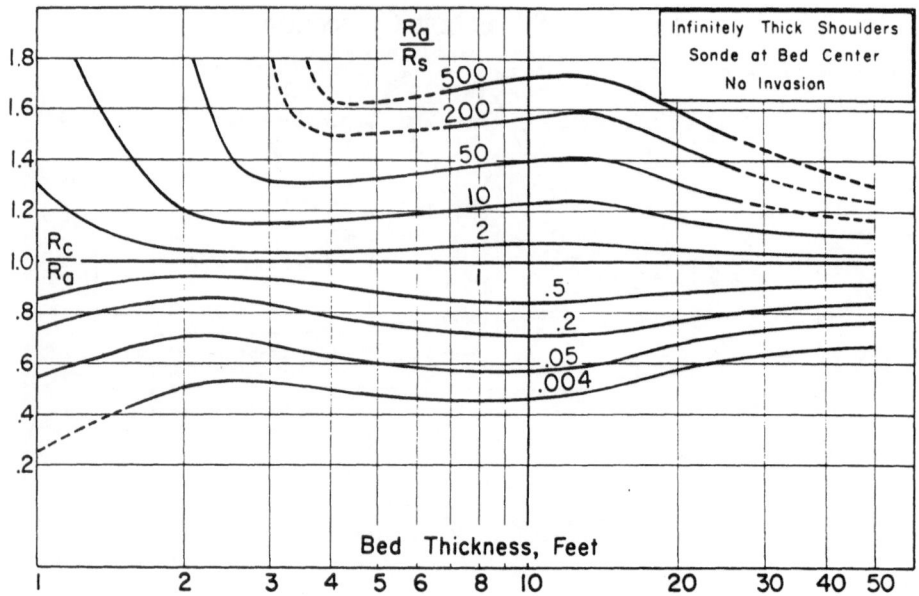

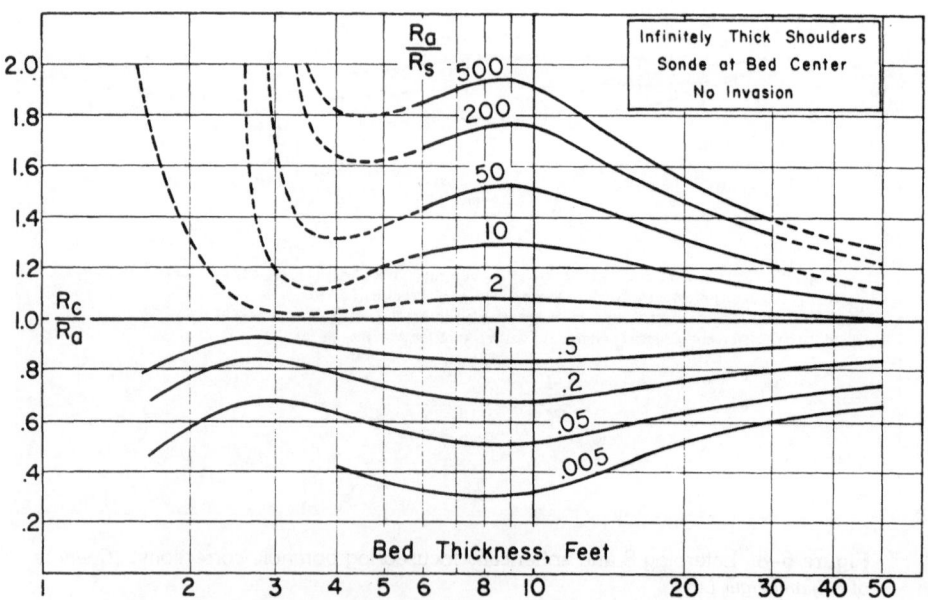

Figure 6–9. Bed-thickness corrections for Laterolog 3 (top) and Laterolog 7 (bottom). *(Courtesy of Schlumberger Ltd.)*

Bed-thickness Corrections

Figure 6–9 illustrates the bed-thickness corrections for the Schlumberger *Laterolog 3* (top) and *Laterolog 7* (bottom). These charts are based on a model where the bed of interest is sandwiched between infinite (or very thick, in practice) beds of equal resistivity R_s and there is no invasion. R_a is the apparent resistivity measured by the tool and R_c is the apparent *correct* resistivity for the bed of interest. For either tool, the correct resistivity may vary from less than half to nearly twice the tool reading, depending on the ratio of the apparent resistivity to the adjacent bed resistivity. For large contrasts, the measured resistivity will depart further from the true resistivity. Note that the *Laterolog 3* performs slightly better than the *Laterolog 7* relative to the range of the ratio of R_c/R_a for the same thickness and range of R_a/R_s. Bed thickness itself does not seem to be the problem for these tools, but rather the ratio of contrasting bed resistivities. Bed-thickness corrections should not be a problem for the *Laterolog 8* or SFL, which are essentially thin-bed tools that respond only to the invaded zone.

Compare Figure 6–9 with Figure 6–10 for the Schlumberger Dual Laterolog (DLS/DE) bed-thickness corrections. Except for beds thinner than 4 ft, there is a significant response improvement for the same resistivity contrasts. For the deep laterolog (top of Fig. 6–10), the range of corrected resistivity is from 0.6 to 1.3 times the measured value over a large range of resistivity contrasts with adjacent beds. Note that the shallow laterolog is not to be compared to the LL3 or LL7 tools since it is an invaded-zone resistivity tool, whereas the LL3 and LL7 are deep resistivity tools like the LLd. For moderate resistivity contrasts, the deep laterolog gives a reading close to the true resistivity without correction, say, when the ratio of resistivity to adjacent bed resistivity is within 10 or .1 for beds at least 3 ft thick. Note the anomaly that for beds of 20 ft thickness the tool apparently does not do quite as well (according to the model on which the chart is based) as for beds of only 5 ft thickness.

Invasion Correction Charts

Figure 6–11 illustrates the invasion corrections for one of the Schlumberger dual laterolog (DLT-B)–R_{xo} combinations. The R_{xo} device is a pad-mounted, micro-SFL tool that is explained shortly.

Unfortunately, we can expect the high resistivity contrast problems to compound invasion corrections for the guard logs and laterologs. Thin-bed effects will lead to apparent invasion profiles and we will find that invasion corrections are routinely applied, both by hand and by computer, without regard to possible thin-bed effects.

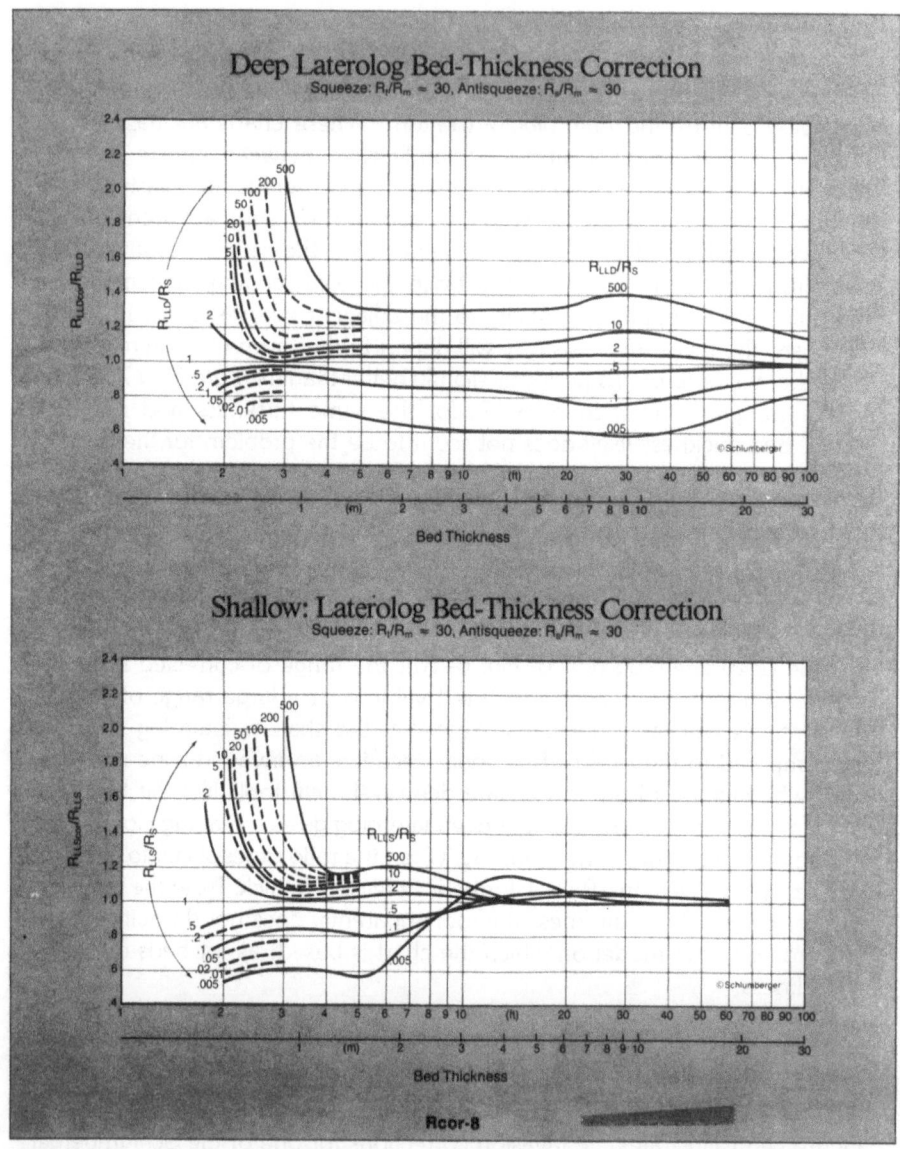

Figure 6–10. Bed-thickness corrections for dual laterolog.

For the guard logs and laterologs, we have seen that the real problem is not so much with the thin beds as the *contrasts in resistivity* between the beds of interest and their adjacent beds. With laterologs and guard logs, we must always keep an eye on the *resistivity contrasts* more than the bed thicknesses to ascertain if the tool is reading too much or too little. If the

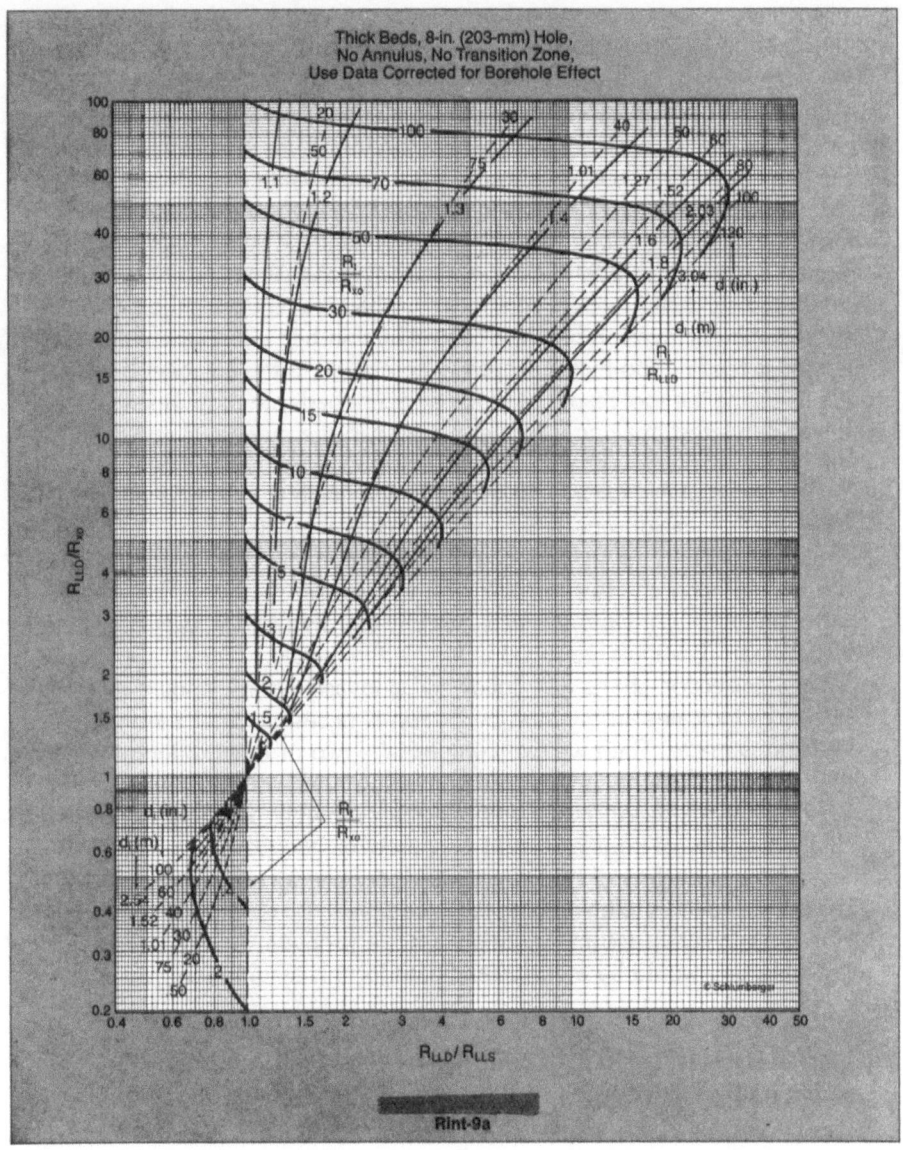

Figure 6–11. Laterolog invasion corrections. *(Courtesy of Schlumberger Ltd.)*

bed is much more resistive than the surrounding beds, the tool reading is likely too low compared to R_t. However, the more common situation with laterologs and guard logs occurs in conductive beds where they may not read low enough. Frequently, in carbonate sedimentary sequences, porous and conductive zones are bounded by tight and very resistive zones. Large resistivity contrasts with the adjacent beds are very likely. The laterolog still offers some advantage over the induction logs with respect to bed effects since we have to be more aware of both bed thicknesses as well as resistivity contrasts with the induction logs.

In Figure 6–11, we can see that in conductive mud systems with the deep laterolog (LLd) reading more than the shallow laterolog (LLs), the true resistivity R_t will be quite a bit larger than the LLd measurement. We can use charts of this type to make invasion corrections if the assumptions listed in the chart are reasonably satisfied. If we see any possibility of a thin-bed effect (large resistivity contrasts between beds), we have to recognize the possibility that the apparent invasion profile is caused by thin-bed effect where the deep laterolog is reading too high a resistivity compared to the adjacent beds that are influencing the tool response. If the adjacent beds are more resistive than the bed of interest, making a further increase in the measured resistivity via the invasion, correction charts will only compound the problem.

In practice, where we are unsure of how much is thin-bed effect and how much is invasion effect, we can calculate a possible range of hydro-carbon saturation. First assume thin-bed effects with no invasion and use the bed correction charts after making any necessary borehole corrections. Then assume only invasion effects without any thin-bed effects but make any necessary borehole corrections first. Alternatively, we can apply bed-thickness correction charts first, followed by invasion corrections if the ratio of LLd/LLs indicates an invasion correction is necessary *after* the thickness correction is made. Then we can compare the two or three different results and make a judgment. In addition, reporting a possible saturation range in this case would seem more practical than reporting a single value. The saturation range, of course, would be based on the possible range in R_t. This same approach could be used with induction logs. The alternative of using the shallow laterolog reading for R_t could also be considered, especially where we can safely assume invasion is not a problem. A very porous and permeable bed would be a likely candidate for this assumption.

PROBLEMS WITH FOCUSED CURRENT RESISTIVITY DEVICES

There are additional problems with laterologs and guard logs, other than bed thickness and invasion profiles, that you should be aware of. With certain laterolog and guard log tools, there may be calibration problems. Also, older tools suffered from what was known as the *Delaware* effect, and modern devices have been described as suffering from the *Groningen* phantom. All these problems are connected with tool design and you should learn how to recognize them.

Delaware Effect

The Delaware effect occurs below thick resistive beds and shows up as an abnormally high resistivity for at least 80 ft below the resistive bed.[4] Tools with a current return electrode placed higher up on the tool rather than at the surface exhibit this problem. Schlumberger states that the *Laterolog 3* is its only tool that still uses this arrangement.[5] However, you should watch for this effect on older as well as modern tools when you are not familiar with the placement of the current return electrode. Figure 6–12 illustrates the Delaware effect. The tool resistivity reading was sometimes as much as 50 times R_t.[6] Note that the effect gradually decreases below the resistive bed. Schlumberger states that a small anti-Delaware effect has been observed

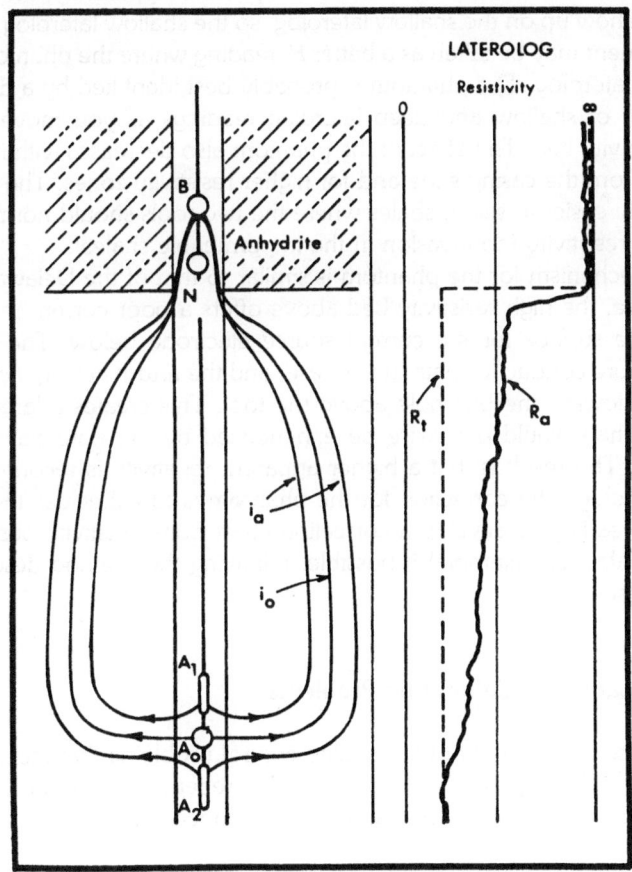

Figure 6–12. Delaware effect. *(Courtesy of Schlumberger Ltd.)*

with *Laterolog 7* and deep laterologs where the measured resistivity is too low just below the resistive beds.[7] Placing the current return electrode up at the surface has apparently solved the problem of the Delaware effect. The anti-Delaware effect has also been reported as solved by the introduction of a nine-electrode tool.[8]

Groningen Phantom

The Groningen phantom is so named because of its apparent discovery in the Groningen gas field in the Netherlands.[9] The Groningen phantom is somewhat similar to the Delaware effect. It is observed as much as 272 ft below a high resistivity bed. The phantom exhibits itself as a resistivity reading that is up to three times what should be expected. Apparently, the phantom does not show up on the shallow laterolog, so the shallow laterolog resistivity measurement may be taken as a better R_t reading where the phantom affects the deep laterolog. The phantom is probably best identified by a decreasing separation of shallow and deep laterolog readings as you move below a high resistivity bed. The effect of the phantom also decreases with increasing distance from the casing shoe and for higher resistivity zones. The phantom is probably easier to see in shales where the two tools should normally read the same resistivity (no invasion in the impermeable shales).

The mechanism for the phantom is similar to that of the Delaware effect. In this case, the high resistivity bed above offers a poor current return path toward the surface for the current source electrode below. The borehole offers a more conductive path in this case, and the return current flow is concentrated toward the borehole above the tool. This creates a larger potential drop than would otherwise be experienced by the potential reference electrode. The result is that a higher apparent resistivity is recorded by the deep laterolog. The correction for the phantom is to either use the shallow laterolog reading or develop a correction chart from resistivity observations (in the shales, for example) if possible, following the method described by Woodhouse.[10]

Possible Laterolog Calibration Problems

Pickett described a study of log quality control problems that included over 500 wells.[11] The logs studied had acceptable repeat sections and calibration data. A wide range of R_{mf}/R_w ratios were included for zones up and down the wells.

The laterolog accuracy studies were prompted by a situation where abnormally low water saturation cutoff values had to be used in a productive area.

For example, it was necessary to use an S_w cutoff of, say, 30%. That is, zones with calculated S_w larger than 30% would produce only water. This implies that the resistivity tool might be reading too high, therefore leading to erroneously low S_w calculations.

The laterolog tools were not specified as to type but from the probable time of this study they were likely *Laterolog 3* or guard type devices since some illustrated example resistivity logs were displayed on hybrid resistivity scales. The anomalous behavior was that the laterolog readings were higher than the R_{xo} tool readings in impermeable beds where they should have read about the same resistivity if both were working all right. It was not stated which R_{xo} devices were studied. From the probable time of the study, the R_{xo} devices would have likely been proximity logs or microlaterolog tools.

Although Pickett found it possible to increase the R_{xo} readings by a factor that would make them equal to the laterolog R_t readings in the impermeable zones, this approach led to too high calculated residual hydrocarbon saturations (too low flushed zone saturation S_{xo}).[12] This would yield a reasonable movable hydrocarbon saturation ($S_{xo} - S_w$), but still compounds the low calculated S_w problem with a low calculated S_{xo} problem.

For these reasons, Pickett concluded the likely culprit was abnormally high R_t readings from the laterolog devices.[13] The corrective action was to multiply the laterolog readings by a factor determined by the reciprocal of their ratio to the R_{xo} tool readings opposite zones with smooth borewall, no hole enlargements, and no invasion.

When the apparently corrected laterolog readings were compared to the readings before correction, it was found from a plot of about a dozen wells that for the laterolog

$$R_t = \text{constant} \times R_a$$

where R_a is the laterolog reading, R_t is the resistivity, and the constant could take on values described as

$$1 < \text{constant} < 3$$

This is an unfortunate result. It means that even with valid R_w information, S_w calculations will be very uncertain.

However, the study reported by Pickett may be too pessimistic when considered in the light of other data reported in the same study. When studying the R_{xo} devices, it was also found that they were not reliable for readings over 10 ohm-meters. When I remove those points from the plot of the dozen wells referred to above, I find that the constant should have a range of something like between 1.0 and 1.5, with maybe 2.0 at the most.

Other possible factors related to this study were not mentioned. For example, a Delaware effect may account for the abnormally high resistivities. Also, low ratios of R_a/R_s in a *Laterolog 3* (see Fig. 6–9) may lead to apparent resistivities as much as twice the actual R_t. On a part of one of the logs shown in Pickett's study, the resistivity log exhibits a possible Delaware effect. Unfortunately, the complete thickness of the resistive bed above (which needs to be known to assess the Delaware effect) is not fully reproduced.

There were large resistivity contrasts for thin beds, possibly contributing to the problem of abnormally high resistivity readings in the more conductive porous zones. I have already mentioned that R_{xo} tool readings over 10 ohm-meters were unreliable. Pickett concluded this from the lack of repeatability of the R_{xo} devices for resistivities over 10 ohm-meters. Even for $R_{xo} < 10$ ohm-meters, Pickett found that the 90% confidence interval from four repeat runs in each well studied was 23% of the average R_a from the four runs.[14] This is to be contrasted with the deep reading R_t devices, which, despite their accuracy problems, showed perfect repeatability.

Pickett's conclusion was that the problems were probably due to difficulties in field calibration. Unfortunately, I know of no similar, modern studies of laterolog calibrations of the same scope and magnitude as the study reported by Pickett. It is hoped that calibration procedures have improved over time and that we see better laterologs today. However, you should always be aware of the possibility of miscalibration on *any* particular log. Nienast and Knox have reported that over *half* of the logs they studied were found to be in error.[15]

Another interesting result from Pickett's log study was a galvanometer inertia effect where the *rate of change* of resistivity was not fully reproduced on the recorded log. The effect increased with increased logging speed (confirming the likely galvanometer effect). For example, Pickett found that at a logging speed of 5,000 ft/hr, only 92% of the actual conductivity change was accounted for in a 3 ft bed. The effect was less severe for thicker beds. Modern logging tools should not exhibit this inertia effect since they are now recorded by computer. However, if you are looking at older laterolog readings in thinner zones (recorded before computer trucks were introduced), you should be aware that the full resistivity contrast may not be achieved on the recorded log.

FLUSHED ZONE LOGGING

In a manner similar to calculating water saturation S_w using the undisturbed zone resistivity, we can use the resistivity of the flushed zone near the well bore to calculate the saturation of the flushed zone S_{xo}. In fact, we could also use an invaded zone resistivity measurement to calculate the saturation in the invaded zone, intermediate between the flushed zone and the undisturbed zones. However, the saturation in this area may vary from the flushed zone value to the undisturbed zone value.

In the flushed zone, the mud filtrate water from the mud system flushes all the original formation water and part of the hydrocarbon from the pore spaces near the well bore. Thus, the resistivity of the flushed zone may be either higher or lower than the undisturbed zone resistivity. Its value will depend on both the flushed zone saturation and the resistivity of the mud filtrate water. Whether or not complete flushing takes place will depend to a great extent on the permeability of the rock formation. Opposite impermeable rocks, the mud filtrate will not invade the formation at all. In some rocks, only partial flushing will take place. You must always watch for this if using flushed zone methods based on the flushed zone resistivity R_{xo}.

Recall that water saturation is calculated from

$$S_w = (R_t/FR_w)^{(-1/n)} \tag{6–1}$$

The formation factor and saturation exponent should remain the same (at least for clean, nonshaly formations) regardless of the fluids in the pore system. If we substitute the resistivity of the mud filtrate R_{mf} for R_w and the flushed zone resistivity R_{xo} for R_t in the above formula, we obtain the formula for the flushed zone saturation.

$$S_{xo} = (R_{xo}/FR_{mf})^{(-1/n)} \tag{6–2}$$

Opposite porous and permeable zones, this flushing by the filtrate water from the mud system constitutes an in-situ pilot water flood test. The hydrocarbon that is moved by the filtrate is essentially the same that will be moved by a water flood. It is also a good qualitative measure of the permeability of the formation since no hydrocarbon will be displaced by the filtrate water in an impermeable formation. The amount of hydrocarbon displaced (movable hydrocarbon) is given by the difference between the initial hydrocarbon saturation before flushing and the residual hydrocarbon saturation after flushing.

$$MHS = S_o - RHS \tag{6–3}$$

Sometimes movable hydrocarbon saturation *MHS* is written as movable oil saturation *MOS*, and the residual hydrocarbon saturation *RHS* is written as residual oil saturation *ROS*. Thus $MOS = S_o - ROS$. In Eqs. 6–1 and 6–2, all saturations are expressed as a fraction of the pore volume, although they may frequently be reported or listed as percentages for convenience.

Since the initial hydrocarbon saturation or initial oil saturation is simply 1 less the water saturation of the undisturbed zone, and the residual hydrocarbon saturation is 1 less the flushed zone saturation

$$
\begin{aligned}
MHS &= S_o - RHS \\
&= (1 - S_w) - (1 - S_{xo}) \\
&= S_{xo} - S_w
\end{aligned}
\tag{6–4}
$$

Although calculation of the movable hydrocarbon saturation is a valuable evaluation method, it requires a knowledge of seven parameters (actually eight parameters if you use the form of the formation factor relation that requires the constant a) rather than the five or six needed for the usual water saturation calculations. In addition to those five unknowns (a, m, n, ϕ, and R_t), R_{mf} and R_{xo} are also needed. Thus, an additional measurement with more possible uncertainty as well as another parameter R_{mf} is needed.

The resistivity of the mud filtrate R_{mf} fortunately is a routine measurement made by the logging service company at the time of logging. There are some pitfalls with this measurement, however. The measurement is subject to errors if the mud sample is collected from some source other than the flowing mud system. Also, this measurement is not usually repeated and some error may be introduced by the temperature conversion process. There must be *complete flushing* to calculate the flushed zone saturation. Although seven values are normally needed to solve for movable hydrocarbon saturation, Pickett demonstrated that cross-plot techniques can eliminate some of the needed parameters (notably, R_{mf}) and provide some compensation for the uncertainties in the other parameters (e.g., a, m, and porosity tool calibration).[16] Also, some of the measurement errors are compensated for by the cross-plot methods.

All resistivity devices used to measure R_{xo} are pad-mounted as opposed to the centralized undisturbed zone and invaded zone tools already examined. The microlog referred to in Chapter 5 was considered by some to be an R_{xo} measuring device, but in practice, it proved more valuable as a good mudcake indicator that could define permeable formations. It was not a quantitative tool. Nevertheless, elaborate interpretation schemes were developed for use with the microlog.

There are three basic R_{xo} devices, all pad-mounted to insure a good chance of measuring resistivity only in the flushed zone. Farther out from the borewall, the rock may not be completely flushed and the tool would be reading the invaded zone resistivity rather than the flushed zone resistivity. The *proximity log* was the first tool used for R_{xo} measurements (it was and still is frequently run in combination with the microlog). Later the microlaterolog was developed, and more recently a pad-mounted version of the spherically focused log was developed: the micro-SFL.

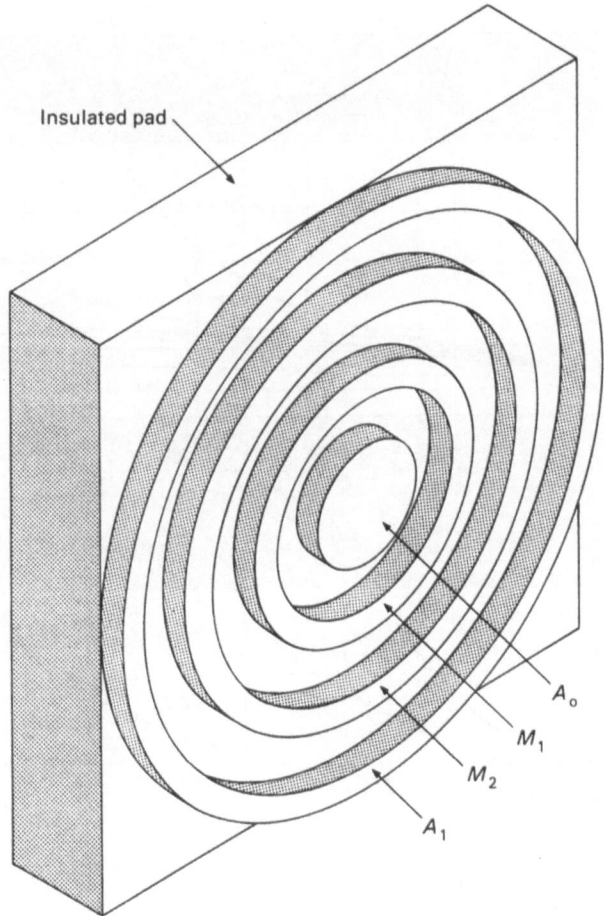

Insulated pad

A_o

M_1

M_2

A_1

Figure 6–13. Concentric ring electrode configuration for the microlaterolog device. (Note: Circular electrodes press against formation.)

Figure 5–8 in Chapter 5 illustrates the pad-mounted devices as well as the centralized devices. All of these pad-mounted R_{xo} devices had miniature electrode systems because of their small size requirements. Figure 6–13 illustrates the concentric ring pattern of the electrodes for the microlaterolog device. The proximity log needs deeper invasion to respond to only the flushed zone because it has a wider path with deeper penetration than the other R_{xo} devices. However it has the smallest mudcake effect of the R_{xo} devices.

Figure 6–14 illustrates the mudcake correction for all three Schlumberger devices. Note that proximity log mudcake correction is virtually nonexistent except for the largest thicknesses (h_{mc}) of mudcake. The microlaterolog operates at the other extreme of invasion from the proximity log. Only a little

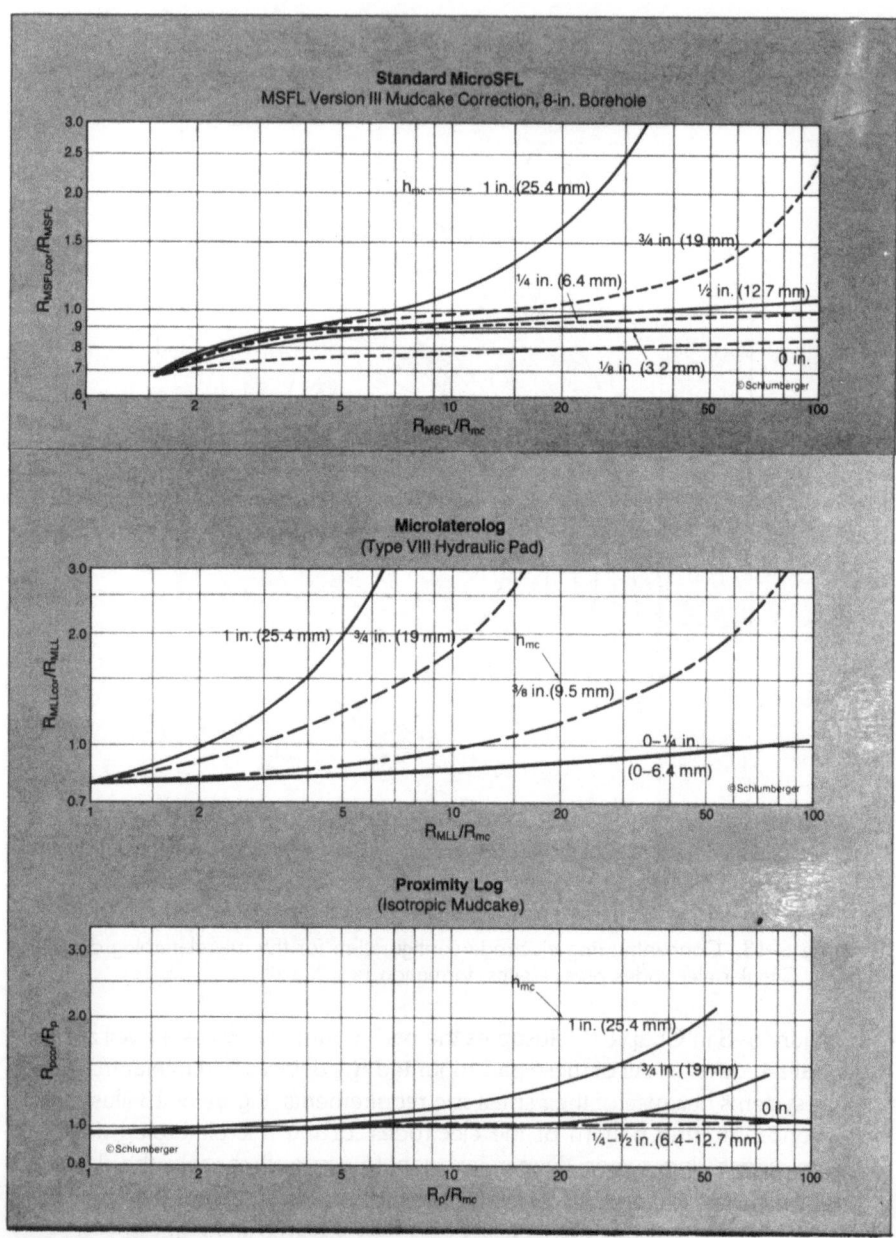

Figure 6–14. Mudcake corrections. *(Courtesy of Schlumberger Ltd.)*

over 4 in. of invasion is required for it to provide an R_{xo} reading. However, it is severely affected by mudcakes thicker than 3/8 in. as can be seen from the mudcake correction chart (Fig. 6–14). It would be the R_{xo} tool of choice if you were not concerned about mudcake in a given situation. The micro-SFL tool is a compromise between the proximity log and the microlaterolog tools. It performs reasonably well for mudcakes that are less than 3/4 in. thick (see the correction chart in Figure 6–14) and does not require as much invasion as the proximity log.[17] However, the actual required invasion depth is not given for this tool. An advantage to the micro-SFL and microlaterolog tools over the proximity log is that they can be run together with a deep reading resistivity tool.

Example of Movable Hydrocarbon Calculation

As an example of a movable hydrocarbon calculation, assume you have the following readings from a deep reading laterolog device and a microlaterolog: $R_{xo} = 0.6$ ohm-meters and $R_t = 7$ ohm-meters. Assume $R_{mf} = 0.02$ ohm-meters (from a salt mud system) and $R_w = 0.10$ ohm-meters. Let $a = 1$ and $m = n = 2$. If the porosity is 30%, what is *MHS* (or *MOS*)? First calculate S_{xo} from:

$$S_{xo} = (R_{xo}/FR_{mf})^{(-1/n)}$$
$$= (R_{xo}\phi^m/R_{mf})^{(-1/n)}$$
$$= (((0.6)(.3)^2)/(.02))^{(-1/2)}$$
$$= 0.609$$
$$= 60.9\%$$

Where I used the relation $F = 1/\phi^m$. Now calculate S_w.

$$S_w = (R_t/FR_w)^{(-1/n)}$$
$$= (R_t\phi^m/R_w)^{(-1/n)}$$
$$= (((7)(.3)^2)/(0.10))^{(-1/2)}$$
$$= 0.398$$
$$= 39.8\%$$

Now take the difference to get *MHS*.

$$MHS = .609 - .398$$
$$= .211$$
$$= 21\%$$

SPONTANEOUS POTENTIAL LOGGING

Spontaneous potential (SP) measurements constituted one of the earliest properties logged in a well. It was found that when a single potential electrode was lowered in a well bore, a potential was measured relative to a ground electrode situated at the surface. This is a so-called *natural* or *spontaneous* potential since no current source is used. The potential varies according to lithology and fluid properties. Explanations for this natural potential include electrochemical potentials, electrokinetic potentials, and cation-selective membrane phenomena. According to Pickett, SP is not well understood, likely being due to a combination of the effects mentioned above.[18]

SP usually develops best in sand–shale sequences in sedimentary rocks in a borehole with a conductive mud system. However, researchers have also recorded SP signals in empty boreholes! In sedimentary rocks, SP is a function of permeability, porosity, shaliness, and the ratio of mud filtrate resistivity to formation water resistivity. One of the primary uses of SP has been to determine formation water resistivity. Since SP is dependent on the ratio of R_{mf}/R_w, we can determine R_w from an SP measurement if R_{mf} is known. R_{mf} can usually be measured at the surface from a sample taken (preferably) from the circulating mud system as close to the time of logging the well as possible. Although the determination R_w has been of primary interest in the petroleum industry, the same measurement can be used in evaluating water quality in ground water studies.

SP is normally measured from a *shale baseline* established by the potential reading opposite the shale zones. SP develops a negative potential relative to the shale base line opposite sands where $R_w < R_{mf}$. In sands with $R_{mf} > R_w$, on the other hand, SP appears as a positive potential. SP is reduced in thinner beds and sands containing silt or clay mineral (shaly sands). It is also reduced in very resistive beds. Correction charts are used to correct the reading for bed thickness and high contrasts of R_t/R_m. Figure 6–15 illustrates SP behavior in a hypothetical sand–shale sequence. In carbonate sequences without shale beds, SP tends to have a more wandering, less definitive appearance. SP can be used to find R_w in some carbonate sequences where the permeable beds are separated by shale beds.

According to Pickett, SP can be described by the equation

$$SP = -K \, \log(R_{mf}/R_w) + E_k \qquad \text{(6–5)}$$

where SP is in millivolts; K is a constant that may vary from one area to another and is a function of temperature (expressed in units of millivolts per logarithmic cycle); R_{mf} is the mud filtrate resistivity expressed in the same units as the formation water resistivity R_w; and E_k is the electrokinetic

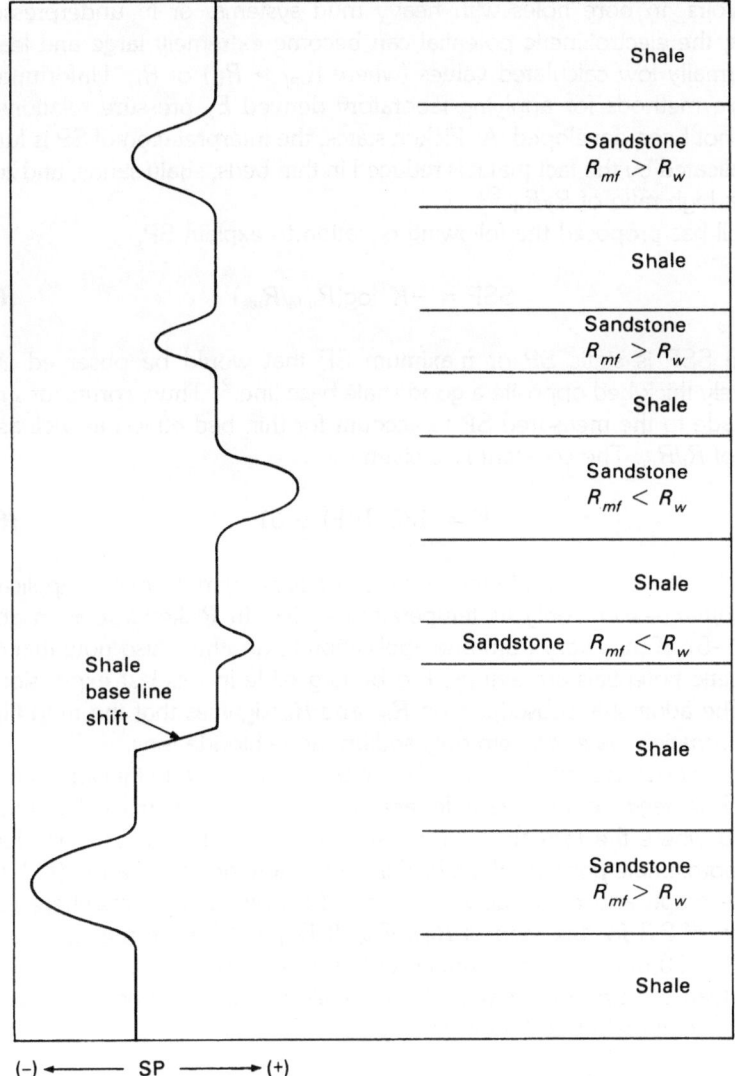

Figure 6–15. Spontaneous potential (hypothetical log).

potential in negative millivolts.[19] I have taken the liberty of re-expressing Pickett's equation (as well as similar equations to follow) using a minus sign in front of the constant K so that SP is expressed as a negative number when $R_{mf} > R_w$. E_k increases negatively with increasing differential pressure measured between the mud weight and the pore fluid pressure. In depleted reservoirs, in bore holes with heavy mud systems, or in underpressured zones, the electrokinetic potential can become extremely large and lead to abnormally low calculated values (where $R_{mf} > R_w$) of R_w. Unfortunately, reliable methods for applying laboratory derived E_k pressure relationships have not been developed. As Pickett states, the interpretation of SP is further complicated by the fact that it is reduced in thin beds, shaly sands, and zones with a high ratio of R_t/R_m.[20]

Doll has proposed the following equation to explain SP.

$$SSP = -K \, \log(R_{mfe}/R_{we}) \tag{6-6}$$

where SSP is static SP or maximum SP that would be observed in an infinitely thick bed opposite a good shale base line.[21] Thus, corrections must be made to the measured SP to account for thin-bed effects as well as the ratio of R_t/R_m. The constant K is given by

$$K = .133 \; T(°F) + 61 \tag{6-7}$$

Note that the constant K in this equation does not vary from one application to another, except only as temperature varies. In Pickett's relation above (Eq. 6-5), K may vary from one application to another. Also note that electrokinetic potentials are assumed to be negligible in this last expression for SP. The additional subscript e on R_{mf} and R_w signifies that the mud filtrate and formation water contain only sodium and chloride ions.

Figure 6-16 is a graph for solving the SP equation. Note that where $R_{mf} > R_w$, SP is *negative*. For example, assume SP reads -30 millivolts opposite a sand where the formation temperature is 170°F. To find R_w, first locate the approximate position of a 170°F line between the 150°F and 200°F lines on the graph and move upward from -30 millivolts to somewhere in the vicinity of 2.3 for the ratio of R_{mfe}/R_{we}. If R_{mf} is .23 ohm-meters, then R_w must be .10 ohm-meters assuming both are NaCl solutions.

Instead of using the graph, which usually requires interpolation, we can use the temperature to find K directly from Eq. 6-7:

$$K = .133 \times 170°F + 61$$
$$= 83.61$$

Then solve for the ratio of R_{mf} to R_w by rearranging Eq. 6-6.

$$R_{mf}/R_w = 10^{-SP/K)}$$
$$= 10^{-(-30mv/83.61)}$$
$$= 10^{(30/83.61)}$$
$$= 2.28$$

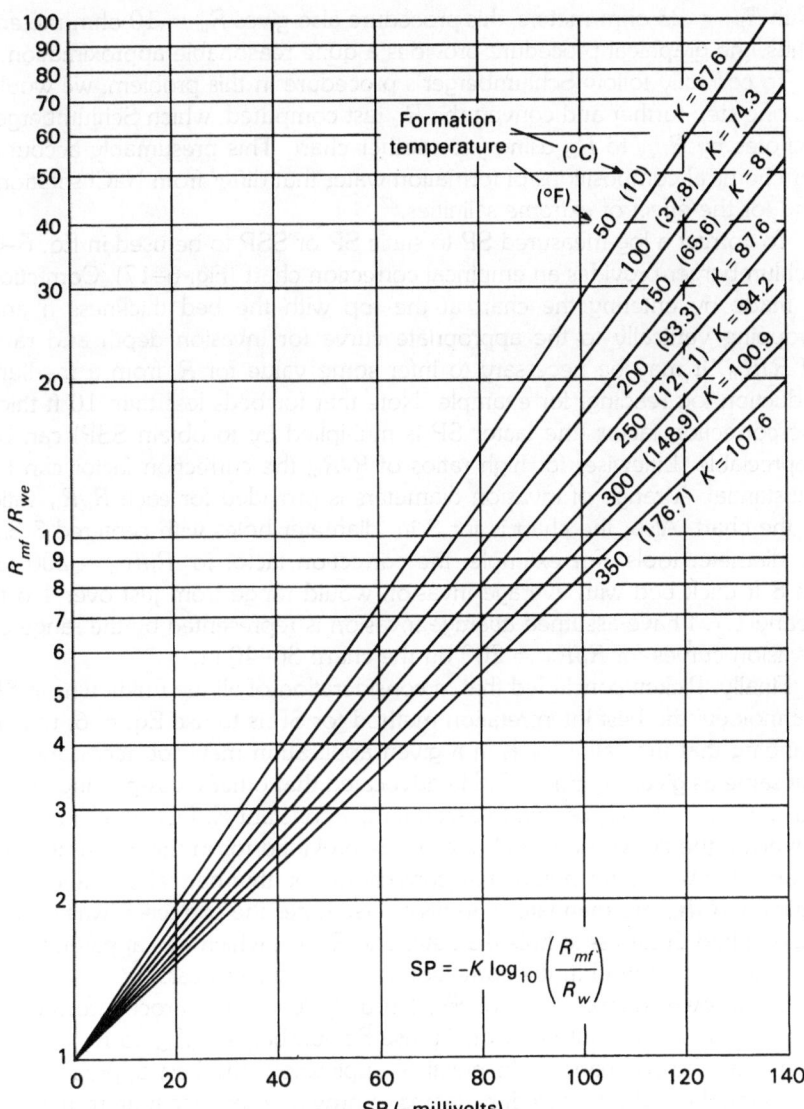

Figure 6–16. Graphic solution of the SP relation for R_w. (Note: For positive SP use this chart, but vertical axis becomes R_w/R_{mf}.)

With $R_{mf} = .23$ ohm-meters, this procedure also gives $R_w = .10$ ohm-meters. Thus, the graphical procedure provides a quite reasonable approximation.

To correctly follow Schlumberger's procedure in this problem, we would go one step further and convert the R_w just computed, which Schlumberger denotes by R_{we}, to R_w using yet another chart. This presumably accounts for chemical compositions of formation water that differ from NaCl solutions and for the effect of *extreme* salinities.[22]

To correct a log-measured SP to static SP or SSP to be used in Eq. 6–6, Schlumberger provides an empirical correction chart (Fig. 6–17). Correction is made by entering the chart at the top with the bed thickness h and dropping vertically to the appropriate curve for invasion depth and ratio of R_i/R_m. It may be necessary to infer some value for R_i from a medium induction log reading, for example. Note that for beds less than 10 ft thick the correction factor (the factor SP is multiplied by to obtain SSP) can be appreciable. Likewise, for high ratios of R_i/R_m the correction factor can be substantial. A range of invasion diameters is provided for each R_i/R_m ratio in the chart. Also, the chart is for 8 in. diameter holes with centered 3 3/8 in. diameter tools. For example, the correction factor for $R_i/R_m = 200$ for an 8 ft thick bed with average invasion would range from just over 1.6 to nearly 1.7. I have assumed *average* invasion is represented by the range of invasion curves for $R_i/R_m = 200$ on the chart: 30–40 in.

Finally, Pickett concluded that in consideration of all the unknowns in SP technology, the best interpretation method for SP is to use Eq. 6–6, understanding that the value for K in a given application may not necessarily be the same as given by Eq. 6–7. He advocated that other cross-plotting methods be used where possible to ascertain the constant K. Pickett also did not advocate the conversion of R_{we} to R_w as provided for in the Schlumberger charts. However, he noted that corrections for thin-bed effects and high ratios of R_i/R_m are mandatory (in his course notes the correction was for the ratio of bed thickness to hole diameter and R_t/R_m, which was apparently the common form found in the correction charts in earlier years).[23]

In my own interpretation of SP, I usually follow the procedure recommended by Pickett but occasionally use Schlumberger's R_{we} to R_w conversion chart for comparison. In practical application, the best approach may be dictated by which procedure seems to provide more accurate results in a given area. For example, you may have R_w samples from produced water or drill stem test in a well for comparison with log-derived SP values. Also, you may have back-calculated R_w values from well log data in apparent water-bearing zones in a well, or perhaps you will be using one of the more advanced cross-plotting techniques to help decide which SP interpretation approach works best. Some of these were outlined by Etnyre[24] in a discussion of an approach advocated by Bassiouni and Matthews.[25]

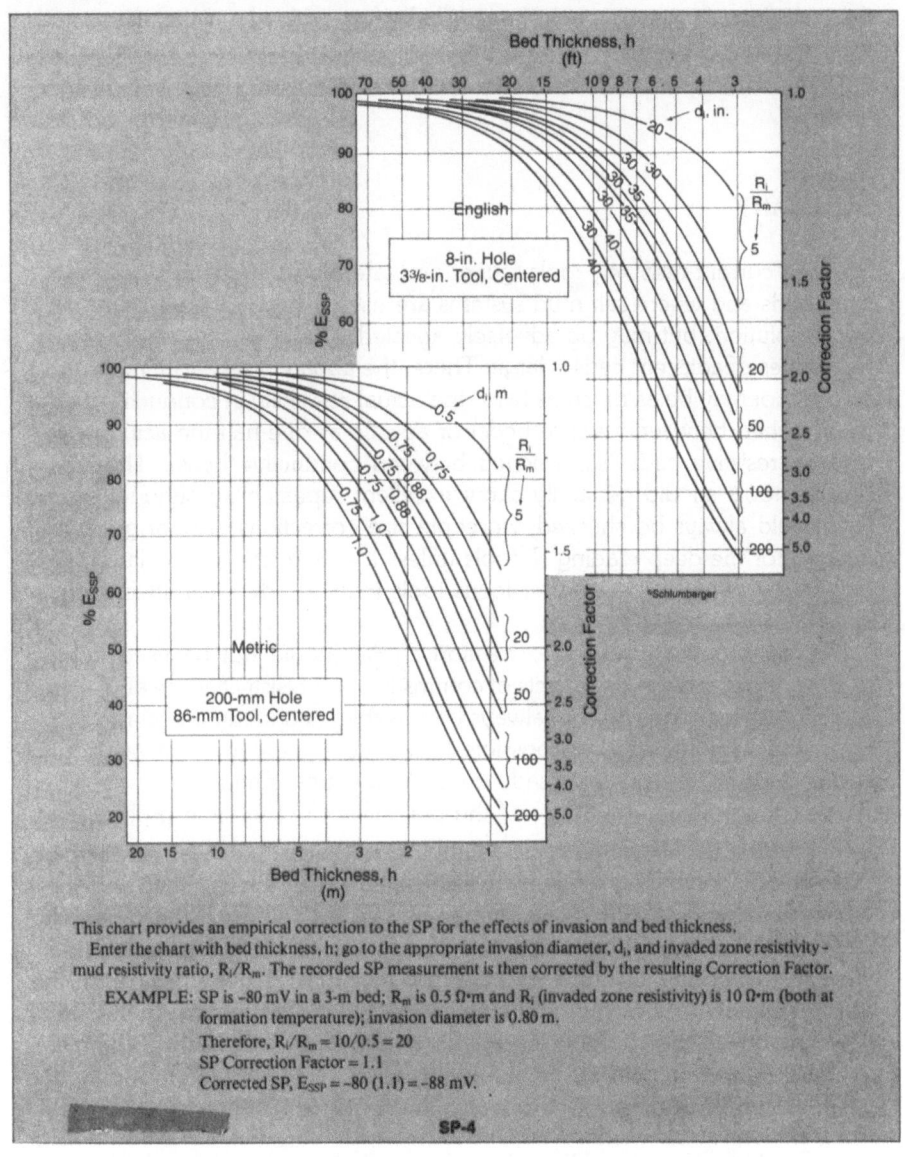

This chart provides an empirical correction to the SP for the effects of invasion and bed thickness.
    Enter the chart with bed thickness, h; go to the appropriate invasion diameter, d_i, and invaded zone resistivity - mud resistivity ratio, R_i/R_m. The recorded SP measurement is then corrected by the resulting Correction Factor.

EXAMPLE: SP is −80 mV in a 3-m bed; R_m is 0.5 Ω•m and R_i (invaded zone resistivity) is 10 Ω•m (both at formation temperature); invasion diameter is 0.80 m.
    Therefore, $R_i/R_m = 10/0.5 = 20$
    SP Correction Factor = 1.1
    Corrected SP, $E_{SSP} = −80 (1.1) = −88$ mV.

SP-4

Figure 6–17. SP correction chart. *(Courtesy of Schlumberger Ltd.)*

SP has also found use as an environmental indicator. In a *fining-upward* sequence (where grain size decreases upward), the SP curve will generally show its largest development in the lower part of the sequence with a steadily decreasing potential as we move upward. The decrease is due presumably to the inclusion of increasing amounts of shaly, fine-grained material associated with the fining-upward sequence. Turbidite sands may show an alternating SP associated with alternately increasing and decreasing grain sizes, and so forth.

SUMMARY

Focused current resistivity tools (guard and laterologs) are best in high resistivity beds and where salt mud systems are used. They have excellent thin-bed resolution, but may be adversely affected where the contrast between resistivities of adjacent beds is large. There, the apparent resistivity measured by the tool may be as much as twice the actual resistivity in conductive beds (sandwiched between resistive beds) or as little as one-half the actual resistivity in resistive beds (sandwiched between conductive beds). The deep reading tools should generally show excellent repeatability although each log should always be checked. Borehole size corrections are not generally needed for the deep reading R_t tools unless the hole size is quite large (see Fig. 6–7). However, the shorter focus tools may be adversely affected and need correction (see Fig. 6–8).

My belief is that invasion correction charts should not be used unless reliable corrections for the effects of contrasting bed resistivities are able to be established first. This may not always be possible, and if invasion corrections are made without regard to possible bed effects, the correction may only compound the error in an already questionable measurement.

We should always be aware of the possibility of calibration problems or other effects (Delaware effect and Groningen phantom). These can best be assessed by comparing resistivity measurements between the deep and shallow measuring devices in shales or other impermeable beds. When analyzing older logs, we should be aware of the possibility of the galvanometer inertia effect when the logging speed was too high. The inertia effect (from the galvanometers of the recording systems on older type logging trucks) may become quite severe for logging speeds greater than 5,000 ft/hr (80 ft/min).

Flushed zone resistivity measurements allow us to calculate not only the saturation of the flushed zone but also the movable hydrocarbon present in the pore spaces (if we also have an R_t measurement for calculation of S_w) from Eq. 6–4. This movable hydrocarbon calculation is very useful in providing information about how much hydrocarbon may be moved by a water drive mechanism. It also provides a qualitative measure of permeability. The disadvantage of movable hydrocarbon calculations is the requirement of finding values for so many unknown parameters. If either R_t or R_{xo} is in error, the calculation will be in error. A benefit is that, if the porosity device is in error, there will be some compensation since the same porosity measurement appears in both S_w and S_{xo}. Since movable hydrocarbon is the difference between these two saturations, the porosity error (or error in porosity exponent) will affect each saturation calculation in the same way, and the error will tend to subtract out when movable hydrocarbon saturation as the difference between S_{xo} and S_w is calculated.

Spontaneous potential (SP) measurements are a function of the ratio of the formation water resistivity to the mud filtrate resistivity (see Eq. 6–6). SP is reduced by shaliness, high resistivities, and thin-bed effects. Once corrections are made for these effects, Eq. 6–6 allows us to determine the formation water resistivity in a formation if R_{mf} is known. The constant K in the equation must be determined either from Eq. 6–7 or from cross-plot methods. The determination of R_w cannot always be made in carbonate zones or in zones where the electrokinetic effect (particularly in abnormally low-pressured zones) cannot be quantified. Shaliness corrections may also be difficult to quantify. Therefore, determination of R_w from SP is best done in relatively clean sands.

PROBLEMS

6–1. According to Figure 6–8, the spherically focused tool will be less affected by the mud system in the borehole than the *Laterolog 8*. Why is this true?

6–2. Many commercial computerized log interpretation systems provide for the routine calculation of resistivities corrected for invasion. Why or why not is this a reasonable practice?

6–3. What will be the effect on the tool reading if a micro-SFL device loses pad contact with the borewall?

6–4. Which of the R_{xo} devices works best for very shallow invasion? Which one is least affected by the larger mudcakes on the borewall?

6–5. In a bed that is 5 ft thick, how much is SP reduced when the ratio of the resistivity of the invaded zone to that of the mud system is 100 and the depth of invasion is 40 in.? How will you determine the resistivity of the invaded zone and the depth of invasion from typical log data?

6–6. In practice, say, in a carbonate evaporite sequence where porous zones appear as low resistivity, thinner beds sandwiched between very high resistivity (tight, low porosity beds) zones, what will be the most common accuracy problem you will encounter with focused current resistivity devices?

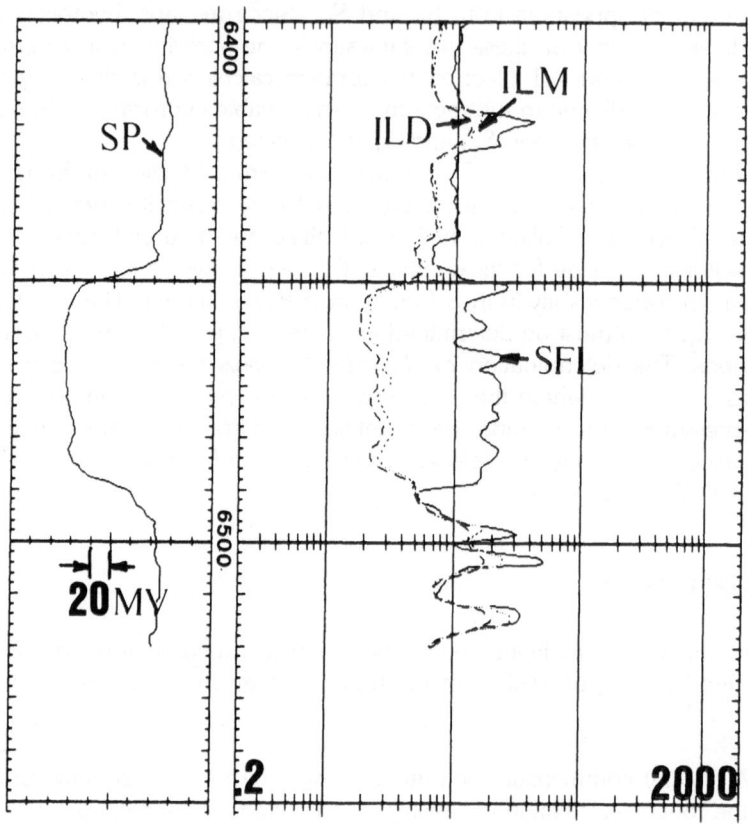

Figure 6–18. SP log for Problem 6–8.

6–7. What is the prerequisite for flushed zone calculations?

6–8. Calculate R_w from the SP log reading in Figure 6–18 for the sand from 6,450 ft to 6,488 ft. In this well, the mud filtrate resistivity R_{mf} is 1.80 ohm-meters at 59°F and the mud resistivity R_m is 2.59 ohm-meters at 63°F. Bottom hole temperature is 131°F at 6,653 ft depth. Thus, it is valid to assume that this is the same as the temperature at the depth of the formation of interest in this problem.

6–9. Correct a shallow laterolog reading of 10 ohm-meters in a 12 ft thick bed with adjacent bed resistivities R_s of 500 ohm-meters and $R_s/R_m = 30$.

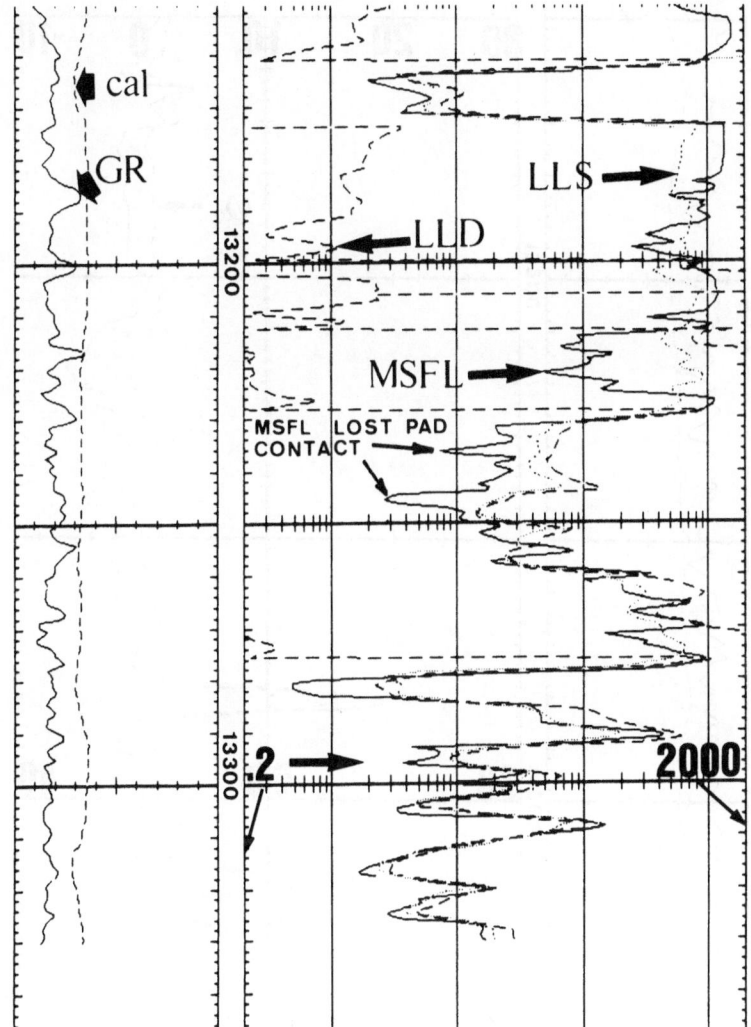

Figure 6–19. Dual laterolog for Problem 6–10.

6–10. Assuming there is no invasion and that $a = 1$, and $m = n = 2$, calculate water saturation, flushed zone saturation, and movable hydrocarbon saturation for the zone from 13,162 ft to 13,171 ft for the well log suite in Figures 6–19 (dual laterolog) and 6–20 (sonic log). R_{mf} is .061

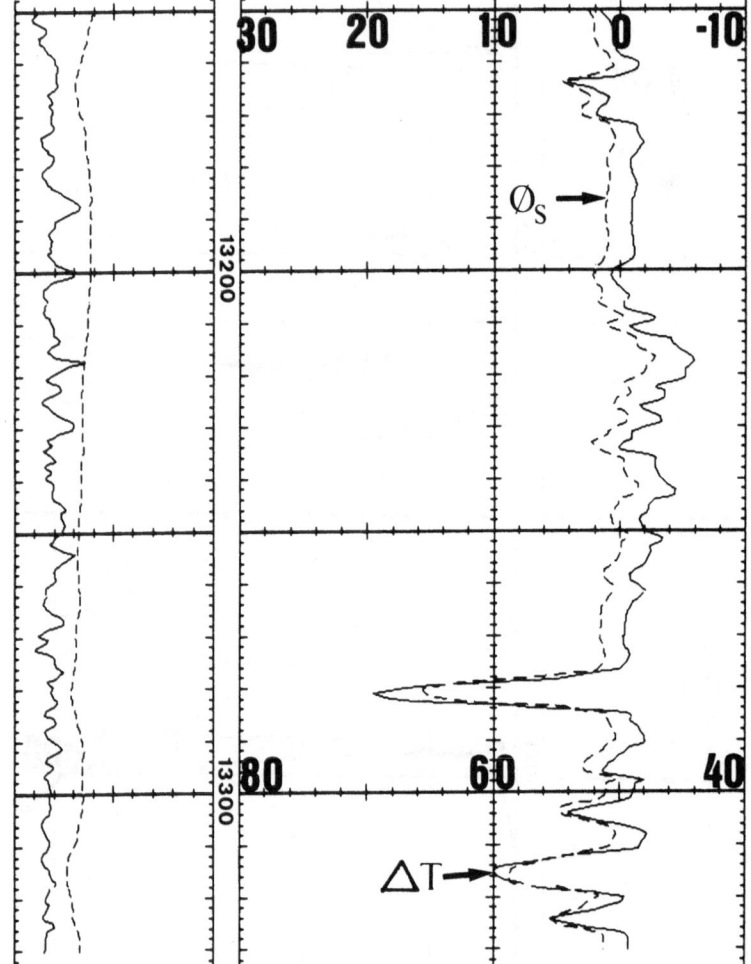

Figure 6–20.   Sonic log for Problem 6–10.

ohm-meters at 60°F and R_w = .015 at formation temperature 244°F. Use ΔTMA = 46μsec/ft (intercept) and slope B = 0.6 μsec/ft/% porosity in the two-parameter, straight-line sonic relation. What are the corresponding parameters for the time-average sonic porosity formula? Repeat your calculations using the shallow laterolog reading for R_t. How would you justify this latter procedure?

6–11. Design a hybrid resistivity scale for track 2 that will record resistivities from zero to infinity ohm-meters without using a backup scale such that a 20 ohm-meters measurement is recorded in the middle of the scale on track 2 (zero–twenty–infinity scale).

REFERENCES

1. Roberto Aguilera, *Naturally Fractured Reservoirs* (Petroleum Publishing Company, Tulsa, Okla., 1980), pp. 200–220.
2. Schlumberger, *Log Interpretation, Volume 1, Principles* (Schlumberger Ltd, New York, 1972), p. 22.
3. Schlumberger, *Log Interpretation Charts* (Schlumberger Ltd, New York, 1986).
4. Schlumberger, *Log Interpretation, Volume 1*, p. 22.
5. Ibid., p. 25.
6. R. Woodhouse, "The Laterolog Groningen Phantom Can Cost You Money," paper presented at the 19th Annual Logging Symposium of the Society of Professional Well Logging Analysts, June 1978.
7. Schlumberger, *Log Interpretation, Volume 1*, p. 25.
8. Woodhouse, "Laterolog Groningen Phantom."
9. Ibid.
10. Ibid.
11. George R. Pickett, "Application of Log Quality Control to Evaluation Problems," paper used with class notes for petrophysics course presented at the Colorado School of Mines, Golden, Colorado, 1976.
12. Ibid.
13. Ibid.
14. Ibid.
15. G. S. Neinast and C. C. Knox, "Normalization of Well Log Digitizing," *The Log Analyst* **15**, no. 2 (1974): 18–25.
16. Author's class notes from course on advanced well log interpretation presented by George R. Pickett at the Colorado School of Mines, Golden, Colorado, 1976.
17. Schlumberger, *Log Interpretation, Volume 1*, pp. 35–36.
18. Author's class notes from a course on well log interpretation presented by George R. Pickett at the Colorado School of Mines, Golden, Colorado, 1975.
19. Ibid.
20. Ibid.
21. H. G. Doll, "The SP Log: Theoretical Analysis and Principles of Interpretation," Pub. No. 2463 (American Institute of Mining and Metallurgical Engineers, New York, 1948).
22. Author's class notes, 1975.
23. Ibid.
24. Lee M. Etnyre, "Discussion" in "Letters to the Editor," *The Log Analyst* **26**, no. 1 (1985): 2–5.
25. Zaki Bassiouni and Diane M. Matthews, "Resistivity–Spontaneous Potential Crossplot for Enhanced Interpretation of Well Logs," *The Log Analyst* **25**, no. 5 (1984): 11–19.

CHAPTER 7

Radioactivity Logging

Radioactivity logging began in the 1940s. Many improvements have been made since then, and several different types of measurements are used in well logging. The atomic nucleus has discrete rather than continuous stable states. When the nucleus changes from one stable state to another, this change is accompanied by the emission of atomic particles or energy. This process of emission of particles or energy is termed *radioactivity*. Radioactivity may be either induced or natural.

Many elements have naturally occurring radioactive isotopes, but radioactive isotopes can also be created in the laboratory. For example, much of the naturally occurring background radiation on the earth's surface is due to gamma rays from a radioactive isotope of potassium that occurs in rocks: potassium-40.

TYPES OF RADIATION

Three basic types of radiated particles or energy are of interest:

1. *Alpha particle emission.* Alpha particles are helium nuclei with considerable mass and a charge of $+2$ (the helium nucleus being the helium atom stripped of its two orbital electrons). Alpha radiation is important to well logging.

2. *Beta radiation.* Beta rays are high-speed (near light velocity), energetic electrons, having a charge of -1 with negligible mass. Beta radiation is not important to well logging.
3. *Gamma radiation.* Gamma rays are photons of very high frequency electromagnetic radiation with energy given by the relation

$$E = h\nu \tag{7-1}$$

where h is Planck's constant and ν is the frequency. They have no mass or charge, and therefore have the greatest penetration of matter compared to other forms of radiation. They are of great significance in well logging.

HALF-LIFE

Each radioactive isotope of any element has a unique half-life. If there are N_o atoms present at time t_o, the number N remaining after time t is given by

$$N = N_o e^{-\lambda t} \tag{7-2}$$

where λ is the decay constant. The time t it takes for half of the atoms present at time t_o to disintegrate can be found from Eq. 7–2 by setting $N = 0.5N_o$. From Eq. 7–2, this time is the half-life T which is found to be $T = 0.693/\lambda$. Most naturally occurring radioactive isotopes have long half-lives, some being thousands of years. On the other hand, most radioactive isotopes created in the laboratory have relatively short half-lives, some being only a fraction of a second.

INDUCED NUCLEAR REACTIONS

Induced nuclear reactions of interest occur when a bombarding particle or photon (such as a gamma ray photon) is scattered or captured when it collides with a *target* nucleus. For example, neutrons are an important bombarding particle in well logging. They have a large mass, nearly the same as a proton, but no charge. They also exhibit a wide range of energies. They do not naturally exist in the free state since they are radioactive and have a very short half-life.

If the bombarding particle exists after collision, it has been scattered. If it no longer exists, it has been captured. Introduction of mass or energy to the target nucleus by the bombarding particle causes the nucleus to become unstable. It will seek a new equilibrium state accompanied by the emission of particles or energy. This process of bombardment and subsequent emission of energy or particles is termed a *nuclear reaction*. Nuclear reactions follow well-defined and observed laws and are therefore predictable. There are three types of nuclear reactions of importance in formation evaluation: elastic scattering, inelastic scattering, and transmutation of a target nucleus to a new nuclear species.

In elastic scattering the *total* kinetic energy of the system, consisting of the bombarding particle and the target nucleus, remains constant. In this type of reaction, the incident particle x usually loses some energy to the target nucleus X in a reaction characterized by Pickett in his logging courses by[1]

$$x + X \rightarrow X + x \qquad (7\text{--}3)$$

Elastic scattering is the principal process for slowing down neutrons used as a source in neutron logging tools.

In inelastic scattering the total kinetic energy of the system decreases with the emission of gamma radiation representing the lost system energy. The target nucleus X comes to an unstable equilibrium (X^*), and then goes to a stable equilibrium with the emission of gamma radiation. Pickett illustrated this as[2]

$$x + X \rightarrow X^* + x \rightarrow X + x + \gamma \qquad (7\text{--}4)$$

Transmutation of a target nucleus to a new nuclear species is illustrated by Pickett as[3]

$$x + X \rightarrow Y + \text{radiation} \qquad (7\text{--}5)$$

where Y is the new nuclear species. For example, this process is used in a neutron source used in logging. An alpha particle source (such as radium or polonium) is used to bombard a beryllium target. This produces an unstable isotope of carbon ($_6C^{13}$) which then emits a neutron to become a stable isotope of carbon: $_6C^{12}$. Each of these nuclear reactions has a relative probability of occurrence that is known as its *cross section*. The probability of any one of the reactions is a function of the energy of the bombarding particle, which should be specified with the cross section. For example, for a given energy, the probability of inelastic scattering may be very small for a given target nucleus, whereas the probability of elastic scattering is large. At a different energy level of the bombarding particle, the relative probabilities may be reversed.

TIME CONSTANT, ACCURACY, AND PRECISION

Associated with well log measurements of radioactivity is the concept of time constant. Electrical smoothing circuits (resistor–capacitor or RC circuits) are used in counting circuits to average out statistical fluctuations. The longer the time constant, the better the smoothing, but the slower the logging tool must be made to move uphole. For the average formation, a logging speed of 30 ft/min with a time constant of 2 sec is the recommended practice. However, in formations where there will be low count rates, a slower speed may improve both the resolution and *accuracy* of the measurement. As we will soon see, this means a density logging tool in low porosity formations (or higher density formations) should be run at a lower speed.

Time constant pertains to all radioactive measurement logging tools: density logs, neutron logs, gamma ray logs, and so forth. The logging service company routinely records the logging speed and time constant somewhere on the log heading. However, it is also a common practice to increase the speed of the logging tool uphole once past the formations of interest. This can be unfortunate if at a later date you decide to use some uphole zone for normalizing log measurements from several wells or an uphole zone proves to be a possible hydrocarbon producing zone.

NATURAL GAMMA RAY LOGGING

A gamma ray logging tool is used to log the natural gamma radiation in the borehole. All rocks, including the sedimentary rocks of interest in petroleum, emit gamma radiation. A scintillation detector is used in the downhole tool to sense the gamma radiation. An electrical signal is sent uphole that is proportional to the count rate of the detector.

In modern logging tools it is also possible to send this information uphole in digital format. The radioactivity is recorded in American Petroleum Institute (API) standard units. On some older tools, other units may be used. For example, a commonly used unit on older tools was micrograms of radium equivalent per ton of formation. Consult service company chart books and documents for unit conversion information for specific tools. Shales have high gamma ray activity, whereas most sedimentary rocks have relatively low gamma ray activity.

Gamma ray logs are used to identify sands and shales, for quantitative evaluation of shale, detection of radioactive minerals (usually potassium, thorium, or uranium), identification of bed boundaries, and correlation. A gamma ray log is also commonly recorded with a cased hole collar locator log (an electromagnetic type device) so that it can be used later to correlate with the open hole logs for correct positioning of perforating guns.

A principal use of the gamma ray tool is to segregate sands and shales on the well logs. However, some carbonates, particularly dolomites, are often radioactive. Shale-free sandstones can also exhibit high gamma ray activity. This is sometimes attributed to the precipitation of uranium-bearing salts in fractures. We must be careful in using the gamma ray for quantitative shale evaluation. A change in clay type may alter the tool response. Gamma ray logs can be run in either open holes or cased holes. They are usually run in combination with other tools. The depth of investigation for a gamma ray tool is about 1 ft. Gamma ray logs must be corrected for large holes or heavy mud systems.

The assumed environment for gamma ray logs is 8 in. holes with 10 lb mud. Use the appropriate logging company charts for corrections if you need to make quantitative measurements from a gamma ray log that was run in an environment other than this assumed standard. In applications such as separating sands and shales, it may be necessary to exclude the uranium contribution to the count rate, since it usually represents the presence of organic material rather than shale.

DENSITY (GAMMA-GAMMA) LOGGING

Density logging tools use both a gamma ray source and two gamma ray detectors. Figure 7–1 is an illustration of modern tool design. Note that the two detectors are mounted on a pad and will be pressed against the borewall by the action of a spring-loaded arm. Together with the detector pad, this arm also provides for a downhole caliper measurement. This caliper will give one of the best available hole size measurements. It is ordinarily better than the three-arm sonic log calipers. However, note that it is essentially a two-dimensional measurement and that in noncircular boreholes this caliper measurement cannot be used for accurate hole volume calculations. Some tool combinations, however, provide for two caliper measurements oriented at right angles by having a second caliper device on the neutron tool run in combination with the density log. However, for many tool combinations, this type of arrangement only works when the neutron tool is also excentered in the hole. Many neutron tools are centered in the hole.

Older tools used only one detector. The two-detector tool allows for compensation for borehole and mudcake effects. The source emits high energy gamma rays (1 Mev and higher). After interacting with the formation, some gamma rays are scattered by the Compton effect and detected by the density logging tool. The count rate at the detectors is proportional to the *electron* density of the formation, which is in turn proportional to the density of the formation.

In the process of Compton effect scattering, a bombarding gamma ray photon strikes a target atom and is scattered with a change in direction

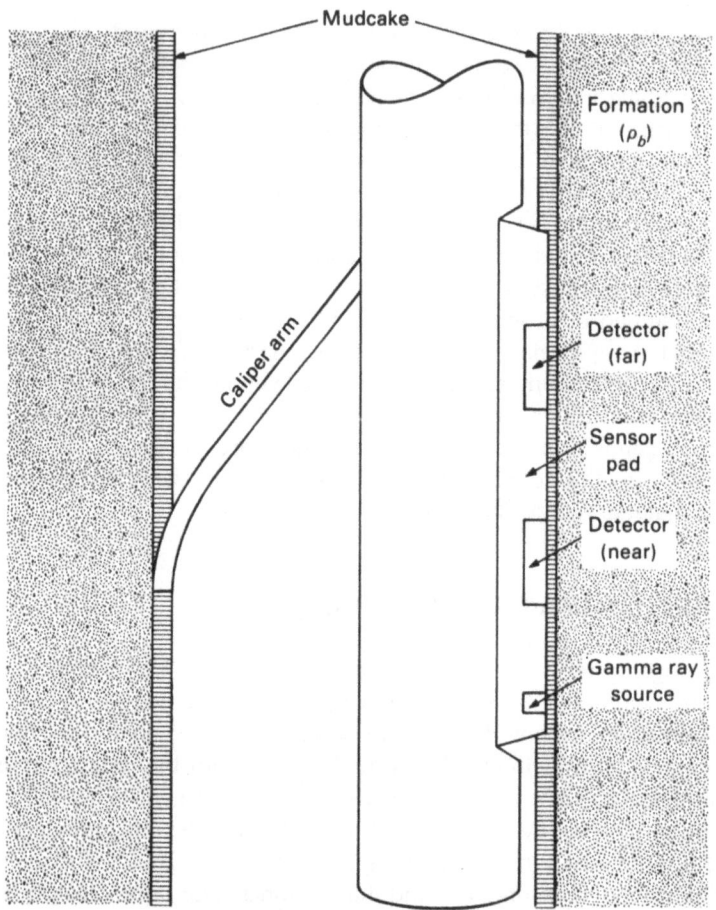

Figure 7–1. Density logging tool concept. *(After Schlumberger's* Log Interpretation, *vol. 1,* Principles, *©1972 by Schlumberger Ltd, New York, 1972, p. 43.)*

accompanied by the emission of a recoil electron (see Fig. 7–2). There are two other gamma ray interactions with matter that will occur but that do not significantly affect the tool response: the photoelectric effect and pair production. In pair production, the gamma ray enters a target nucleus and disappears with the emission of an electron–positron pair. This is a very high energy phenomenon (at least 1.02 Mev). In the photoelectric effect, the gamma ray also disappears, but with the emission of a high-speed electron of energy $h\nu = E_b$, where E_b is the binding energy of the electron in the target atom. This is a low-energy phenomenon (less than 150 Kev). By selection of source energy and tool design, the tool response is primarily from Compton scattering. However, in some modern density logging tools, the photoelectric

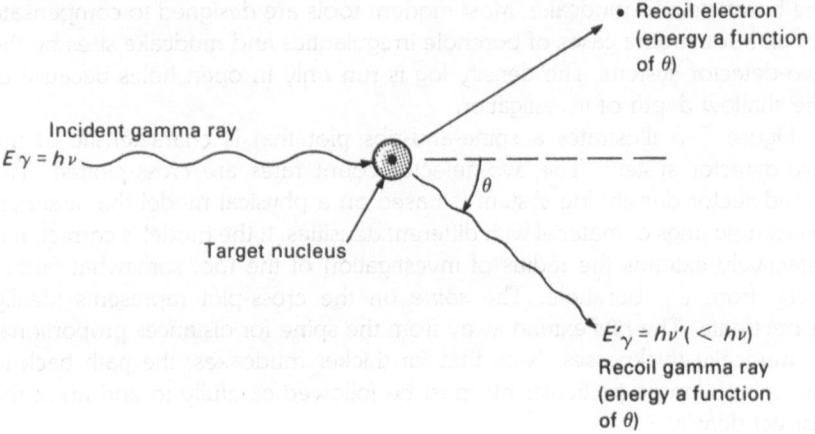

Figure 7–2. Compton effect.

effect is utilized to provide lithology information in addition to the formation density. This application of the photoelectric effect is discussed later in this chapter.

The number of electrons returning to the tool by Compton scattering is controlled by the electron density of the formation[4]

$$n_e = N_o(Z/A)\rho$$

where

n_e = electron density (number of electrons per cc)
N_o = Avogadro's number (6.02×10^{23} molecules [or atoms] per gram molecular [atomic] weight)
Z = atomic number (number of electrons per molecule [atom])
A = molecular (atomic) mass number (grams per mole)
ρ = bulk density (gm/cc) of the scattering material

For many sedimentary rocks of interest in formation evaluation the ratio $N_o(Z/A)\rho$ is constant. Thus the counts at the detector(s) is proportional to the formation bulk density ρ_b.

According to Pickett, detector spacing is critical in the design of the density tool. Small spacings yield more scattered gamma rays with increasing density of the scattering medium, and large spacings give fewer scattered gamma rays.[5] The latter is probably due to gamma ray absorption by photoelectric effects. The depth of investigation of the density log is a few inches at most due to the relatively low energy gamma radiation. Thus the tool is affected by the borehole and mudcake. Most modern tools are designed to compensate for all but extreme cases of borehole irregularities and mudcake sizes by the two-detector system. The density log is run only in open holes because of the shallow depth of investigation.

Figure 7–3 illustrates a spine-and-ribs plot that is characteristic of the two-detector system. The two-detector count rates are cross-plotted. The two-detector density log system is based on a physical model that assumes concentric rings of material with different densities. If the model is correct, this effectively extends the radius of investigation of the tool somewhat farther away from the borehole. The *spine* on the cross-plot represents ideally correct data. The *ribs* extend away from the spine for distances proportional to mudcake thicknesses. Note that for thicker mudcakes, the path back to the spine along a particular rib must be followed carefully to end up at the correct density.

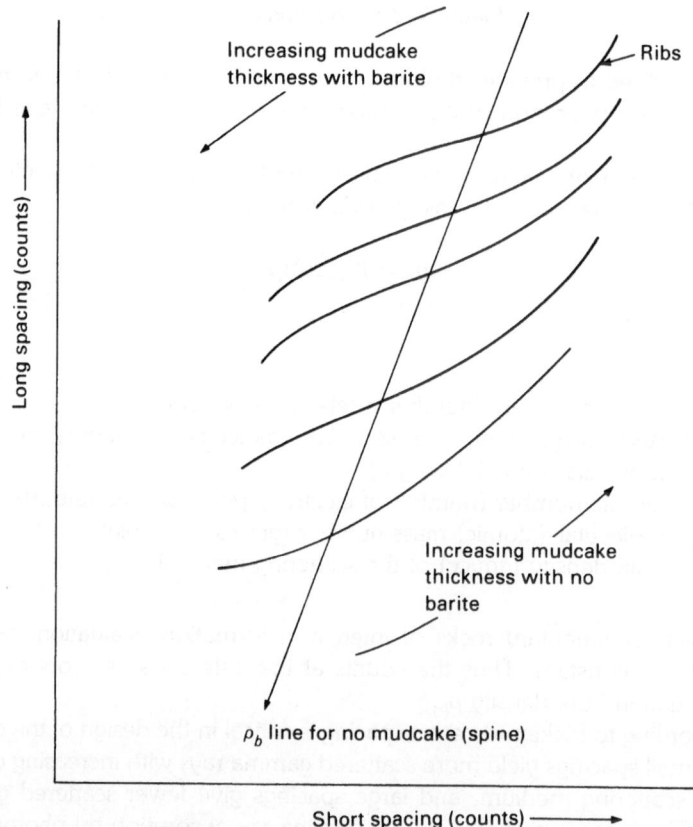

Figure 7–3. Spine and rib plots. *(After Schlumberger's* Log Interpretation, *vol. 1, Principles.* ©*1972 by Schlumberger Ltd, New York, 1972, p. 44.)*

The distance moved on a rib back to the spine corresponds to the delta rho ($\Delta\rho$) correction. This correction may be either positive or negative, depending on the density of the mud system. Lower density mud systems result in positive corrections, whereas sufficiently dense mud systems can lead to negative corrections. If this correction is more than .05 gm/cc, the log accuracy is questionable and the measurement at that point probably should not be used for any calculations. Some analysts accept corrections up to .10 gm/cc and higher, but I feel this may be an excessive tolerance. In some unusual cases, larger corrections may be acceptable, depending on judgment and understanding of the risks in using such data. Although the spine-and-ribs plot suggests that larger corrections would apply for large mudcakes, corrections of over 0.05 gm/cc frequently appear when borehole irregularities (rugosity) and borehole enlargements result in loss of pad contact. Note that an enlarged hole does not of itself mean bad density data. The density data are invalid only if the pad contact is lost, which is usually accompanied by an excessive $\Delta\rho$ value.

Figure 7–4 is an example of a typical density log presentation. This is the density log that accompanies the resistivity and sonic logs suite presented in Figures 6–19 and 6–20 of Chapter 6. Note that the scale used on density logs is standard with 2.00 gm/cc at the left of track 2 and 3.00 gm/cc at the right of track 3. Note that each one of the 20 scale divisions across tracks 2 and 3 represents 0.05 gm/cc.

A gamma ray log is presented in track 1 (gamma ray scale not labeled) along with a caliper (caliper scale not indicated but usually from 6 in. on the left of track 1 to 16 in. on the right of the track). The $\Delta\rho$ curve is shown with its zero just one division to the right of the left side of track 2. The scale will normally be listed as -0.05 at the left of track 2 and $+0.45$ at the right of track 2. Another common location for the correction, or $\Delta\rho$ curve zero, is five scale divisions from the right side of track 3 (identical to the 2.75 gm/cc density scale point).

There is an important pitfall in limiting the amount of correction on the $\Delta\rho$ curve to one scale division, assuming this allows a maximum correction of 0.05. Although the standard scale for modern density logs is from 2.00 gm/cc to 3.00 gm/cc, the $\Delta\rho$ *scales are not always presented on a scale with 0.05 gm/cc correction per scale division.* Therefore, when comparing one density log correction scale to another or making a quick visual estimate of density log quality, be sure to verify the scaling of the $\Delta\rho$ curves.

On older density logs, scales were listed in some arbitrary count rate measurement. It was hoped that the logging service company provided a density calibration scale in addition to the count rate scale. Perhaps they also provided some necessary correction information for this type of density log in their literature. Sometimes it will be necessary to establish your own scale by recognizing that the count rate is a logarithmic function of porosity. You can

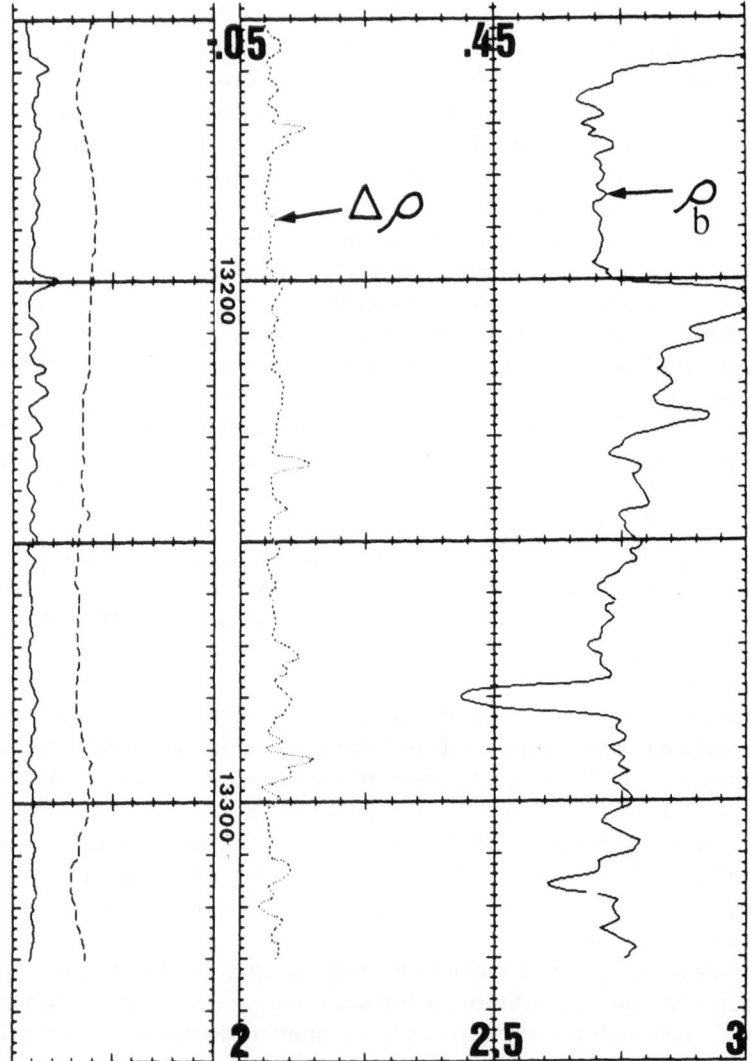

Figure 7–4.   Density log example.

look for a maximum reading on the log (highest count rate) and assign some low expected density reading to this point, and try to assign some expected high density reading at a low count rate point. This procedure is somewhat similar to what is done with the older neutron tools (also scaled in count rate but with low porosity associated with high count rates). In any event, use of these older density logs will likely be more qualitative than quantitative. Not too many were used before the compensated tools appeared, but you may run into one if you are looking over logging suites from wells drilled before the 1960s.

A porosity scale is frequently presented across tracks 2 and 3 with a density log when it is not run in conjunction with a neutron log. A neutron log *was* run in combination with the density log of Figure 7–4, so the porosity scale is shown on a combination presentation of both the neutron and density logs in porosity units (to be illustrated later). At this point, you should recognize that the density log is the recorded data and that any density porosity recording will be *derived* from the bulk density recording.

For fluid-filled porosity, the bulk density of a rock formation may be expressed as

$$\rho_b = (1 - \phi)\rho_{ma} + \phi\rho_f \tag{7–6}$$

where

ϕ = rock porosity
ρ_f = density of the fluid in the pores
ρ_{ma} = grain density of the solid rock matrix

If the rock matrix density and fluid densities are both known, we can re-express Eq. 7–6 to solve for the formation porosity.

$$\phi = (\rho_{ma} - \rho_b)/(\rho_{ma} - \rho_f) \tag{7–7}$$

Now it is easily seen why I earlier referred to the *density* porosity as *derived* information from the recorded bulk density. In practice, the density porosity is calculated based on *assumed* values for pore fluid and rock matrix densities.

The common practice is to assume the pore fluid is water and use $\rho_f =$ 1.00 gm/cc for the fluid density. In salt mud systems, the practice is to use 1.10 gm/cc for fluid density of the rock's pore system, following the assumption that near the well bore, the mud filtrate will have flushed or at least significantly invaded the pore system. In many areas of interest it turns out that where salt mud systems are used, the formation waters are also very saline. Where there is a density contrast between the formation water and the mud filtrate accompanied by incomplete flushing, there may be a combination of the two in the pore system. The fluid density will be somewhere between 1.00 gm/cc and 1.10 gm/cc. If no flushing occurs, as might happen in an impermeable formation with some porosity, the fluid density will be that of the formation water. In any event, the error introduced into density porosity calculations will not be too severe. For example, a rock with a matrix density of 2.65 gm/cc and fluid density of 1.00 gm/cc with 10% porosity will calculate to have 10.6% porosity if you use the incorrect fluid density of 1.10.

On the other hand, low density hydrocarbon in the pore spaces can lead to significant errors in porosity calculations. If the fluid in the pores is 70% gas with a density of only 0.10 gm/cc, the same rock above will be calculated to have a porosity of 13.8%. This magnitude of error can be quite severe.

Typical matrix densities for some sedimentary rocks of interest are: limestone, 2.71 gm/cc, pure quartz sandstone, 2.65 gm/cc, and dolomite, 2.87 gm/cc. However, limestone is the only one that does not really vary much in practice. Few sandstones are pure quartz sandstones, although it is a common practice to assume the quartz matrix value of 2.65 gm/cc or to use 2.68 gm/cc for sandstones with the assumption of some small percentage of more dense minerals or calcite cement in the sandstone. Dolomites have even more variable matrix densities, usually anywhere from 2.83 gm/cc to 2.87 gm/cc. Many carbonate rocks are also composed of varying ratios of calcite and dolomite.

The extreme case for gas effect on the density log may be examined by setting ρ_f equal to zero in Eq. 7–6, then solving for porosity.

$$\phi = (\rho_{ma} - \rho_b)/\rho_{ma} \qquad\qquad (7\text{–}8)$$

If we designate the porosity calculated from the density log using Eq. 7–7 by the symbol ϕ_D, and multiply the expression on the right side of Eq. 7–7 by a factor necessary to make it equivalent to the right side of Eq. 7–8, we obtain

$$\phi = \phi_D \times (\rho_{ma} - 1)/(\rho_{ma}) \qquad\qquad (7\text{–}9)$$

This expression assumes a fluid density in the pores of 1.00 gm/cc, but there is little difference in the result using 1.10 gm/cc for fluid density.

For a rock with a grain density of 2.7 gm/cc with a zero density gas in the pore system, Eq. 7–9 tells us that we must multiply the apparent porosity calculated from the density log by 0.63 to obtain an accurate porosity estimate. In practice, there should be something less than 100% gas in the pores near the borewall after flushing, so we might choose to multiply by 0.7 or 0.75 in gas-bearing reservoirs. Even if there is no flushing, there should be some irreducible water trapped by capillary forces in the pore spaces. This could be 20% or 30% for rocks with large pores and good permeability to 50% or even higher in rocks with small pore sizes and poor permeability. In both oil- and gas-bearing reservoirs, we may use a larger factor, say, 0.85 or 0.9, depending on the assumed relative oil and gas saturations. Of course, we could always choose the practical alternative of calculating a range of porosity.

Since density porosity is derived from bulk density measurements, it is essential to establish the fluid and matrix densities used for any log presentations of a density–porosity curve. Also, the curve may be derived assuming a limestone matrix, yet the entire interval logged may include sands as well as other lithologies.

Another problem encountered with density logs is the vertical variation of grain density in the rocks of interest. Many sandstones will exhibit a variable grain density, for example, the density may vary from a low of 2.63 gm/cc to a high of 2.72 gm/cc because of a variable mineral content in the sand. Using a single-matrix density value for the porosity calculations is going to result in some significantly erroneous porosity numbers. If we know from previous experience or core measurements that we have a rock type with significant matrix density variations, this problem can be handled only by calculating a *range* of possible values using the range of matrix densities. We may be able to use a histogram of the grain density range and find that, for example, most matrix densities fall between 2.74 gm/cc and 2.68 gm/cc, though there are a few isolated values above and below this range. In this case we may be justified in making our calculations with the more restricted range where most of the grain densities lie. The ultimate situation would be where the distribution of grain densities follows some well-behaved statistical distribution that would allow us to establish some quantitative confidence in our calculations.

The grain densities may sometimes follow a pattern of increasing grain density with decreasing porosity. This pattern will usually be discovered from core data. For example, this might occur in a rock where the loss of porosity is accompanied by some dense, pore-filling mineral. To calculate porosity from density logs in this type of rock, we may have to use a variable expression for grain density as a function of porosity. This usually leads to a quadratic equation for porosity. An alternative is to use the lower grain density associated with the more porous rocks and permit overly pessimistic calculated porosity results for the less porous rocks where the lower porosity rocks would not be of commercial interest anyway.

Patchett and Coalson published an excellent discussion of porosity logging problems in sandstones that provides some excellent solutions and recommendations.[6] Although Patchett and Coalson include applications of advanced statistical methods to formation evaluation, I highly recommend it for the many practical insights they provide regarding porosity logging and pitfalls. For those who are serious about formation evaluation using well logs, further study in the use of statistical methods (such as those exemplified by Patchett and Coalson) and their power in both evaluating log data quality and extracting information from log data is also highly recommended.

An important conclusion in Patchett and Coalson's paper is that the density log is probably the best porosity tool in sandstone and shaly sandstone reservoirs where there are good hole conditions and the actual grain densities of the rocks can be reliably established rather than using assumed values.[7] Establishing the correct grain density is particularly important when the sand contains accessory minerals in addition to quartz. Very small amounts of some minerals (e.g., pyrite) can seriously alter the grain density. The unfortunate aspect of these grain density alterations is that they may vary both vertically and laterally) in the same well. In this case, it may be impossible to derive accurate porosity information from any well log. Where the borehole is not in good condition due to hole size changes that affect the density log response, the sonic log can often be used for porosity estimates. However, it is often adversely affected by the same mineral constituents that make porosity estimates from the density log difficult. Finally, Patchett and Coalson conclude that, although density log-derived porosity measurements can be improved in carbonate sedimentary sequences by combining the density log readings in some manner with another porosity tool measurement, such combination of porosity tool readings usually leads to a worse estimate of porosity in sandstones comprised of some accessory minerals.

Despite the possible problems in interpretation of density logs, they have many applications in identification of lithologies and pore fluids as well as in calculating porosity. With an independent measurement of porosity, gas-bearing zones can be identified by the fluid density effect, or different lithologies can be identified by the grain density effect. The density log-derived porosity will be too high for gas zones, and for different lithologies it may be either too high or too low, depending on the difference between the assumed and actual grain density.

The density log is best in higher porosity (lower density) formations since the count rate is highest for the lower densities. With the higher count rate, the statistical effect will be less. Density logs measure grain density with an accuracy of $\pm$ .01 gm/cc or .02 gm/cc, depending on the statistical effects. Porosities calculated from the density measurement are total porosities.

PHOTOELECTRIC EFFECT LOGGING

Most major companies now have density logging tools that provide a measurement of the photoelectric effect (PE). Schlumberger's tool provides a measurement of PE from one of the two density log detectors, and the Welex tool provides two PE measurements, one from each detector. Presumably, the dual detector PE measurement provides a check on PE measurement quality, with separation between the two curves indicating the measurement may be affected too much by the borehole conditions.[8] Other companies employ similar, but not identical designs. The theoretical concepts of PE measurements are discussed by Bertozzi et al.[9]

The usual density log tool is designed to minimize the effects of both PE and pair production. In the density log with PE measurement capability, count rates from the lower energy gamma ray spectrum are used to provide a PE measurement as well as enhance the density measurement.[10] Welex's tool is the *Spectral Density Log* and Schlumberger's tool is the *Lithodensity* tool. Dresser Atlas has introduced the *Z-Density* tool. GO Wireline designates their new PE device the *Compensated Spectral Lithodensity* tool.

The measurement is presented in units of *barns per electron* (one barn = 10^{-24} cm^2). The PE measurement has no *physical* significance since the photoelectric absorption cross section is determined by a whole atom. The arbitrary unit of barns/electron allows comparison with other effects (e.g., Compton effect).

In the Schlumberger single-detector concept, both a high energy window counter and a low energy window counter are used. The high energy count rate is sensitive only to formation density, but the low energy window count rate is sensitive to both formation density and atomic number Z. The ratio of the two energy window count rates essentially has the density dependency removed. Since the photoelectric absorption cross section for gamma rays varies exponentially with the atomic number, this ratio provides a PE measurement that can be related directly to the atomic number Z, which uniquely identifies an element (or combination of elements).[11]

The PE measurements for some materials of interest to formation evaluation are: calcite = 5.084, dolomite = 3.142, quartz = 1.806, anhydrite = 5.05, fresh water = 0.358, salt water = 0.807, and oil = 0.119. The PE value for rocks with liquid-filled porosity varies little from the rock matrix value. This makes lithology identification possible. Gardner and Dumanoir describe methods for using PE measurements with the neutron–density log combination to identify the presence of gas as well as lithology, using both two- and three-mineral models.[12] The so-called dolomite–shale ambiguity still cannot be resolved with this approach, but at least one source lists a method of doing so that incorporates sonic log measurements.[13] A problem with the three-mineral method is that the sensitivity of the method depends on the combination of minerals selected. For example, a dolomite–calcite–quartz triad seems to work out well in practice, but a dolomite–quartz–anhydrite triad (such as found in the Minnelusa formation in the Powder River Basin) can yield misleading results. The problem arises because the appropriate equations are ill-behaved with this combination of three minerals. That is, small changes in the input data (such as tool statistical error) produce large changes in the results. This problem can be examined visually by noting the location of the three selected minerals on the three-mineral PE cross-plot (see Schlumberger's chart CP-21[14]).

Figure 7–5 is an example of Schlumberger's PE log. Note that the PE curve is denoted by a long dash. It is displayed across tracks 2 and 3 with a scale from 0 PE units on the left to 10 PE units on the right. A bulk density curve, scaled from 2.00 gm/cc on the left to 3.00 gm/cc on the right, across

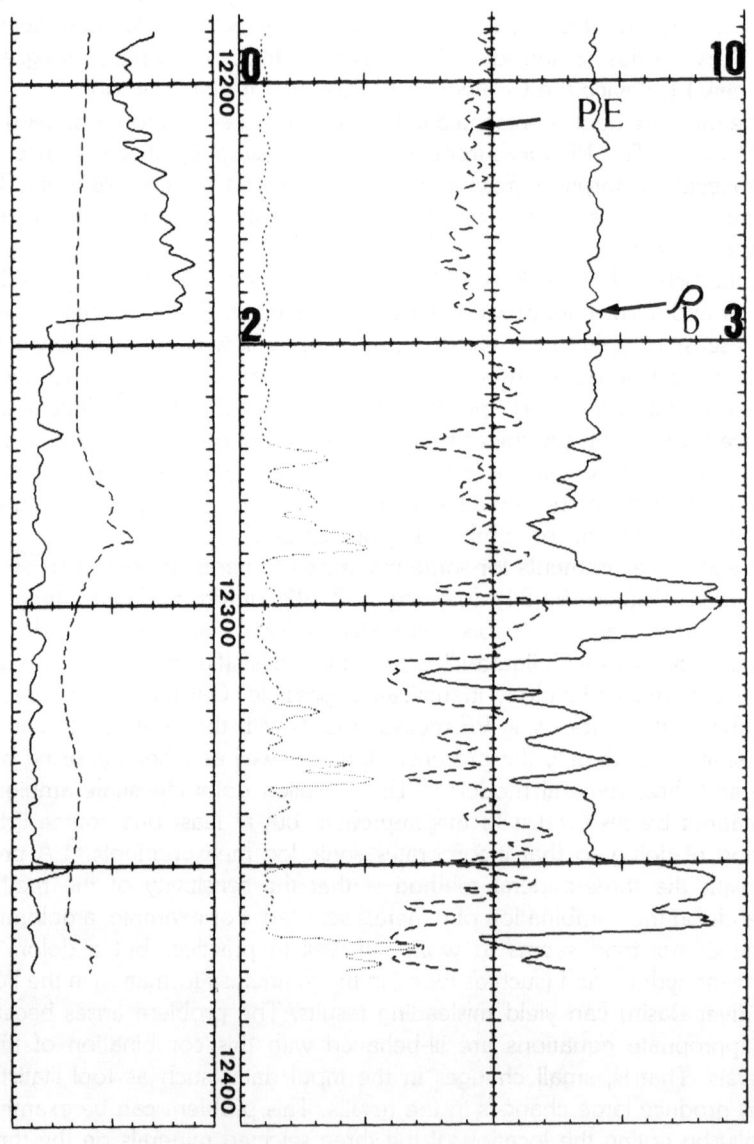

Figure 7–5. Photoelectric effect log example.

Table 7–1. Useful Rock Properties for Photoelectric Effect Log Interpretation

Mineral	Apparent Log Density (gm/cc)	PE (barns/electron)	ΔT (μsec/ft)	Gamma Ray (API units)	Neutron Porosity
Calcite	2.710	5.084	45–50	0	0
Dolomite	2.876	3.142	42–48	0	4
Quartz	2.648	1.806	50–55	0	−4
Halite	2.032	4.65	67	0	0
Anhydrite	2.977	5.055	50	0	0
Coal	0.7–1.8	.161–.180	110–170	0	50+
Barite	4.5	266.8	—	—	—

Sources: Acoustic: Author's class notes from course on well log interpretation presented by George R. Pickett at the Colorado School of Mines, Golden, Colorado, 1975; photoelectric effect: W. Bertozzi, D.V. Ellis, and J.S. Wahl, "The Physical Foundations of Formation Lithology Logging with Gamma Rays," *Geophysics* **46**, no. 10 (1981): 1451; density, neutron, and gamma ray: Schlumberger, *Log Interpretation, Volume 1, Principles* (Schlumberger Ltd, New York, 1972), p. 108.

tracks 2 and 3 is also shown. Note the PE curve reading of almost 6 units at 12,300 ft depth. This could be due to either a calcite or anhydrite, although the reading is a little high. The bulk density measurement at the same depth is 2.95 gm/cc. This is much too high for a calcite, which, with a grain density of 2.71 gm/cc, would have a bulk density reading of no more than this even with zero porosity. Therefore, if we consider the possibility of some data scatter due to statistics, the most likely mineral is anhydrite, which has a grain density close to 3.00 gm/cc. A neutron tool reading of zero porosity units or slightly less or an acoustic wave travel time of 50 μsec/ft at the same depth would also confirm this conclusion. Just below, at 12,314 feet the PE reading is 3 units. This is diagnostic of a dolomite.

In some sequences, this might represent a mix of quartz (about 1.8 PE units) and calcite (about 5 PE units). The bulk density reading of 2.36 gm/cc at the same depth suggests that, if this is dolomite, it must have a very high porosity (grain density for dolomite is 2.87 gm/cc). By using the PE measurement in conjunction with other tool responses, it is often possible to make reasonable lithology identification. Table 7–1 illustrates the PE and other parameters for some sedimentary rocks of interest in logging.

NEUTRON POROSITY LOGGING

Another radioactivity logging method is to bombard the formation with neutrons of a selected energy range and measure the responsive properties of the target nuclei in the formation. These response properties can often be related to properties of interest in formation evaluation. In this section I discuss neutron logging with thermal and epithermal neutrons to obtain information about the hydrogen concentration in the formation, which can, in turn, be related to porosity. Figure 7–6 is a schematic of the older type of thermal neutron logging tool. It had a neutron source for thermal neutrons and a gamma ray detector (scintillation type). More modern tools use two detectors and employ other enhancements designed to improve tool response. As the neutrons from the source lose energy to the bombarding target nuclei by elastic scattering, their energy approaches the thermal range where it is comparable to the energy of vibration of the atoms in the media. At the thermal energy level, the probability of capture of a neutron by a media nucleus becomes relatively large. When a neutron is captured, a *capture*

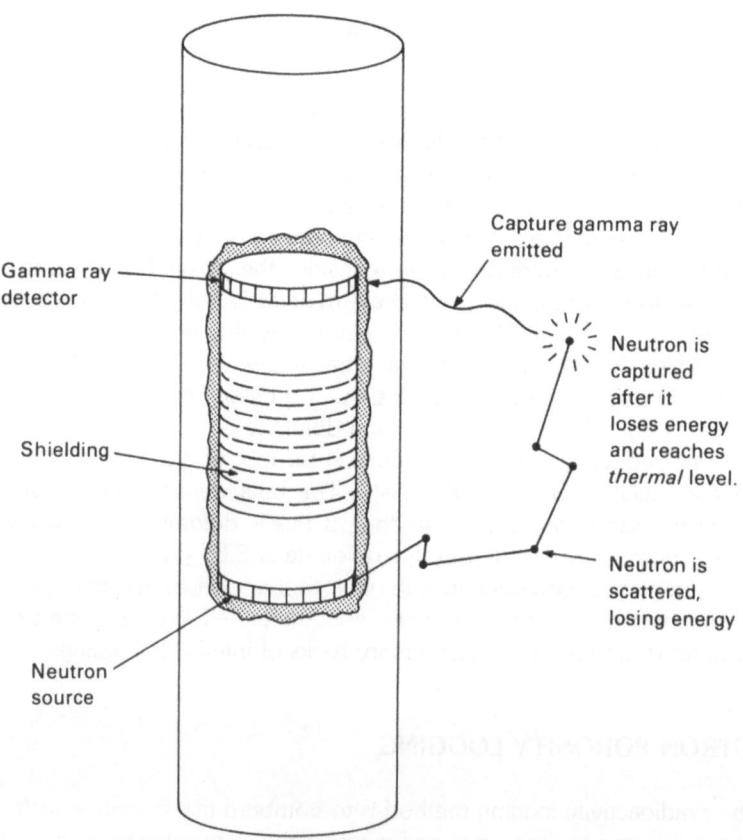

Capture gamma ray
emitted

Gamma ray
detector

Neutron is
captured
after it
loses energy
and reaches
thermal level.

Shielding

Neutron is
scattered,
losing energy

Neutron
source

Figure 7–6.   Thermal neutron logging tool concept.

gamma ray is emitted by the nucleus absorbing the neutron. Some fraction of these capture gamma rays reach the gamma ray detector to be counted.

Some versions of neutron tools use direct detection of the neutrons (e.g., using a boron trifluoride proportional counter). The closer to the source the neutron is captured, the less is the probability of the capture gamma ray reaching the detector. Hydrogen is the most effective element usually found in the formation that causes neutrons to lose energy by elastic scattering. There are some trace elements that have extremely high capture cross sections for thermal neutron absorption that sometimes occur in sufficient concentrations to cause problems in neutron log interpretation. Boron is one such element that has been credited with causing anomalously low neutron tool count rates in some areas.

Since the thermal neutron tool is especially sensitive to hydrogen concentration, we might expect that it will be especially responsive to the hydrogen in water and oil contained in rock pores in shale-free rocks. Its response will be proportional to the rock porosity in these shale-free rocks. In shaly rocks, there will be a high proportion of hydrogen in the chemically bound water of the clay minerals as well as some hydrogen in the *dry* part of the clay mineral. The increase in hydrogen concentration in the clay minerals will cause an anomalously low count rate at the detector. Therefore, the apparent porosity as seen by the neutron tool in a shaly rock will be higher than the true porosity. If gas is present in the rock's pore system, the hydrogen concentration will be much less in the low density gas than if only water or oil were present. There will be a significantly higher count rate at the detector, resulting in an abnormally low porosity measurement. That is, the neutron porosity will be less than the true porosity. This fact is used in conjunction with another porosity measurement that is not sensitive to hydrogen (such as a density log, which would actually read a little high in this situation) to sense the presence of gas in the formation. Borehole size can have an adverse effect on neutron tool response due to the hydrogen present in the borehole fluids. If the hole is too large, the neutron tool response may come primarily from the borehole. Some modern neutron tools are designed to minimize the borehole effects. Tool design and operation also play a significant part in neutron tool response. Source strength is constantly changing because of the finite half-life of the source. The shorter the half-life of the source, the more frequently the source must be calibrated.

Logging speed affects this tool as it affects all radioactive devices. Optimum logging speed is 30 ft/min. Slower speeds will give a better response, particularly in *higher*-porosity, thin-bed formations. The presence of casing and cement will also reduce the count rate, and slower speeds will be necessary when logging behind casing. The performance of neutron logging tools is very sensitive to the neutron tool source-to-detector spacing.[15] For the single-detector type thermal neutron tool, the response follows an approximate *S* curve as in Figure 7–7. The usable portion of the curve is the approximately linear portion between some low porosity around 2% to an upper limit of 20% to 30%. This portion of the curve may be approximated by the linear relation

$$ND = C + D \log \phi \qquad\qquad (7\text{–}10)$$

where *C* and *D* are parameters that are a function of the tool design and operating environment. Older versions of the single-detector neutron tool had the count rate scaled in some arbitrary unit, usually expressed simply as the *counts per second*. Modern or later versions of the tools have output scaled in API units. The constants *C* and *D* of Eq. 7–10 are obtained from service company charts, which also provide some environmental corrections. When you are uncertain about the tool type (which is needed to use the

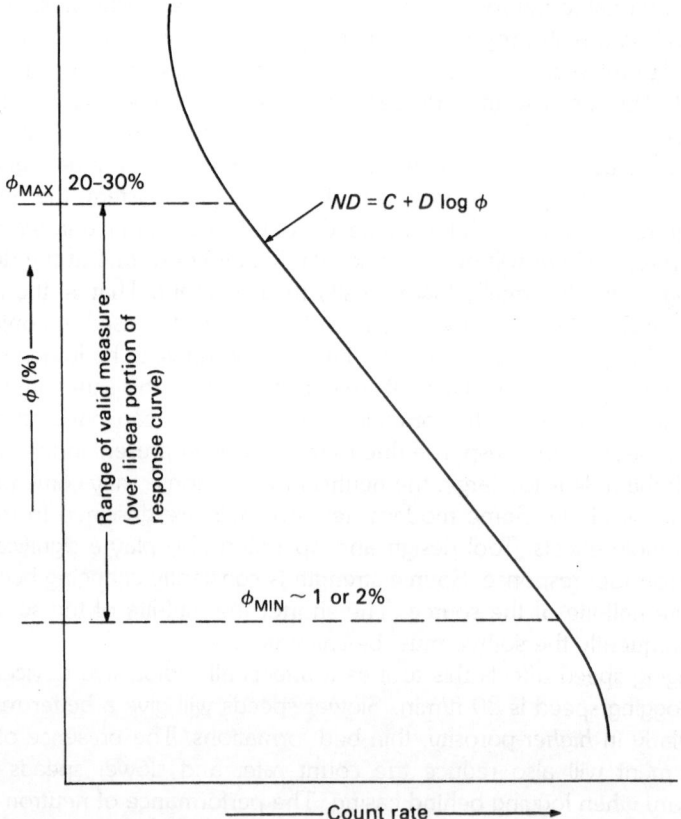

Figure 7–7.   Thermal neutron log response.

appropriate charts), the log can be scaled by plotting neutron count rate (in whatever units are stated on the log or even use the log scale divisions) as an *X* axis and the logarithm of porosity as the *Y* axis. This is best done on a semilogarithmic plotting grid as in Figure 7–8. You can assign a value of 1% or 2% porosity to the highest count rate observed on the neutron log. This should be somewhere in a tight, very resistive zone, or maybe an

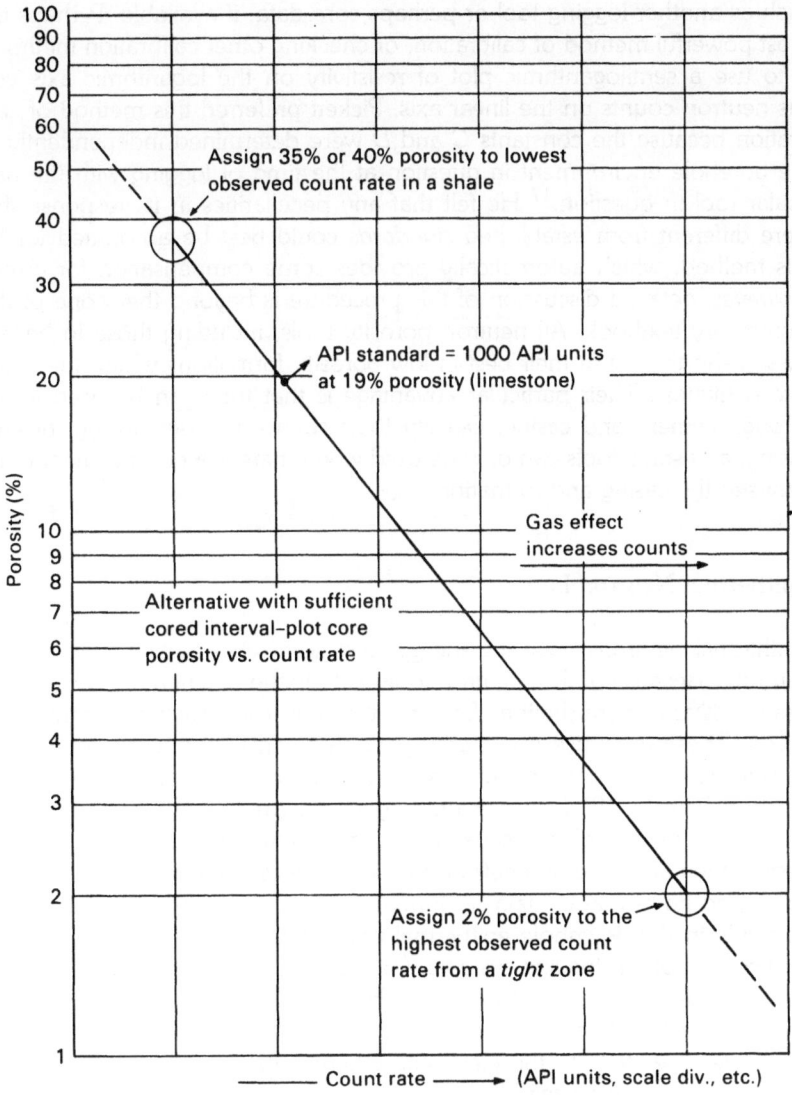

Figure 7–8. Thermal neutron tool calibration.

anhydrite. Then find a shale with a low count rate and assign 35% or 40% porosity to this point. The calibration is completed by extending a straight line between the two. Another calibration point that can be used to confirm your calibration is the API-defined standard of 1,000 API units at 19% porosity for a water-filled limestone.[16] However, there will be some lithology effect for formations other than limestone. Another alternative to older neutron tool calibration is to calibrate the neutron against another porosity source such as another logging tool or perhaps core data, if available. Perhaps the most powerful method of calibration, or checking other calibration methods, is to use a semilogarithmic plot of resistivity on the logarithmic axis versus neutron counts on the linear axis. Pickett preferred this method of calibration because the constants C and D were determined independently in the borehole environment in question at the time of logging with the particular tool in question.[17] He felt that any peculiarities in its response that were different from established *standards* could best be accounted for by this method, which automatically provides some compensation for errors. However, detailed discussion of this procedure is beyond the scope of this elementary textbook. All neutron porosity tools, including those to be discussed shortly, are at their best in low porosity formations where the count rate is highest. Their particular advantage is that they can be used to log through cement and casing. Density tools cannot be used to log through casing, and sonic tools can only be used when there is a good cement bond between the casing and formation.

Epithermal Neutron Log

Epithermal neutrons have an energy level somewhat higher than thermal neutrons. Because of this, neutron tools with epithermal neutron sources are less sensitive to perturbations from neutron absorbers such as chlorine and boron. These tools use excentered pads that contain both the source and detector to reduce borehole effects. Although the epithermal tools (Sidewall Neutron Porosity [SNP] is Schlumberger's designation for its tool) cannot be used in cased holes because of the pad-mounted source and detector design, they are the only neutron porosity devices that can be run in air- or gas-drilled holes that contain no formation fluid. Automatic corrections can be made in the downhole epithermal neutron tool (or by computer at the surface in modern logging systems) for mud weight, borehole fluid salinity, borehole diameter, and temperature. However, we should recall Pickett's observation: "System output is no better than the selection of appropriate parameters needed for the electronic schemes used or than the response fidelity of the system itself."[18]

The SNP-type tool is designed to cut into the mudcake as much as possible to reduce mudcake effect on the tool. Any remaining mudcake will affect the tool. According to Schlumberger's correction chart for its SNP tool, a quarter inch mudcake will result in correcting an apparent SNP neutron tool porosity from 13% down to 11%.[19] Schlumberger's instruction is to use the caliper measurement minus the bit size for the mudcake thickness. The only problem with this is that some boreholes are enlarged. This may make routine application of the correction difficult. The lithology effect for Schlumberger's epithermal neutron (SNP) tool is illustrated in Figure 7–9. Note that the lithology effects for the SNP tool (designated by the curves closest to the line) are significantly less than that for the compensated neutron logging tool, which will be discussed shortly.

Epithermal neutron logs and other neutron tools whose output is in porosity units are presented on tracks 2 and 3 of a well log with varying scales. Some common scales are 30% on the left of track 2 and −10% on the right of track 3. Another common scale ranges from 45% on the left to −15% on the right. Of course, minus porosity values have no physical meaning, but apparent negative porosity values occur frequently because of lithology effects.

The common lithology setting for neutron logs is limestone (corresponding to API calibration standards), but neutron logs may be recorded on any matrix scale. Be sure to check log headings for the matrix setting. The matrix setting is also usually included on the scale heading at the top and bottom of the recorded log. From Figure 7–9, an SNP porosity reading of 15% on a limestone matrix setting is corrected to 18% for a sandstone or 12% for a dolomite. To convert from limestone settings to another matrix, you simply enter the bottom of the graph with the apparent limestone porosity, move vertically upward to the appropriate lithology curve, then move horizontally to the vertical porosity scale and read the corrected porosity for the indicated matrix material. You use this same scale to convert from a matrix setting of other than limestone. To convert from a *sandstone* matrix setting neutron reading to a limestone reading, enter the left side on the vertical *true* porosity scale and move horizontally to the right to the *sandstone* curve. Then move vertically down to the matrix of interest (limestone in this case). Finally, we move horizontally back to the true porosity axis and read the porosity. If you have one of the unusual neutron logs scaled in dolomite porosity units, enter the vertical porosity scale with the apparent dolomite porosity and move horizontally to the right to the dolomite curve. Then move vertically to the curve for the matrix of interest. Finally, move horizontally back to the true porosity, vertical axis. In using the neutron porosity lithology effect graph, keep in mind that there is no such thing as a standard dolomite and that sandstones can have a varying mineral content (limestone cement and some clay minerals). These properties may change the effect on the neutron tool.

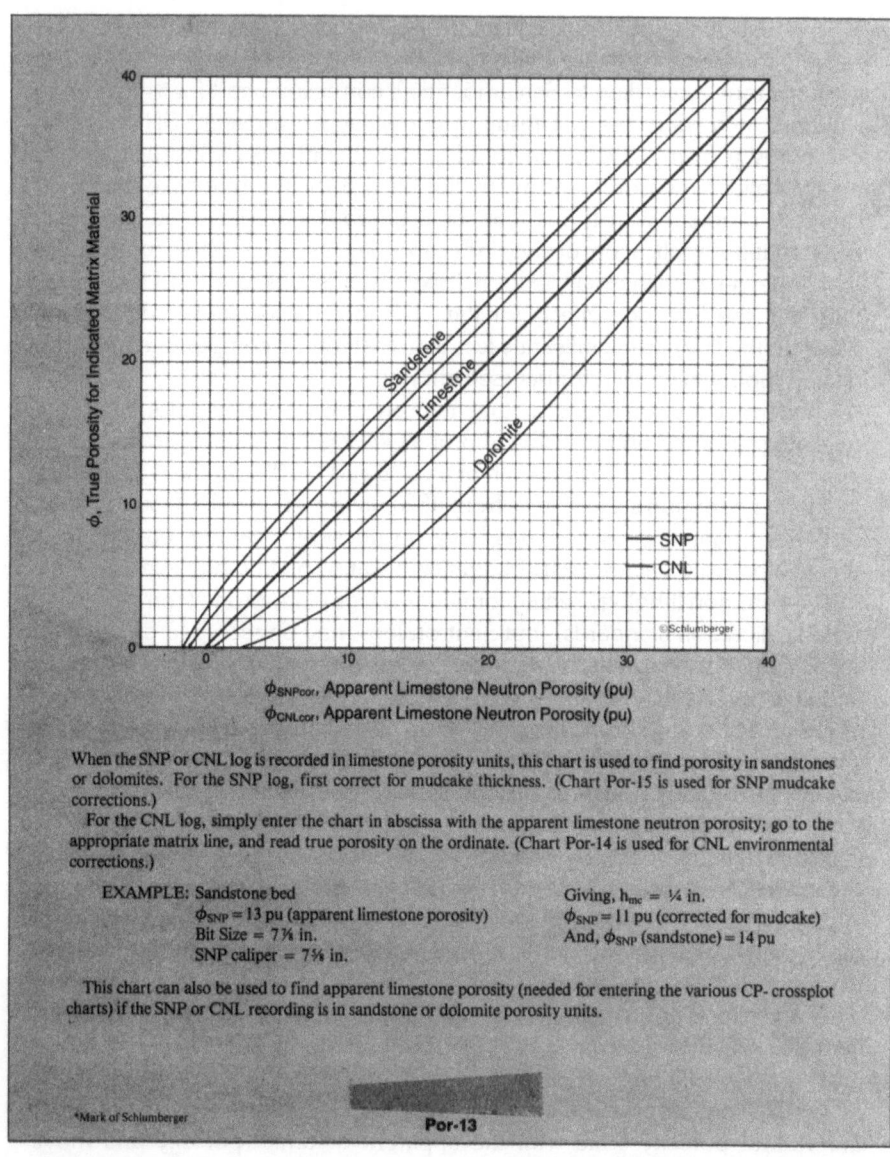

When the SNP or CNL log is recorded in limestone porosity units, this chart is used to find porosity in sandstones or dolomites. For the SNP log, first correct for mudcake thickness. (Chart Por-15 is used for SNP mudcake corrections.)

For the CNL log, simply enter the chart in abscissa with the apparent limestone neutron porosity; go to the appropriate matrix line, and read true porosity on the ordinate. (Chart Por-14 is used for CNL environmental corrections.)

EXAMPLE: Sandstone bed
        ϕ_{SNP} = 13 pu (apparent limestone porosity)
        Bit Size = 7⅞ in.
        SNP caliper = 7⅝ in.

Giving, h_{mc} = ¼ in.
ϕ_{SNP} = 11 pu (corrected for mudcake)
And, ϕ_{SNP} (sandstone) = 14 pu

This chart can also be used to find apparent limestone porosity (needed for entering the various CP- crossplot charts) if the SNP or CNL recording is in sandstone or dolomite porosity units.

*Mark of Schlumberger

Por-13

Figure 7–9. Neutron porosity equivalence for common lithologies. *(Courtesy of Schlumberger Ltd.)*

Compensated Neutron Tool

Most service companies have a compensated neutron logging tool in addition to single-detector thermal and epithermal devices in their repertoire of logging tools. The compensated neutron logging tool (CNL is Schlumberger's designation) is a two-detector device that uses a source of thermal neutrons. The two detectors extend the radius of investigation and reduce borehole effects. The compensated neutron tool can be run only in liquid-filled holes. It can be used behind casing. Porosity has been found to be proportional to the ratio of the two detector count rates in this tool. It has a greater radial investigation than the epithermal neutron device.[20] For all neutron tools, the radial investigation is probably somewhat less than 1 ft, decreasing with increasing porosity. The vertical investigation resolution of the tool is roughly the same as the source-to-detector spacing (15 in. to 20 in. for most neutron tools). The lithology effect is much larger for the compensated neutron tool (see Fig. 7–9). Allen et al. have attributed this large dolomite departure for the compensated neutron log to small but significant concentrations of trace elements (particularly gadolinium and boron) with large thermal neutron absorption cross sections.[21] From data included in their discussion, I suspect that the dolomite lithology curve in Figure 7–9 is an *average* curve about which there are significant departures for individual porosity measurements. Whether these departures from the average are due to variations in trace elements or some other effect are not clear. The important thing to keep in mind is that, regardless of the cause, the departures observed on a given well log can be significantly different than indicated by the average lithology curve. Although borehole and environmental effects are minimized in the compensated neutron tool, Schlumberger provides for automatic caliper correction in its device.[22] Charts are also provided for other effects that can add or subtract 10% of the measured porosity. For example, a 30% porosity may be adjusted up or down by as much as 3%. These corrections are tedious and not used routinely since they do not alter the tool response too much. Also, they tend to cancel for some applications. These corrections include: borehole size, mudcake thickness, borehole and formation salinities, mud weight, standoff distance, borehole temperature and pressure, and casing and cement corrections in cased holes.

Excavation Effect

Excavation effect on a neutron log occurs opposite gas-bearing zones. The neutron porosity is lower than one would predict on the basis of the lower density of hydrogen in the rock pores alone. Apparently, the formation has a smaller neutron slowing-down effect when the rock matrix is *excavated* and replaced by the gas-filled portion of the porosity than if the gas-filled portion of the porosity is assumed to be replaced by the rock matrix.[23] The maximum error occurs when the saturation of the flushed zone near the well bore is 50%. The maximum error is about -1% at 10% porosity, -3% at 20% porosity, and -6% at 30% porosity. For saturations significantly above and below 50%, the excavation effect decreases rapidly for the higher porosities and declines more slowly for lower porosities (say, 10% or less). Schlumberger provides corrections that can be used for excavation effect on the neutron log.[24]

Gobran et al. provide an algorithm for analysis of shaly gas sands that includes corrections for excavation effect.[25] The problem with these corrections is that S_{xo} must be known or estimated to make the correction. If the formation is too tight for flushing, S_{xo} will be the same or nearly the same as S_w. Whenever there is an unusually large apparent gas effect on the neutron log, evidenced by the neutron porosity reading more than 5% less than a density porosity tool (after making any necessary allowances for lithology effects), the possibility of either a tight formation or excavation effect, or both, should be considered.

NEUTRON–DENSITY LOG COMBINATION

Any combination of available porosity tools may be made: sonic–density logs, sonic–neutron logs, neutron–density logs, or even triple log combinations. I discuss only the most commonly used combination here, the neutron–density log, the principles of which apply to other log combinations as well. Often, porosity estimates can be improved, gas-bearing zones can be identified, lithology can be determined, and log quality control problems can be identified and corrected by using porosity log combinations. Two common methods of using porosity log combinations are log overlays and cross-plots.

A log overlay is made when any combination of logs is displayed in the same log track using the same scales. A related technique is to record several logs, juxtaposed to enhance lithology or other rock property identification. Cross-plots are usually two-dimensional with one log plotted as the Y axis and the other plotted as the X axis. Multidimensional plots are also possible, often using a Z axis, whereby the third axis is presented by plotting some numerical value instead of an x or some other plot symbol. The Z axis is represented by making the numerical plotted value proportional to the value of the third axis variable. A fourth axis can be represented by using some variable color spectrum. A particular shade of color corresponds to a numerical value for the fourth axis.

Figure 7–10 is a neutron–density cross-plot for the Schlumberger CNL and formation density log in a fresh water-filled hole. Figure 7–11 is the

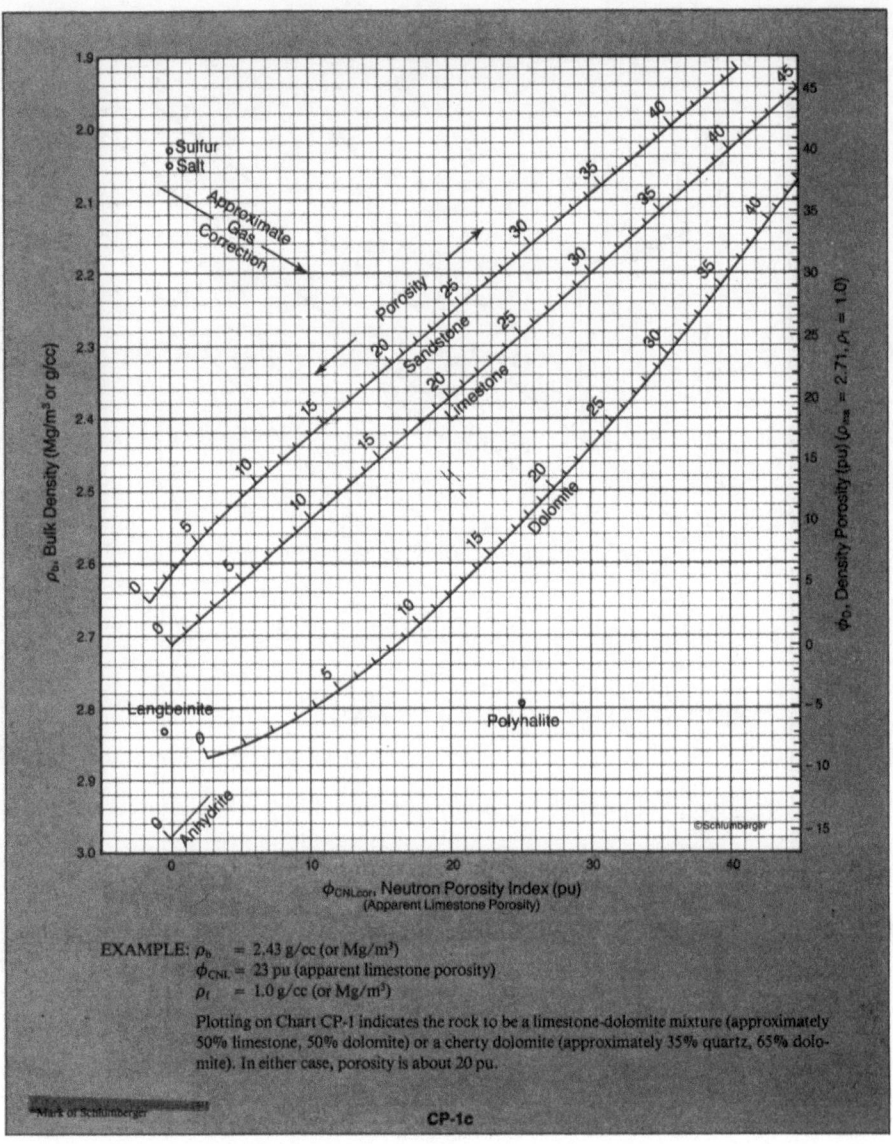

EXAMPLE: ρ_b = 2.43 g/cc (or Mg/m³)
ϕ_{CNL} = 23 pu (apparent limestone porosity)
ρ_f = 1.0 g/cc (or Mg/m³)

Plotting on Chart CP-1 indicates the rock to be a limestone-dolomite mixture (approximately 50% limestone, 50% dolomite) or a cherty dolomite (approximately 35% quartz, 65% dolomite). In either case, porosity is about 20 pu.

CP-1c

Figure 7–10. Compensated neutron log–density log cross-plot (fresh muds). *(Courtesy of Schlumberger Ltd.)*

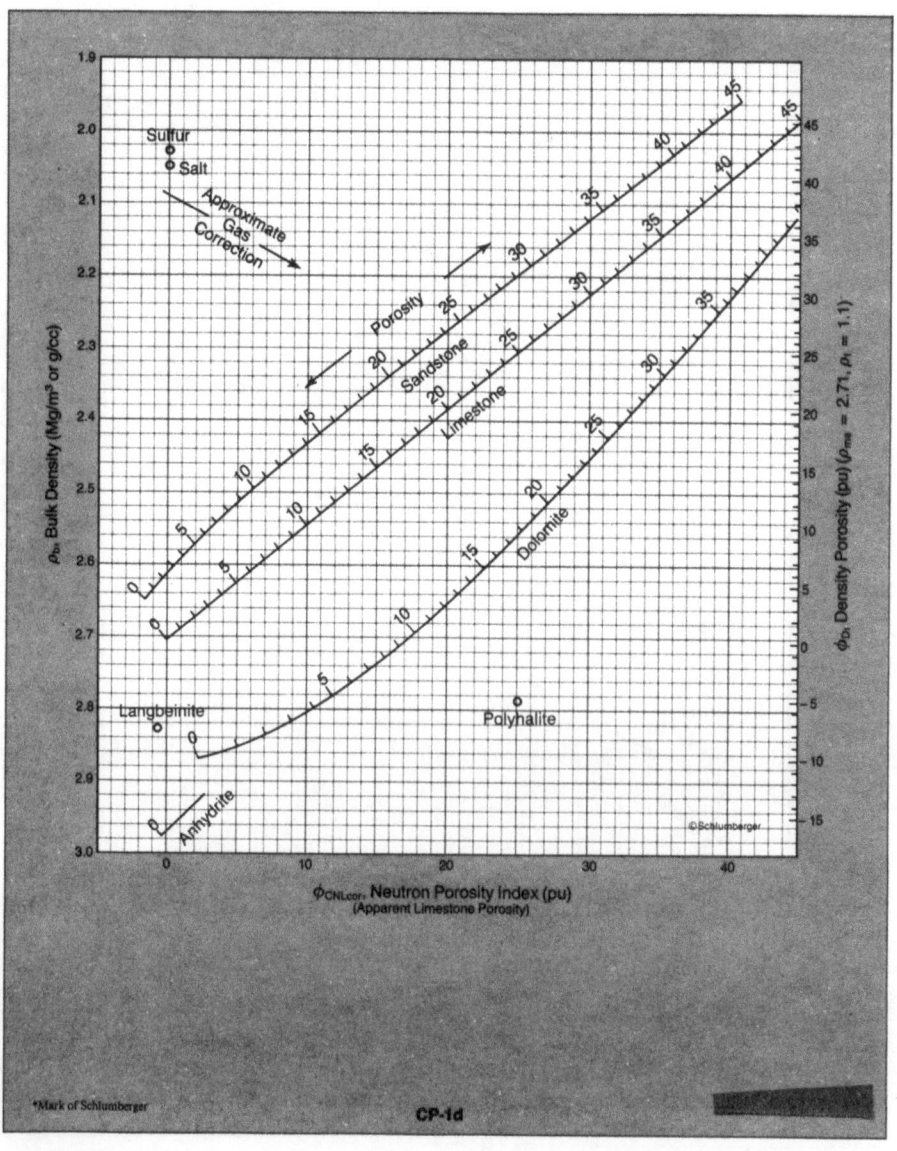

Figure 7–11. Compensated neutron log–density log cross-plot (salt muds). *(Courtesy of Schlumberger Ltd.)*

same plot for salt water-filled holes. Other cross-plot charts must be used for sidewall neutron porosity tools. The illustrated charts apply only to the CNL for Schlumberger's tools. Appropriate charts for a particular service company's tool must be consulted when that company's tools are used. The plots are constructed using the respective tool response characteristics of the neutron and density logging tools. Note that both the neutron log and the density log must be scaled in limestone porosity units. The bulk density can be used directly on the vertical axis instead of the density–porosity calculated using limestone units. The location of the lithology curves on the cross-plot is due to the lithology effect on the neutron tool and the differing grain density values for different lithologies.

On the chart, the indicated direction to compensate for gas effect must be noted. This means that one cannot distinguish between a mixture of calcite and dolomite and a pure dolomite with a gas effect. The PE measurement can sometimes be used to resolve this problem.[26] Figure 7–12 is a cross-plot porosity chart for a sidewall neutron log–density log combination. The closer spacing of the quartz–calcite–dolomite lines on this plot, as compared to the compensated neutron log–density log cross-plot, reflects the smaller lithology effect on the sidewall neutron device. In some applications, such as shaly sands, this could be an advantage. In other applications, the lithology effect can readily be used to advantage (such as complex lithology and carbonate reservoirs).

Use of these cross-plot porosity charts is rather straightforward. For example, in the chart in Figure 7–10 for the CNL in fresh muds, we would enter the chart with a neutron porosity (converted first to limestone units, if necessary) of 30% with either a bulk density reading of 2.44 gm/cc on the left or a calculated density porosity using limestone units (2.71 gm/cc grain density and 1.0 gm/cc fluid density) of 16% on the right side and find that we had a dolomite (point falls exactly on the dolomite line in the chart) with 23% porosity and no gas effect apparent. For points that fall between the lines, we construct iso-porosity lines by joining adjacent corresponding porosity scale points on the various lithology curves on the plot. A point with a bulk density of 2.5 gm/cc and 20% neutron porosity (limestone setting, of course) falls between the limestone and dolomite curves, approximately on a line joining the 16% points of the two curves. This point could be *either* an approximately 50% mixture of calcite and dolomite *or* a dolomite (or even slightly limey dolomite) with gas in the pores and a porosity somewhat higher than 16% (the direction of the approximate gas correction is not quite parallel to the iso-porosity line).

We have already seen how gas affects neutron tool response and how its presence in a rock's pore system affects density log calculations of porosity. Other than using the cross-plot porosity charts, we can use an average of

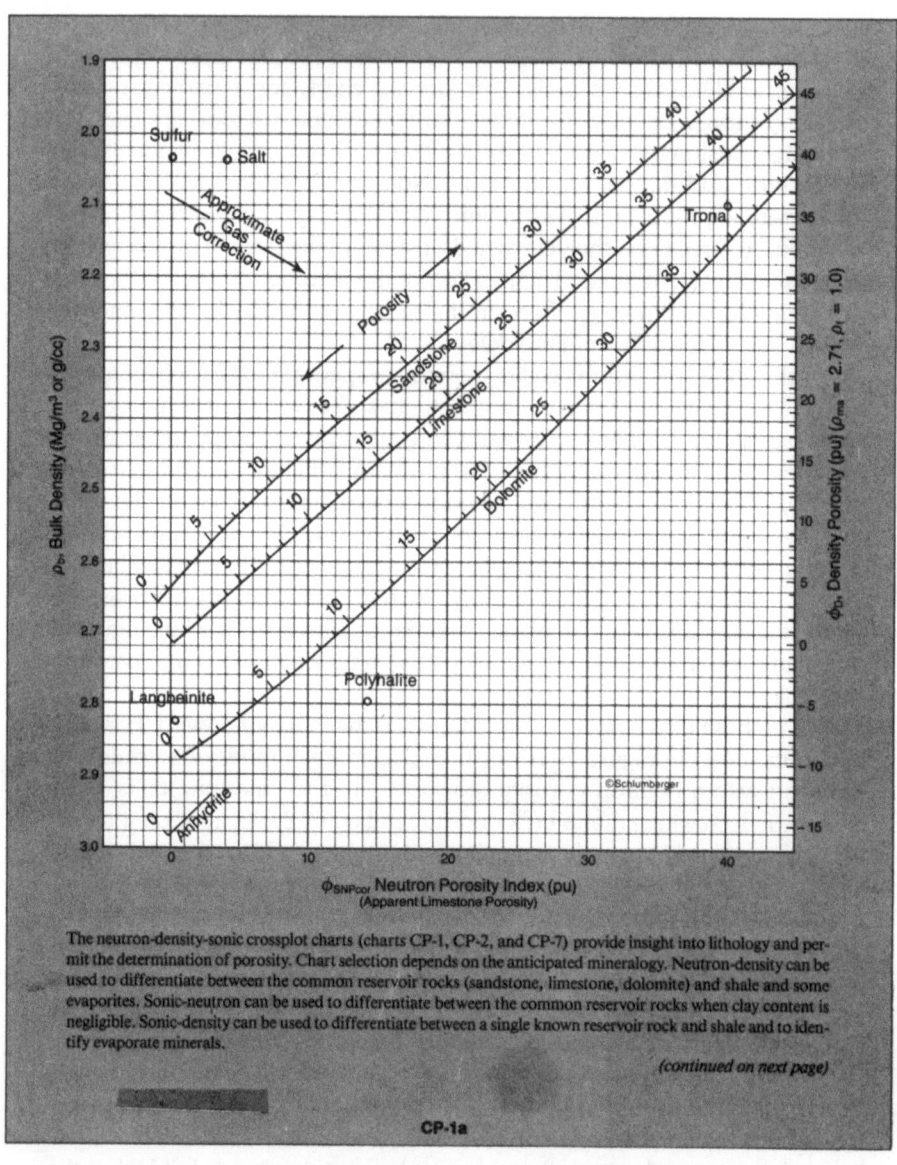

The neutron-density-sonic crossplot charts (charts CP-1, CP-2, and CP-7) provide insight into lithology and permit the determination of porosity. Chart selection depends on the anticipated mineralogy. Neutron-density can be used to differentiate between the common reservoir rocks (sandstone, limestone, dolomite) and shale and some evaporites. Sonic-neutron can be used to differentiate between the common reservoir rocks when clay content is negligible. Sonic-density can be used to differentiate between a single known reservoir rock and shale and to identify evaporate minerals.

(continued on next page)

CP-1a

Figure 7–12. Sidewall neutron log–density log cross-plot (fresh muds). *(Courtesy of Schlumberger Ltd.)*

the neutron and density porosities or the square root of the average of the sum of the squares of the two measurements. Pickett proposed the following gas effect porosity equation using a sidewall neutron porosity reading with a bulk density measurement.[27] The equation is developed from the bulk density relation

$$\rho_b = (1 - \phi)\rho_{ma} + \phi(1 - S_{xo})\rho_g + \phi S_{xo}\rho_f \qquad (7\text{--}11)$$

If the gas density ρ_g is assumed to be zero, or at least very small compared to the density of other fluids in the pore spaces, the middle term disappears and we solve for porosity.

$$\phi = (\rho_{ma} - \rho_b + \phi_N)/\rho_{ma} \qquad (7\text{--}12)$$

where the neutron porosity has been substituted for the ϕS_{xo} product term of Eq. 7–11. This is the liquid-filled porosity that essentially represents the response of a neutron device. Note that this equation is to be used in gas reservoirs, not reservoirs with both oil and gas present in the pore system. In some *tight* or low permeability rocks there is very little flushing. In that case, the S_{xo} term above is more likely S_w (which is ordinarily less than S_{xo}). This means that there will be some small error introduced into the porosity estimate based on Eq. 7–11 since the term with the gas saturation $(1 - S_{xo})$ may not be quite as small with S_w introduced instead of S_{xo}. We might also use this equation in gas reservoirs with the compensated neutron log, but *only* after correcting it to the appropriate lithology.

From studies in *shaly sand* and *sandstone reservoirs* conducted by Coalson and Patchett[28] and my own observations in similar reservoirs in other areas, it appears that use of the neutron–density cross-plot porosity may actually degrade the quality of the porosity estimate that can be had from the density log alone if the correct grain density is used. Juhasz also prefers the use of the density porosity alone in the *normalized Qv* shaly sand method and compares this application to shaly sand methods that use combinations of neutron–density porosity log measurements.[29] However, if the cross-plot porosity is used in complex carbonates, not only is the porosity estimate improved over that calculated from a density log alone,[30] but lithologies can also be identified. Performance of the cross-plot porosity in complex carbonates is further often enhanced with a PE measurement.[31] Patchett and Coalson also suggest statistical methods that may be used to evaluate the accuracy of using any cross-plot porosity chart for a particular reservoir if core measurements are available.[32]

Figure 7–13 is an overlay or presentation of a neutron–porosity log and a porosity curve calculated from the corresponding bulk density log. This is the neutron log that accompanies the PE log and bulk density log of Figure 7–5. Both logs are presented across tracks 2 and 3 on a scale of 30% porosity on the left to −10% porosity on the right. A gamma ray and caliper curve are presented in track 1. The bulk density correction curve is also repeated on the neutron–density overlay in track 2 with a scale from −.05 gm/cc on the left to +.45 gm/cc on the right. Although the overlay may be presented using any lithology (sometimes repeat log sections are presented with different lithology settings/calculations), limestone is the usual lithology because the cross-plot porosity charts are constructed for neutron readings with limestone settings and density porosity calculations using limestone parameters.

Always check the log headings for the listed lithology settings and grain/fluid densities used for the calculated density porosity. The anhydrite bed at 12,300 ft shows up with 0% neutron porosity and a calculated density porosity of less than −10%. Three apparent dolomite beds (or dolomitic calcite beds) show up between 12,310 ft and 12,336 ft with apparent neutron porosities reading several porosity units higher than the calculated density porosity in each zone. In zones where the neutron porosity agrees with the calculated density porosity, we would likely conclude the presence of pure calcite such as the zone immediately below 12,336 ft, which appears to be a tight calcite, perhaps with a trace of dolomite since the neutron tool reads slightly higher. Note that the density correction curve here shows no correction, indicating that the density tool reading should be quite accurate. Looking back at the photoelectric curve of Figure 7–5, there is some variability of the PE measurement in this zone, indicating the possibility of something other than pure calcite, although the mixture should still be predominantly calcite. Perhaps there is some dolomite at the very top of the zone as indicated by the slight excursion of the PE curve below 5 barns/electron at 12,341 ft. Where the bulk density correction curve exceeds one or two scale divisions (better stated as .05 gm/cc or .10 gm/cc), this means the PE curve as well as the bulk density reading may be unreliable. There are several points on the example of Figure 7–5 where the correction curve exceeds these limits (e.g., 12,271 ft).

A commonly observed anomaly is the pseudogas effect observable in sandstones when the density—and neutron—porosity curves are recorded in limestone units. Because the density—porosity curve will be calculated using a 2.71 gm/cc grain density rather than a more likely 2.65 gm/cc for a quartz, the density–porosity curve based on limestone parameters will measure about 1.5 porosity units too high in a sandstone. At the same time, a compensated neutron log will exhibit the lithology effect illustrated in Figure 7–9. The neutron log recorded on a limestone matrix setting will read a porosity that is approximately 4 units too low for the sandstone.

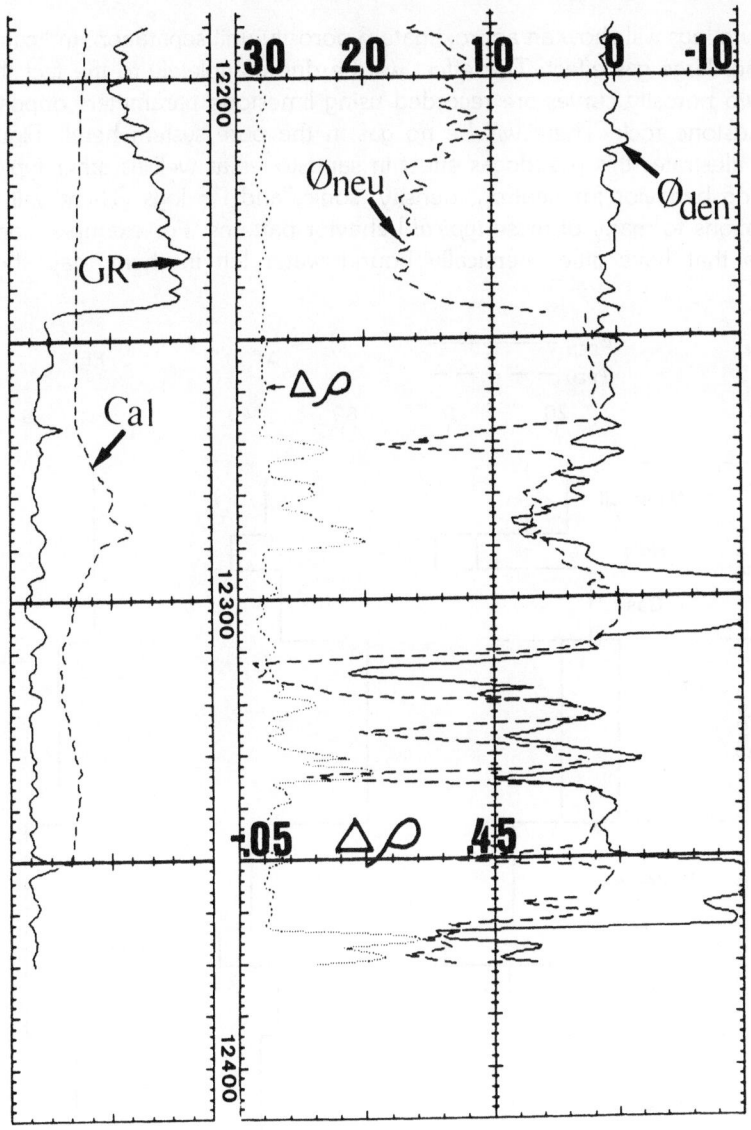

Figure 7–13.   Neutron–density log presentation (overlay).

The two logs will show an approximate 6 porosity unit separation, indicating a rather large gas effect. This effect will be due completely to the fact that the two porosity curves are recorded using limestone parameters opposite a sandstone rock. There will be no gas in the pore system here! Figure 7–14 illustrates this pseudogas effect in sandstones as well as other typical well log behavior for neutron, density, sonic, and PE logs. There will be exceptions to many of these *typical* behavior patterns. For example, some shales that have little chemically bound water left in them may show

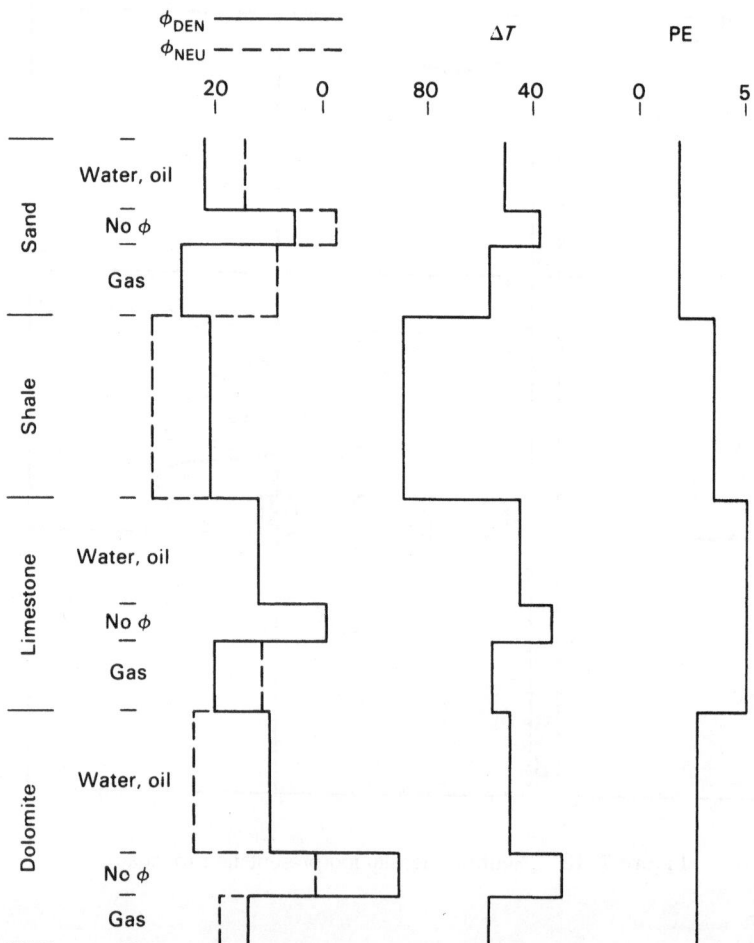

Figure 7–14. Typical lithology effects on selected well logs. (Note: Values are approximate and will vary according to actual porosity and fluid saturations.)

faster (smaller) travel times than the travel time in adjacent sands! Some may question whether such rocks are still really shales. At least, some may think of shales as composed of clay minerals among other constituents. Normally, a clay mineral is composed of some chemically bound water. This type of anomaly is not often seen, but remember that it does exist. Usually, someone experienced in a given area can advise you about such anomalous log behaviors. The danger always is that you may find one of these anomalies in a new area (as in wildcat wells).

OTHER RADIOACTIVE PROPERTY LOGGING DEVICES

In this chapter on the elements of radioactive property well logging, I do not discuss some of the other less commonly used radioactivity logging devices. However, many of them are of importance to well logging and should be mentioned. The thermal neutron capture cross-section devices can be used in cased holes as well as open holes to obtain both porosity and saturation measurements. Natural gamma ray spectral logs break down the received gamma ray energy into relative quantities of its potassium, thorium, and uranium constituents, which can help identify clay types. Neutron activation devices can bombard a formation with neutrons, and the resulting capture gamma rays can be classified by predictable energy levels to produce spectral data of the elements making up the activated rock. Other less frequently used radioactive property logs will probably become increasingly important as more difficult to find reservoirs and shaly sandstone reservoirs are explored.

SUMMARY

In radioactivity logging, most properties of interest in formation evaluation are deduced from actual properties measured. Neutron logs essentially respond to hydrogen content in the formation, whether included as water or hydrocarbons in the pore system or as chemically bound water in clay minerals. This hydrogen concentration must be converted to porosity through appropriate equations to relate count rates to porosity. High hydrogen concentration in the formation is associated with *low count* rates at the neutron tool detector(s). Low count rates are associated with *high porosities*. On the other hand, the neutron tool is best in *low porosity* formations where the *higher count rate* means the porosity relation is less sensitive to the usual statistical fluctuations that occur with any radioactivity measurement.

Compensated neutron logs and the older, single-detector thermal neutron devices can be run in open or cased holes containing liquid but cannot be run in air- or gas-drilled holes. Sidewall neutron tools cannot be run in cased holes but can be run in air- and gas-drilled holes. Sidewall neutron tools have the smallest lithology effect and will be a better device in many sandstone and shaly sandstone reservoirs if the borehole is relatively smooth and not enlarged. In complex carbonates, on the other hand, we can take advantage of the larger lithology effect on the compensated neutron devices to help identify lithologies and improve porosity estimates.

A density log responds to the electron density in the formation. This can, in turn, be related to formation bulk density. If we have knowledge of the correct grain density of the rocks of interest, it is possible to obtain reliable porosity estimates from a density log. The main problem will be uncertainty in the grain density. Commonly used values for quartz (2.65 gm/cc), calcite (2.71 gm/cc), or dolomite (2.87 gm/cc) work well for many applications, particularly the carbonates. However, there can be significant problems in attempting to use the typical quartz values in some sandstones and shaly sandstones. In some areas with variable grain density distributions, it may be necessary to calculate a *range* of possible porosity values that will, in turn, result in a range of possible water saturation or hydrocarbon saturation values. The density log is best in medium to high porosity rocks. It is also the best overall porosity device where hole conditions are reasonably good. It can be run in air- or gas-drilled holes as well as liquid-filled holes but not behind casing.

Natural gamma ray logs may be used to discriminate sands from shales and to calculate shale volumes. The gamma ray response often must be adjusted for uranium content before shale volume calculations. Natural gamma ray logs are also commonly used as a correlation log and to locate the correct depths for perforating zones of interest. They can be run in open holes or cased holes.

PE logs provide lithology information and can be used in conjunction with neutron–density log combinations to identify gas-bearing zones and provide better porosity estimates, as well. However, some combinations of three different minerals cannot be reliably discriminated with the PE measurement.

Neutron and density log cross-plots are very valuable for improving porosity estimates and lithology identification, particularly in complex carbonate rocks. However, their use in sandstone and shaly sandstone reservoirs may lead to unreliable porosity estimates, depending on the types and variety of clay minerals, feldspars, and iron-bearing rock present and the variability in grain densities. It is possible that some of these problems in shaly sandstones can be resolved by newer methods based on PE measurements and gamma ray spectral data.

PROBLEMS

7–1. A sidewall neutron tool measures a porosity of 20% in a sandstone when it is set to read data for a limestone matrix. The hole has been drilled using a fresh mud system and no corrections are necessary. What is the apparently correct porosity? Give a reason (other than any gas effect) why this corrected porosity measurement might be in error by 1% or 2% even though the tool is correctly calibrated and working properly.

7–2. A sidewall neutron tool measures a porosity of 15% in a dolomite when it is set to read data for a sandstone matrix. The hole has been drilled using a fresh mud system. What is the apparently correct porosity for this dolomite? Give a reason (other than any gas effect) why this corrected porosity measurement may be in error even though the tool is correctly calibrated and working properly.

7–3. Refer to the density log of Figure 7–4 and calculate the porosity for the zone at 13,278 ft to 13,282 ft. Assume that the interval is a limestone and that a salt mud (fluid density = 1.10 gm/cc) system was used to drill the well. How much error in the porosity calculation is made if the rock is actually 50% dolomite and 50% limestone (use an assumed dolomite grain density of 2.87 gm/cc)? From the correction curve and caliper logs, would you assume that the bulk density data in this problem is valid? Why or why not? What would the porosity be in this zone if the pores contained only zero density gas? What might be a more practical porosity calculation if there is some irreducible water and possibly some residual oil in the pore system?

7–4. Porosity and bulk density are not directly measured using neutron and density logging tools. What properties do these tools actually measure? What effect does the presence of gas have on these measurements?

7–5. Which neutron tool can be run in air- or gas-drilled holes? Which one can be run in cased holes?

7–6. A gamma ray tool can be used to separate sands from shales. Why can you not be sure of separating shales from dolomites with the gamma ray measurement?

7–7. How might you use a gamma ray measurement to correctly position a perforating gun in a well bore?

7–8. Is there any reason why neutron or density logging tools designed by different companies should not read the same in all formations? Suggest a

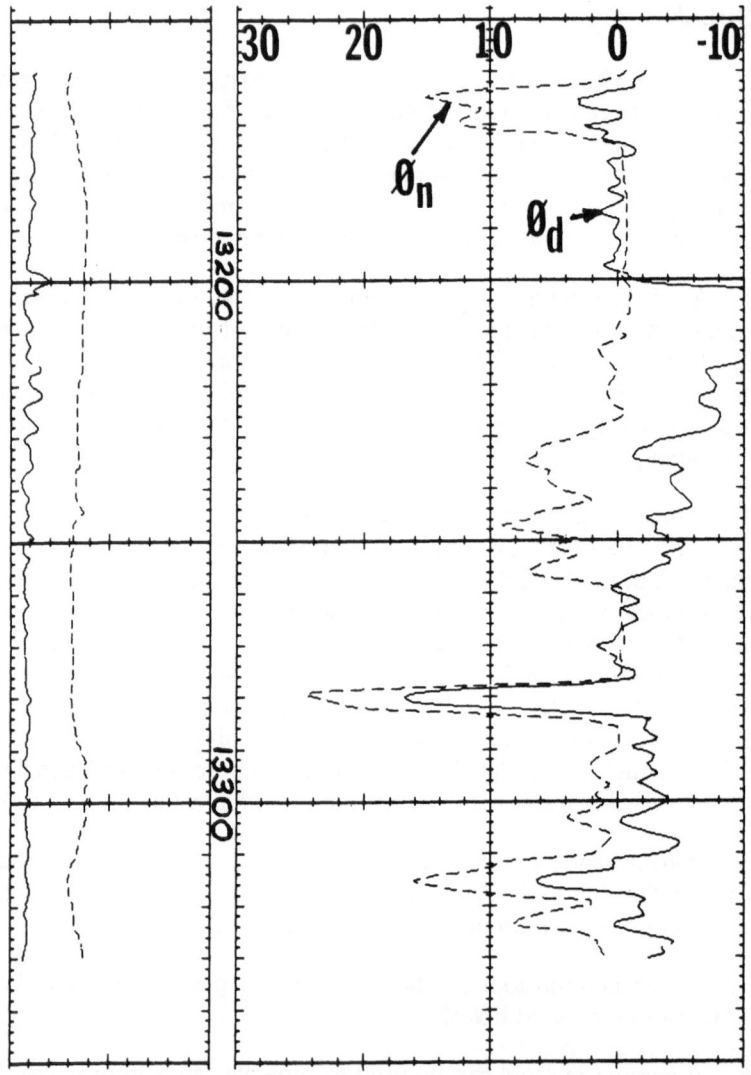

Figure 7–15. Neutron–density logs for Problem 7–13.

way that you might *normalize* the readings from different tools. What pitfalls should you be aware of when using a normalization procedure?

7–9. You are using a density tool to calculate porosities in sands where the grain density varies uniformly from 2.64 gm/cc to 2.7 gm/cc. What can you do to provide more usable results in these sands?

7–10. What would be a good logging practice with all radioactivity logging devices in holes where there are many thin beds with relatively low porosities? Which porosity tool would be more accurate in this situation?

7–11. Can you think of a situation where the sidewall neutron device would be preferable to the compensated neutron device because of the smaller lithology effects? In what depositional environment can you take advantage of the larger lithology effect on the compensated neutron device?

7–12. In Figure 7–5, what lithology is indicated by the photoelectric effect curve at 12,310 ft through 12,320 ft?

7–13(a). What lithology is indicated by the neutron log and density log responses (assuming this is a compensated neutron log recorded with a density tool calculated porosity in a salt mud system) for the interval from 13,162 ft to 13,171 ft in Figure 7–15? Is there a gas effect apparent, or is the zone a mixture of two lithologies? What log might be used to help resolve this uncertainty about the presence of gas here?

7–13(b). Using the neutron density log combination of Figure 7–15 with the logs from Figure 6–19 (dual laterolog) in Chapter 6, calculate the porosity and water saturation for the interval from 13,162 ft to 13,171 ft assuming a salt mud system with no invasion, $a = 1$, $m = n = 2$, and $R_w = .015$ ohm-meters at formation temperature 244°F. The porosity logs of Figure 7–15 are presented on a limestone matrix so that you can use the crossplot porosity from the chart of Figure 7–11. Use both the deep laterolog and shallow laterolog measurements to calculate two water saturations (with no invasion assumed; the justification for this was provided in the answer to problem 10 of Chapter 6).

REFERENCES

1. Author's class notes from course on well log interpretation presented by George R. Pickett at the Colorado School of Mines, Golden, Colorado, 1975.
2. Ibid.
3. Ibid.
4. J. Tittman and J. S. Wahl, "The Physical Foundations of Formation Density Logging (Gamma-Gamma)," *Geophysics* **30**, no. 2 (1965): 284–294.
5. Author's class notes, 1975.
6. J. G. Patchett and E. B. Coalson, "The Determination of Porosity in Sandstone and Shaly Sandstones, Part 2: Effects of Complex Mineralogy and Hydrocarbons," paper presented at the 23rd Annual Logging Symposium of the Society of Professional Well Log Analysts, Corpus Christi, July 1982.

7. Ibid.
8. W. E. Shultz, A. Nunley, J. G. Kampfer, and H. D. Smith, Jr., "Dual Detector Lithology Measurements with a New Spectral Density Log," paper presented at the 26th Annual Logging Symposium of the Society of Professional Well Log Analysts, Dallas, June 1985.
9. W. Bertozzi, D. V. Ellis, and J. S. Wahl, "The Physical Foundations of Formation Lithology Logging with Gamma Rays," *Geophysics* **46**, no. 10 (1981): 1439–1455.
10. Ibid.
11. Ibid.
12. J. S. Gardner and J. L. Dumanoir, "Litho-Density Log Interpretation," paper presented at the 21st Annual Logging Symposium of the Professional Society of Well Log Analysts, Lafayette, July 1980.
13. T. R. McGinley and T. M. McKnight, Jr., "Log-Derived Stratigraphic Reservoir Description," SPE Paper 12897, Permian Basin Oil and Gas Recovery Conference, Midland, Texas, March 1984.
14. Schlumberger, *Log Interpretation Charts* (Schlumberger Ltd, New York, 1986).
15. Author's class notes, 1975.
16. Schlumberger, *Log Interpretation, Volume 1*, p. 51.
17. Author's class notes, 1975.
18. Ibid.
19. Schlumberger, *Log Interpretation Charts* (Schlumberger Ltd, New York, 1986), p. 19.
20. Schlumberger, *Log Interpretation, Volume 1*, p. 50.
21. L. S. Allen, W. R. Mills, K. P. Desai, and R. L. Caldwell, "Some Features of Dual-Spaced Neutron Porosity Logging," *The Log Analyst* **13**, no. 4 (1972): pp. 22–28.
22. Schlumberger, *Log Interpretation, Volume 1*, p. 54.
23. Ibid., p. 52.
24. Ibid., p. 53.
25. Brian D. Gobran, Miguel A. Saldana, Susan L. Brown, and Subir K. Sanyal, "A Comprehensive Mathematical Approach and a Hand-held Calculator Program for Analysis of Shaly Gas Sands," *The Log Analyst* **21**, no. 5 (1980): 11–21.
26. Gardner and Dumanoir, "Litho-Density Log Interpretation."
27. Author's class notes, 1975.
28. Patchett and Coalson, "Determination of Porosity."
29. Istvan Juhasz, "Normalized Qv—the Key to Shaly Sand Evaluation Using the Waxman-Smits Equation in the Absence of Core Data," paper presented at the 22nd Annual Logging Symposium of the Professional Society of Well Log Analysts, June 1981.
30. Patchett and Coalson, "Determination of Porosity."
31. Gardner and Dumanoir, "Litho-Density Log Interpretation."
32. Patchett and Coalson, "Determination of Porosity."

Answers to Problems

CHAPTER 1

1–1. 310,320 bbl of oil. By setting recovery factor equal to 1.00 in Eq. 1–2, you can calculate hydrocarbons in place.

$$HIP = (7758 \text{ bbl/acre-ft}) \times (80 \text{ a}) \times (1) \times (0.5) \times (0.10) \times (10)$$
$$= 310,320 \text{ bbl}$$

1–2. 341,352 bbl. In this problem it is necessary to sum $S_o \phi h$ for each zone. Note that if the zones had different drainage areas (DAs) or recovery factors, you would have to include these quantities along with the separate $S_o \phi h$ summations for each zone. That is, you would form the summation of $DA \times S_o \phi h$ rather than just $S_o \phi h$ if DA was different, for example, for each zone. In this problem

$$RH = (7758 \text{ bbl/acre-ft}) \times (0.20) \times (80 \text{ a})$$
$$= (0.6 \times 0.2 \times 20 \text{ ft} + 0.7 \times 0.1 \times 5 \text{ ft})$$
$$= 341,352 \text{ bbl}$$

1–3. 14.554 Bcf < RH < 20.376 Bcf. In this problem you first compute the recoverable hydrocarbon (gas) at reservoir conditions of temperature and pressure. Since the recovery factor is not given in the problem, you must

273

compute a range of recoverable gas volume based on the possible range of recovery factors quoted in the chapter, that is, from 50% to 70% for typical gas reservoirs.

At reservoir conditions of temperature and pressure, you have a recovery factor of 0.5

$$RH = (43,560 \text{ ft}^3/\text{acre-ft}) \times (0.5) \times (640 \text{ a}) \times (0.7) \times (0.3) \times (20 \text{ ft})$$
$$= 58,544,640 \text{ ft}^3$$

For a recovery factor of 0.70

$$RH = (43,560) \times (0.7) \times (640) \times (0.7) \times (0.3) \times (20)$$
$$= 81,962,496 \text{ ft}^3$$

Thus, at formation temperature and pressure (reservoir conditions)

$$58,544,640 \text{ ft}^3 < RH < 81,962,496 \text{ ft}^3$$

Now, you convert a volume at reservoir conditions to a volume at surface temperature and pressure by multiplying the volume at reservoir conditions by the factor

$$(0.433 \text{ psi/ft} \times 10,000 \text{ ft} + 14.7 \text{ psi})$$
$$\times (70°F + 460°)/(14.7 \text{ psi}) \times (170°F + 460°)$$
$$= 248.6$$

After multiplying the above range RH by this factor, you should obtain the correct answer.

CHAPTER 2

2–1. No. The assumption that hydrocarbon saturation is equal to 1 less the water saturation is not always valid. However, it is necessary to *assume* this is true to calculate hydrocarbon saturation. This necessity arises because you can only calculate water saturation from most common well log tool responses.

2–2. Yes. The average water saturation is given by using Eq. 2–9.

$$\bar{S}_w = \sum_{i=1}^{2} \phi_i S_{w_i} h_i / \sum_{i=1}^{2} \phi_i h_i$$
$$= (\phi_1 S_{w_1} h_1 + \phi_2 S_{w_2} h_2)/(\phi_1 h_1 + \phi_2 h_2)$$

$$= ((0.10)(0.60)(100 \text{ ft}) + (0.40)(0.20)(100 \text{ ft}))/$$
$$((0.10)(100 \text{ ft}) + (0.40)(100 \text{ ft}))$$
$$= (14/50)$$
$$= 0.28$$

Substituting this result into Eq. 2–10 for S_w

$$S_h = 1 - S_w$$
$$= 1 - 0.28$$
$$= 0.72$$

which is the correct figure for average hydrocarbon saturation calculated in the example. Putting 0.72 into Eq. 1–2 with $RF = 1$, along with the average porosity of 0.25

$$HIP = (7758 \text{ bbl/acre-ft})(1.0) \times (1 \text{ a})(0.72)(0.25)(200 \text{ ft})$$
$$= 279,288 \text{ bbl}$$

which is what you set out to show.

Remember that in this problem the correct average porosity is the arithmetic average of the two porosities (10% and 40%) *only* because the two depth intervals (100 ft each) are the same. In general, the correct average is the weighted-by-thickness average porosity given by Eq. 2–7.

2–3. $\bar{\phi} = 0.14$, $\bar{S}_w = 0.717$, $\bar{S}_h = 0.283$. However, 0.50 is not a realistic value for porosity. You might see nearly this value (48%) for ideal, equal-sized, spherical grains packed together in the least compact way. Some unusual rocks, such as pumice, might actually have higher porosities, but the common sedimentary rocks usually have porosities smaller than 50%.

The average porosity is calculated from Eq. 2–7.

$$\bar{\phi} = ((1 \text{ ft})(0.50) + (9 \text{ ft})(0.10))/(1 \text{ ft} + 9 \text{ ft})$$
$$= (0.5 + 0.9)/10$$
$$= 0.14$$

The average water saturation is calculated from Eq. 2–9.

$$\bar{S}_w = ((1 \text{ ft})(0.50)(1.00) + (9 \text{ ft})(0.10)(0.56))/((1 \text{ ft})(0.50) + (9 \text{ ft})(0.10))$$
$$= (0.5 + 0.504)/(0.5 + 0.9)$$
$$= 1.004/1.4$$
$$= 0.717$$

Average hydrocarbon saturation is found as 1 less average water saturation $(1 - 0.717 = 0.283)$. You could confirm this the hard way by using Eq. 2–9, but substituting hydrocarbon saturations for the two zones $(1 - 1 = 0$ for Core 1 and $1 - 0.56 = 0.44$ for Core 2).

$$\text{average hydrocarbon saturation} = \frac{((1\ \text{ft})(0.50)(0) + (9\ \text{ft})(0.10)(0.44))}{((1\ \text{ft})(0.50) + (9\ \text{ft})(0.10))}$$
$$= (0 + 0.396)/1.4$$
$$= 0.283$$

2–4. Bulk volume oil equals 22.5%. It is the product of porosity (0.30) with hydrocarbon saturation $(1 - 0.25)$. You *cannot* calculate bulk volume oil as 1 less bulk volume water. Recall that bulk volume water is the fraction of the total rock volume that contains water. The remainder must be not only oil but the rest of the rock matrix material. However, the bulk volume hydrocarbon and the bulk volume water must add up to the total porosity

$$\phi = \phi S_w + \phi(1 - S_w)$$

You can use this relation to calculate bulk volume hydrocarbon as *porosity* less the bulk volume water. In this problem, $\phi S_w = 0.25 \times 0.30 = 0.075$. Thus, $0.30 - 0.075 = 0.225$.

2–5. You will be 24,825.6 bbl too low (approximately 24,850 bbl is okay, too) if you use straight average porosities and saturations in this problem. When you use the straight average porosity of $(0.08 + 0.18)/2 = 0.13$, you will be all right since, fortuitously, the interval of rock with 8% porosity is the same thickness as the interval with 18% porosity. Thus, you have accidentally weighted the porosities by the appropriate thicknesses if you use the straight average porosity. However, the straight average oil saturation of $(0.40 + 0.80)/2 = 0.60$ is going to create problems. Using the recovery factor of 0.20 with the 80 a drainage area in Eq. 1–2

$$RH = (7758\ \text{bbl/acre-ft})(80\ \text{a}) \times (0.20)(0.13)(0.60)(20\ \text{ft})$$
$$= 193,639.68\ \text{bbl}$$

You should actually use a correctly weighted average hydrocarbon saturation. In this problem, you can conveniently use the ϕS_o column where the product is computed for each depth. Note that this is actually an $h\phi S_o$ column with $h = 1$ ft for each entry. Using Eq. 2–9, but substituting S_o values for S_w values

S_o = ((10 ft)(0.032) + (10 ft)(0.144))/((10 ft)(0.08) + (10 ft)(0.18))
    = 1.76/2.6
    = 0.677

If you use this correct average value for hydrocarbon saturation in Eq. 1-2 instead of the straight average of 0.60, you should obtain 218,490.11 bbl of recovered oil. This correct value is 24,850 bbl higher than the incorrect value. It is possible to get a more accurate estimate by using the column with the products of ϕS_o directly in Eq. 1–2 without calculating average porosity and hydrocarbon saturation to use for ϕ and S_o. The accumulation of the ϕS_o column is 1.76. So,

$$RH = (7758 \text{ bbl/acre-ft})(80 \text{ a})(0.2) \times (1.76)$$
$$= 218,465.28 \text{ bbl}$$

This is 24,825.6 bbl higher than the estimate based on the incorrect arithmetic average S_o value of 0.60. This increased accuracy in the estimate comes from using the accumulation of ϕS_o directly rather than first dividing the accumulation of 1.76 by 2.6 above to get the correct average S_o. The correct average of 0.677 is a rounded off answer. You avoid the round-off error by using the product of ϕS_o directly in Eq. 1–2.

CHAPTER 3

3–1. .075 ohm-meters; 31,000 ppm NaCl. This is a straightforward application of the graph for NaCl solutions in Figure 3–2.

3–2. 18,519 millimhos/meter. Be sure to convert the concentration given for only *chlorides* to ppm NaCl by multiplying by the factor 1.65 before entering the NaCl graph. There you should have found the resistivity to be .054 ohm-meters. Dividing 1,000 by this resistivity (Eq. 3–2) should give you the correct answer for conductivity.

3–3. If you simply use Eq. 3–4 to make the temperature conversion, you get .071 ohm-meters for the answer.

3–4. 0.30 ohm-meters at 68°F. In this problem there are significant quantities of ions other than Na and Cl listed in the water sample analysis. It is necessary to convert the concentrations of these other ions to their NaCl equivalents using the graph in Figure 3–4 (multipliers for ion concentrations) before entering the graph in Figure 3–2 for resistivities of NaCl solutions. Before using either of these charts, you need to convert the listed ion concentrations in units of mg/1 to ppm using Eq. 3–10. Then, the ppm numbers for each ion are converted to ppm NaCl using the appropriate multiplier from Figure 3–4.

Ion	(ppm)	Multiplier	NaCl equivalent (ppm)
Ca	1,321	0.96	1,268
Mg	122	1.50	183
Na	7,156	1.00	7,156
HCO_3	464	0.33	153
Cl	12,006	1.00	12,006
SO_4	1,885	0.64	1,206
Totals	22,954		21,972

Note that you use the total solids concentration 22,954 when entering the graph in Figure 3–4 on its horizontal axis. Finally you arrive at the equivalent NaCl concentration of 21,972 ppm, which can be used in the graph in Figure 3–2 to obtain the solution resistivity at 68°F. If only the chloride concentration had been reported in this analysis and you had multiplied the chloride concentration of 12,006 ppm by 1.65 and used the resultant 19,810 ppm in the graph in Figure 3–2 to find the resistivity, you would have found a somewhat higher resistivity. It is also interesting to note that in the analysis of this water sample the actual measured resistivity was 0.32 ohm-meters at 68°F. This is not an unusual circumstance. Recall from the discussion in the text that the resistivity calculated from the ion concentrations can be more accurate.

3–5(a). $F = 46$. Use Eq. 3–15 and a hand calculator or Figure 3–9.

3–5(b). $F' = 23$. Follow the same procedure as in **(a)** to find the apparent F, assuming $a = 1$ ($F = 46$), then multiply by $a = 0.5$.

3–5(c). $\phi = 6.7\%$ (or 0.067). Use Figure 3–9 or Eq. 3–15 and solve for ϕ. If using the figure, find the formation factor on the horizontal axis and move up to the appropriate line with a slope of 2.2 to find ϕ.

3–5(d). $\phi = 2.4\%$ (or 0.024). Since $F' = 200$ and $a = 0.5$, from the relation $F' = aF$, where F without the prime refers to formation factor from the graph assuming $a = 1$, we have

$$F = F'/a$$
$$= 200/0.5$$
$$= 400$$

Entering the graph with $F = 400$ and going up the line for $m = 1.6$: $\phi = 2.4\%$.

3–6(a). $S_w = 45\%$. Use Eq. 3–16 to find $I = 20/4 = 5$, then enter Figure 3–13 or use Eq. 3–17 with a hand calculator.

3–6(b). $S_w = 61\%$. Find $I = 200/80 = 2.5$ and use the graph in Figure 3–13 or a hand calculator to solve Eq. 3–17.

3–7(a). 8.4 ohm-meters. From the graph in Figure 3–9, $F = 42$. Using Eq. 3–14: $R_o = 42 \times 0.2$ ohm-meters $= 8.4$ ohm-meters.

3–7(b). At least 33.6 ohm-meters. From Figure 3–13, we see that for $n = 2$, $S_w < 50\%$ implies $I > 4$. Rearranging Eq. 3–16

$$R_t = IR_o$$

So, for $S_w < 50\%$ (or $S_{hc} > 50\%$), $R_t \geq 4R_o \geq 4 \times 8.4$ ohm-meters ≥ 33.6 ohm-meters.

3–7(c). 336 ohm-meters. First, $R_o = FR_w = 42 \times 2$ ohm-meters $= 84$ ohm-meters. As above for $S_w \leq 50\%$ with $n = 2$, $R_t \geq 4R_o$ and ≥ 336 ohm-meters.

3–7(d). 3.4 ohm-meters. First, $R_o = FR_w = 42 \times 0.02$ ohm-meters $= 0.84$ ohm-meters. As above, $R_t \geq 3.36$ ohm-meters.

3–8(a). $S_{hc} = 38\%$. Since the lower part of the formation is assumed to be wet, its resistivity (5 ohm-meters) is R_o for the upper part of the formation as long as both upper and lower parts have the same porosity. From Eq. 3–16

$$
\begin{aligned}
I &= R_t/R_o \\
&= 11 \text{ ohm-meters/5 ohm-meters} \\
&= 2.2
\end{aligned}
$$

From the graph in Figure 3–13, for $n = 1.7$, $S_w = 62\%$, thus, $S_{hc} = 38\%$.

3–8(b). $30\% \leq S_{hc} \leq 40\%$. Using the graph in Figure 3–13 for $n = 1.6$, $S_w = 60\%$. For $n = 2.2$, $S_w = 70\%$. Therefore, $60\% \leq S_w \leq 70\%$, or $30\% \leq S_{hc} \leq 40\%$.

3–8(c). $12\% \leq S_{hc} \leq 56\%$. This is getting closer to what a practical problem looks like. Many older logging suites did not include a porosity log, and sometimes only a resistivity log is available. There are two ways to work this problem. First, find the possible range for R_o. From the graph in Figure 3–9 for $m = 1.8$, $F = 30$ if $\phi = 15\%$, and $F = 18$ if $\phi = 20\%$. Therefore, $18 \leq F \leq 30$. Since porosity can vary from 15% to 20%, consider two possible extreme cases:

Case 1. Upper part of formation: $\phi = 20\%$ $(F = 18)$
         Lower part of formation: $\phi = 15\%$ $(F = 30)$
Case 2. Upper part of formation: $\phi = 15\%$ $(F = 30)$
         Lower part of formation: $\phi = 20\%$ $(F = 18)$

From Eq. 3–14, we can solve for R_w in the lower part, knowing $R_o = 5$ ohm-meters for the lower part, which is assumed 100% wet. For Case 1, $R_w = R_o/F = 5/30 = 0.17$ ohm-meters and for Case 2, $R_w = R_o/F = 5/18 = 0.28$ ohm-meters. For the upper part, $R_o = FR_w$, and for the two extreme cases, in Case 1 $R_o = 18 \times 0.17 = 3$ ohm-meters and in Case 2 $R_o = 30 \times 0.28 = 8.4$ ohm-meters.

Since R_t for the upper part is 11 ohm-meters, for Case 1 $I = R_t/R_o = 11/3 = 3.7$ and for Case 2 $I = R_t/R_o = 11/8.4 = 1.3$. Since n can vary from 1.6 to 2.2; from the graph in Figure 3–13, for Case 1 $44\% \leq S_w \leq 56\%$, and for Case 2 $83\% \leq S_w \leq 88\%$. Thus, $44\% \leq S_w \leq 88\%$, or $12\% \leq S_{hc} \leq 56\%$.

A second, and more elegant solution is to use the relation

$$R_t = IFR_w$$

with subscripts 1 for the upper part of the formation and 2 for the lower part. The ratio of resistivities is then

$$R_{t_1}/R_{t_2} = I_1F_1R_w/I_2F_2R_w$$
$$= I_1F_1/F_2$$

since R_w is known to remain constant for both the upper and lower parts of the formation, and $I_2 = 1$ for the lower part if $S_w = 100\%$ for the lower part. Solving this expression for resistivity index in the upper part

$$I_1 = R_{t_1}F_2/R_{t_2}F_1$$
$$= (11/5)(F_2/F_1)$$
$$= 2.2\, F_2/F_1$$

Since both F_1 and F_2 can vary from 18 to 30, you can state that

$$I_1(\text{maximum}) = 2.2 \times (30/18) = 3.7,$$

and for $1.6 \le n \le 2.2$, we have $44\% \le S_w \le 56\%$. Further,

$$I_1(\text{minimum}) = 2.2 \times (18/30) = 1.3,$$

and for $1.6 \le n \le 2.2$, we have $82\% \le S_w \le 88\%$. Thus, from the two possible extremes

$$44\% \le S_w \le 88\%$$
$$12\% \le S_{hc} \le 56\%$$

3–8(d). All calculated hydrocarbon saturations will be too low (or S_w too high). From the solution to part **(c)**:

$$I_1 = R_{t_1}F_2/R_{t_2}F_1$$

I_1 will be too low if $R_{t_2} > R_o$ in the lower part. This means S_w will be too high (see Fig. 3–13).

Author's Note: Problem 3–8 illustrates a typical practical problem you may encounter. In particular, part **(c)** illustrates two common techniques used to handle the many uncertainties in both data and parameters used in interpretation:

1. Calculate a possible range in results corresponding to the known uncertainties in data or parameters.
2. Use ratios to eliminate unknowns such as R_w in **(c)**.

3–9(a). $F = 20$. From Eq. 3–11

$$F = R_o/R_w$$
$$= 10 \text{ ohm-meters}/0.5 \text{ ohm-meters}$$
$$= 20$$

3–9(b). 22%. Solving Eq. 3–15

$$\phi = (a/F)^{(1/m)}$$
$$= (1/20)^{(1/2)}$$
$$= 0.22$$
$$= 22\%$$

Also, you can use the graph in Figure 3–9 to solve this problem.

3–9(c).   7.1%. First, convert the measured rock resistivity and R_w to a common temperature. Using the temperature at which the rock resistivity is given, 150°F, and Figure 3–2, the water resistivity of 0.5 ohm-meters at 75°F = 0.25 ohm-meters at 150°F. Thus

$$F = R_o/R_w$$
$$= 50 \text{ ohm-meters} /0.25 \text{ ohm-meters}$$
$$= 200$$

and

$$\phi = (a/F)^{(1/m)}$$
$$= (1/200)^{(1/2)}$$
$$= 0.071$$
$$= 7.1\%$$

3–9(d).   Too low. If the rock's pores contain hydrocarbons, the apparent R_o will be too high, hence F will be too high, and ϕ from rearranged Eq. 3–15 will be too low.

3–9(e). S_{hc} = 75%. Two methods can be used to solve this problem. First, if ϕ = 28.4%, then

$$R_o = FR_w$$
$$= a\phi^{-m}R_w$$
$$= (0.284)^{-2} \times 0.25 \text{ ohm-meters}$$
$$= 3.1 \text{ ohm-meters at 150°F}$$

(since R_w = 0.25 ohm-meters at 150°F). You can use either a hand calculator or Figure 3–9 to solve ϕ^{-m} = F = 12.4. Then

$$I = R_t/R_o$$
$$= 50 \text{ ohm-meters/3.1 ohm-meters}$$
$$= 16$$

and from Figure 3–13 (or a calculator and Eq. 3–17) S_w = 25% and S_h = 75%.
    Second, use the ratio of actual porosity to calculated porosity as discussed in the section at the end of the chapter on *quick look* methods:

$$I = (\phi/\phi')^m$$
$$= (0.284/0.071)^2$$
$$= 4^2$$
$$= 16$$

and finish the same as for the first method.

CHAPTER 4

4–1(a). 60 μsec/ft $\times$ 3.28 ft/m = 196.8 μsec/m, 110 μsec/ft $\times$ 3.28 ft/m = 360.8 μsec/m.

4–1(b). 195 μsec/ft $\times$ 1 m/3.28 ft = 59.45 μsec/ft.

4–1(c). 7,500 ft/sec $\times$ 1 m/3.28 ft = 2,286.6 m/sec, or 1/7,500 ft/sec $\times$ $10^6\mu$sec/sec $\times$ 3.28 ft/m = 437.3 μsec/m and 1/(437.3 $\times$ 10^{-6} sec/m) = 2,286.6 m/sec.

4–2. The likely cause could be thin-bed effects. However, you have appropriately used a running average of the core data, which should have alleviated the thin-bed problems to some extent. Another cause could be heterogeneity of the rock, which might result in core plug samples not reflecting a good average porosity for the same volume of rock as that sampled by the acoustic logging tool. If the problem is thin-bed effects, you could try using a different acoustic logging tool (if available) with a shorter spacing between receivers. If the problem is heterogeneity of rock type, you may need to take core plugs every few inches (as frequently as possible) from the core or use a whole core analysis to obtain better agreement between the core data and the acoustic log data.

4–3. A vuggy carbonate will have a faster velocity (shorter travel time) since the rock matrix will offer an effective acoustic path that bypasses the vugs to some extent. The constant B in Eq. 4–5 is smallest for vuggy rocks. Also, the acoustic log is sometimes said to *ignore* vuggy type porosity when interpreted in accordance with the time-average relation.

4–4. From the equivalence relations discussed after Eq. 4–3

$$B = C_p(\Delta TF - \Delta TMA)$$
$$= (1)(189\mu\text{sec/ft} - 51.3\ \mu\text{sec/ft})$$
$$= 137.7\mu\text{sec/ft/fraction porosity}$$
$$= 1.377\mu\text{sec/ft/percent porosity}$$
$$A = \Delta TMA = 51.3\ \mu\text{sec/ft}$$

Either set of parameters provides equally accurate answers since they both describe the same straight-line relation of porosity with travel time. However, an extra parameter is required to use the time-average relation. Although it expresses the same straight-line equation, it does so in a more complicated fashion.

4–5. From the same equivalence relations used in problem **4–4**

$$(\Delta TF - \Delta TMA)C_p = 60 \ \mu sec/ft$$
$$(189 \ \mu sec/ft - 42 \ \mu sec/ft) \ C_p = 60 \ \mu sec/ft$$
$$147 \ C_p = 60 \ \mu sec/ft$$
$$C_p = 0.41 (\text{with } \Delta TF = 189 \ \mu sec/ft)$$

or

$$(\Delta TF - \Delta TMA) = 60 \ \mu sec/ft$$
$$\Delta TF = 60 \ \mu sec/ft + \Delta TMA$$
$$= 60 \ \mu sec/ft + 42 \ \mu sec/ft$$
$$= 102 \ \mu sec/ft \ (\text{with } C_p = 1)$$

Neither choice leads to physically realistic values for either ΔTF or C_p. C_p should be at least 1, and it is unlikely that ΔTF would be as low as 102 $\mu sec/ft$.

4–6. First read ΔT values for each depth. Then use Eq. 4–5 to find porosity ϕ for each depth.

Depth (ft)	ΔT ($\mu sec/ft$)	ϕ
8,415	74	0.20
8,416	75.5	0.21
8,417	74	0.20
8,418	71.5	0.18

You might elect to use the peak value at 8,416 ft to represent the whole interval because the bed is relatively thin for a 2 ft (or 3 ft) spacing log. Only the peak value may have approached the correct value for the entire bed. Also note that at the depths 8,415 ft, 8,417 ft, and 8,418 ft, the acoustic log signature is in a *transition* from lower to higher values (or vice versa). These transition readings are normally useless for quantitative calculations. It would not make much sense to *average* over the interval 8,415 ft to 8,418 ft since the bed was so thin (relative to the tool spacing) that only the peak at 8,416 ft likely represents anything close to *valid* information. The average in this case would only contaminate the *available* information.

CHAPTER 5

5–1. The corrected resistivity is *higher* after you have made any necessary borehole signal corrections. The borehole signal contributes an additional *conductivity* to the received signal, which lowers the tool resistivity reading. The tool reading must then be corrected to a higher resistivity reading.

5–2. To read the resistivity as close to the borewall as possible you must use the shortest electrode spacing possible. This provides the best thin-bed resolution also, but the tool will not be able to sense the resistivity in the undisturbed volume of rock removed far enough from the borewall so that invading mud system fluids have not affected the rock's resistivity.

5–3. Resistivity devices other than the induction log depend on propagating an electrical current into the formation. This current requires a conductive path to flow into the rock from the electrodes in the borehole. A conductive mud system provides the necessary current path, whereas a nonconductive, oil-base mud system will not conduct current. The induction logs function without a conductive mud because their functioning does not require that currents propagate directly into the formation. They propagate an *electromagnetic field* into the formation that *induces* a current flow in the rock. The electromagnetic field will propagate even through insulators or dielectrics.

5–4. Your calculated values for the foot-by-foot analysis may differ somewhat from mine depending on your picks of the resistivity at each depth and numerical roundoff.

Depth (ft)	R_t (ohm-meters)	Porosity (%)	Water Saturation (%)
7,151	35	6	84
7,152	60	7	55
7,153	120	5.7	48
7,154	160	5.5	43
7,155	150	6.2	40
7,156	140	5	51
7,157	110	3	95

This saturation profile is not realistic. It would be unusual to see water saturations going from above 80% down to 40% as depth increases through continuous, permeable, and porous beds. Likewise, the 95% saturation at 7,157 ft seems more likely due to a bed boundary effect whereby the resistivity reading is being affected by the low resistivity bed below. Probably the water saturation numbers in the middle of the interval are the most realistic. There, the resistivity tool will be affected the least by the adjacent beds. An alternate approach to working the problem would be to use the maximum resistivity of 160 ohm-meters for the entire bed on the assumption that the peak value is probably the only value that *approached* the actual resistivity during the entire interval from 7,151 ft to 7,157 ft. This could be coupled with a central average of the porosities or the porosities for each foot could be used as they stand. Using a single peak resistivity for this entire thin interval may provide an improved answer.

Using an *eyeball* average porosity for the interval of 6% with the peak resistivity of 160 ohm-meters gives a water saturation of 40% for the whole interval. From the porosity tool response, you might have to break the interval into two zones: one from 7,151 ft through 7,155 ft and one below from 7,156 ft through 7,157 ft since the porosities are lower in the lower part. This will give a somewhat higher water saturation for the lower part that may or may not be true. A thin-bed correction is indicated since there is a large contrast between the bed of interest and the adjacent beds, but the correction charts do not apply for adjacent bed resistivities as high as those in this problem. Therefore, you cannot make a quantitative thin-bed correction. Note that the medium induction log reads higher in the lower part of the interval as the porosity decreases. You might expect that the true resistivity should also follow such a pattern if water saturation remains relatively constant. Therefore, it might not be totally unreasonable to use the medium induction log reading for R_t in this case. There could also be some invasion effect, but note that you *cannot* use the invasion charts because they assume thick beds (or corrected thin-bed resistivity readings).

Finally, in a practical application you would probably assume no significant invasion and use the medium induction log readings to analyze the interval as follows:

Depth (ft)	R_t (ohm-meters)	Porosity (%)	Water Saturation (%)
7,151–7,155	160	6	40
7,155–7,157	250	4	47

This probably represents a good practical compromise since the problem here is that the resistivity tool does not have the same bed resolution as the porosity device.

5–5.   You could use the 500 ohm-meter reading as is. However, you must remember that the reading is probably not as precise as you might think. Five hundred ohm-meters represent only 2 millimhos/meter of conductivity, which is in the range of zeroing error of the induction tool. It might still be a good bet to truncate the reading at a maximum of 200 (or possibly 250) ohm-meters and use that number. That is about the highest resistivity the induction log can reliably indicate, even in relatively thick beds.

5–6.   The true resistivity would be at least as high if not higher than the tool reading. As any filtrate from the saline mud system invades far enough from the borewall, it will provide a more conductive path for electrical current in the rock than would be the case in the undisturbed rock. Of course, in induction log applications the mud systems will usually be *fresh*, and the reverse will be true: in a fresh mud system with some invasion the true resistivity will be somewhat lower than the tool reading. In practice, this may be more than offset in thin beds by the shoulder-bed effects and skin effects.

CHAPTER 6

6–1.   The spherically focused tool is not affected as much by the mud system in the borehole as the *Laterolog 8* because the spherical focusing electrode configuration maintains an approximate spherical shape of the equipotential surfaces near the tool for a much wider range of hole size and mud resistivities than the focusing electrode configuration of the *Laterolog 8* device.

6–2.   It is not a reasonable practice to routinely correct laterologs for invasion if it is not possible to first correct for the shoulder effects from high resistivity contrasts between adjacent beds. Invasion corrections should only be applied for relatively thick beds (preferably tens of feet thick or more). Unfortunately, it is not possible to correct for bed effects in many practical applications because the bed-thickness correction charts are based on assumed infinitely thick beds of the same resistivity both above and below the bed of interest.

6–3.   When the micro-SFL or any other pad-mounted device loses contact with the borewall, the tool will respond only to the fluid in the borehole. It will essentially measure the mud system resistivity.

6–4.   The microlaterolog is the best R_{xo} tool for shallow invasion. It requires only a few inches invasion to provide a valid R_{xo} reading. The proximity log, on the other hand, requires the deepest invasion of all the R_{xo} devices but is insensitive to all but the largest mudcakes.

6–5. The SP has been reduced by 50%. Use the correction chart in Figure 6–17. Move down the 5 ft thickness vertical line to one of the two curves bracketed by the ratio $R_i/R_m = 100$. The upper curve is for invasion of 30 in. and the lower curve for invasion of 40 in. For 40 in. invasion, the correction factor (read from the right margin of the chart) is 2.00. Therefore the SP has been reduced by 50%. Note that SP corrections, in general, are quite empirical. Always note the stated conditions for the use of each SP correction chart. It may not always be possible for you to make an exact correction. The invaded zone resistivity may be read (at least approximately) from the shallow laterolog. Otherwise, you will have to infer the invaded zone resistivity from the deep reading resistivity tool. In a salt mud system, the invaded zone resistivity should be somewhat less than the deep reading resistivity. If you also have an R_{xo} reading, the invaded zone resistivity should be somewhere between the deep laterolog reading and the R_{xo} tool reading. The depth of invasion may be more difficult to assess. It can be done from invasion charts if the situation justifies their use. However this implies relatively thick beds, and you probably would not be concerned about correcting an SP reading for bed thickness in this case. Perhaps you will have to consider a possible range of invasion values and find a corresponding range of corrected SP values. Rocks with larger pore sizes and good permeability should, as a rule, have more shallow invasion than rocks with smaller pore spaces and poor permeability.

6–6. In thinner zones with high resistivity contrasts with adjacent beds, the common accuracy problem comes from the inability to make a usable thin-bed correction from the correction charts that assumes infinitely thick adjacent beds above and below the bed of interest. The problem can be compounded if invasion corrections are applied in this situation of alternating beds with large resistivity contrasts. It may be possible, where a dual laterolog combination tool is used, to use the shallow laterolog reading for R_t if invasion effects can be assured to be negligible. This is strictly a judgment call that comes from experience. In some instances, you may assume that the invasion effects are offset by the bed-thickness (contrasting resistivities) effects and use the resistivity reading without correction.

6–7. The prerequisite for flushed zone calculations is complete flushing of the zone near the borehole. Otherwise, the R_{xo} tool will be measuring something more like the invaded zone resistivity.

6–8. R_w calculated from the SP measurement is 0.06 ohm-meters. To solve this problem, first convert the mud and mud filtrate resistivity measurements using the equation (or consult Fig. 3–2 for the resistivities of NaCl solutions):

$$R_m = 2.59 \text{ ohm-meters} \times (63°F + 6.77)/(131°F + 6.77)$$
$$= 1.31 \text{ ohm-meters at } 131°F$$

Likewise,

$$R_{mf} = 0.86 \text{ ohm-meters at } 131°F$$

Examining the log in Figure 6–18 a little distance above and below the interval of interest, you can establish the shale baseline at about three scale divisions to the left of the right side of track 1. You may possibly have established the baseline to the right of this third scale division. There is some arbitrary judgment involved here, and different people may judge this position slightly differently. For the most part, the SP in the interval 6,450 ft to 6,488 ft develops about 4 1/2 scale divisions of deflection to the left of this baseline. This gives SP $= -90$ millivolts (mv) if the scale is 20 mv per division with negative going SP to the left. Note a slight reduction in the SP to about -80 mv at 6,479 ft. This may be due to a slight shaly streak in the otherwise relatively clean sand. In any event, I would elect to use the -90 mv SP figure for the entire interval for the purposes of determining R_w.

Examine Figure 6–17 for bed-thickness corrections. Using the SFL tool reading for R_i, you can see that it varies from about 10 ohm-meters to just over 20 ohm-meters throughout the interval of SP development. Comparing this to the mud resistivity of 1.31 ohm-meters at formation temperature, the ratio of R_i/R_m is then from a low of about 7 to a high of about 16. In the chart in Figure 6–17, for a bed of 38 ft thickness with this range in R_i/R_m, you can see that the SP develops at least 97% or 98% of the *static* SP. No correction is necessary in this case, and you can see from the curves on the correction chart that the resolution for different ratios of R_i/R_m and invasion depths is nonexistent. A correction based on a specified R_i/R_m would be meaningless.

You can use either the graphical solution from Figure 6–16 or the equations as follows to solve for R_w:

First, solve for K from Eq. 6–7

$$K = .133 \times (131°F) + 61$$
$$= 78.4$$

Then use Eq. 6–6

$$SP = -K \log (R_{mfe}/R_{we})$$

and solve for the ratio of R_{mfe}/R_{we}

$$(R_{mfe}/R_{we}) = 10^{-(SP/K)}$$
$$= 10^{-(-90mv/78.4)}$$
$$= 14.1$$

Then use the value for R_{mfe} to solve for R_{we}

$$R_{we} = R_{mfe}/14.1$$
$$= 0.06 \text{ ohm-meters at } 131°F$$

For those interested, you could go one step further and, calculating R_w from R_{we} using the Schlumberger charts, find $R_w = 0.07$ ohm-meters. Whether or not this extra step has any practical significance, considering the inherent lack of precision (due to the involvement of individual judgment and the actual tool response characteristics) in the entire procedure as well as practical problems inherent in application of SP theory, may be subject to debate. I have elected to follow George R. Pickett's lead in this matter for the reasons already discussed in Chapter 6 and report the answer as $R_w = 0.06$ ohm-meters. With this approach, there is no distinction between R_w and R_{we}, and Doll's equation is used to solve for R_w directly.

6–9. The corrected shallow laterolog reading is anywhere from 10 to 11 ohm-meters. With the adjacent bed resistivity of 500 ohm-meters, $R_{LLS}/R_s = .02$. Using the 12 ft bed thickness in the graph in Figure 6–10 (note that this puts you at a point where there is little resolution on the graph), you find that the ratio of the corrected reading to the measured reading could be anywhere from slightly less than 1 to maybe a little over 1. In practice, you would probably assume correction here would be meaningless and use the *LLS* measurement without correction.

6–10. Place your data and answers in tabular format

Depth (ft)	R_{LLD} (ohm-meters)	R_{LLS} (ohm-meters)	R_{MSFL} (ohm-meters)	ΔT (μsec/ft)	ϕ_s (%)	S_{w_1} (%)	S_{w_2} (%)
13,162–13,165	4	3	2	54	13.3	46	53
13,165–13,168	11	11	6	51	8.3	44	44
13,168–13,171	7.2	5	4.2	52	10.0	46	55

Continuing with the flushed zone data and answers

Depth (ft)	S_{xo} (%)	MHS_1 (%)	MHS_2 (%)
13,162–13,165	67	21	14
13,165–13,168	62	18	18
13,168–13,171	62	16	7

You may not have read the data exactly the same as in my answer. However, the values should be reasonably close. There do seem to be three rather well-defined zones: two peaks (with high ΔT or low resistivities) with a valley (lower ΔT and a higher resistivity) sandwiched between them in the interval of interest. I have elected to designate the zone boundaries at the apparent inflection points: 13,165 ft and 13,168 ft. The inflections on the recorded log curves at the top and bottom of the interval of interest complete the zone boundaries. It is probably best to use a porosity tool (the sonic log in this problem) to identify your bed boundaries. The deep reading resistivity tools (especially induction logs) do not have as good a thin-bed resolution as the porosity tools and frequently do not define the bed boundaries too well. You can easily verify this by comparing the porosity tool readings with a short focus tool such as the MSFL in this problem. In contrast to the deeper reading R_t devices, the shorter focus resistivity logs also define the beds fairly well. Also, note in this problem that there is a slight depth discrepancy between the resistivity tools and the porosity tool: the MSFL depths are 1 or 2 ft deeper than the sonic depths for the same events on the recorded log. Always check the depths of the various recorded logs for agreement and adjust any accordingly. I prefer to use a porosity tool recorded with a gamma ray log because perforating logs (gamma ray–drilling collar locating logs) will use gamma ray correlation to locate the perforating guns at the correct depth opposite any zones to be *shot*.

Once you have read the data, it should be straightforward to calculate porosity from the sonic log using the slope-intercept equation from Chapter 4. To use the time-average formula for porosity, you use the same value for ΔTMA as the value A in the slope-intercept relation. From the slope B in μsec/ft/% porosity, you can solve for either the *lack of compaction* correction C_p or fluid travel time ΔTF.

$$100\,B = C_p\,(\Delta TF - \Delta TMA)$$

Remember to multiply B by 100 if you used the form of the slope-intercept equation that gives results in percent porosity rather than porosity as a decimal fraction. Now you can use $\Delta TMA = 46\ \mu$sec/ft and solve for C_p by fixing the value of ΔTF (typically at 189 μsec/ft); or solve for an apparent

ΔTF by fixing C_p (usually at 1.00, although other values could be used). Fixing ΔTF at 189

$$100\, B = C_p\, (189\ \mu sec/ft - 46\ \mu sec/ft)$$

and

$$C_p = 0.42$$

Once you solve for porosity, then use the saturation relationships with both the deep laterolog (LLD) and shallow laterolog (LLS) resistivity readings to calculate two water saturations: S_{w_1} and S_{w_2}. In the problem statement, you have been told to assume no invasion, so your justification in using the shallow laterolog for R_t would be that the observed separation on the resistivity logs must be due to bed effects (contrasts in resistivities). The shoulder-bed correction charts would be difficult to use here because there are three low resistivity beds sandwiched between very high resistivity beds. Since these beds are of appreciable thickness, you might consider trying some type of bed-thickness correction, but it seems simpler to use the shallow laterolog readings here. Note that there is no difference between the deep and shallow laterolog readings in the middle zone: 13,165 ft to 13,168 ft. If the shallow laterolog readings are a reasonable indication of R_t, this also implies no invasion.

There is always the possibility of *both* shoulder-bed and invasion effects. However, with the high resistivity beds of appreciable thickness both above and below the interval of interest, it seems likely there would be *some* shoulder effect on the R_t readings. This would also likely account for all the small difference between the shallow laterolog readings and the deep laterolog readings.

You can use the MSFL readings directly for R_{xo}. There is a sizeable mud-cake indicated by the caliper on the resistivity logs (the dashed line in track 1 is the MSFL caliper), but unless it is 1 in. or more in thickness with high ratios of R_{MSFL}/R_{mc}, the correction is negligible according to the correction chart in Figure 6–14. Use Eq. 6–2 with $F = \phi^{-m}$ to solve for S_{xo}. Then use Eq. 6–4 to solve for movable hydrocarbon saturation, using both of the calculated S_w values. With either value for S_w, note that there is a significant movable hydrocarbon indicated. However, the water saturation (S_w) numbers based on the higher reading LLD are more optimistic about potential producibility since they are all less than 50%. If the more pessimistic S_w numbers based on using the shallow laterolog for R_t are correct, production from this zone would probably include water as well as oil since you may be near a cutoff water saturation. Note that another zone below in this well

(13,312 ft to 13,318 ft) was completed and produced *both* oil and water. You might make calculations on this interval for comparison, assuming that both zones have the same R_w *and* cutoff saturations.

Finally, note on these logs where the MSFL tool has apparently lost pad contact and is reading the mud resistivity, for example, at 13,235 ft and 13,245 ft. Moreover, in neither place does the MSFL follow the same pattern as the deeper tools. The sharp decrease in reading is indicative of loss of pad contact. I have placed a note on the log to the effect that you could also compare the $\Delta\rho$ curve from a density log to verify the possible loss of pad contact. If this log was also included, you could look to see if a rugose hole was indicated, which might lead to loss of pad contact. A caliper anomaly would also be diagnostic. In fact, the calipers on the logs shown are both indicating a rather rugose hole, but I would suggest checking the $\Delta\rho$ curve also when it is available.

6–11. The hybrid scale will be linear from zero on the left to 20 ohm-meters in the middle and then each of the remaining five divisions would read progressively: 25, 33, 50, 100, and infinity (all in units of ohm-meters). If the midscale is 20 ohm-meters, the conductivity must be 1,000/20 = 50 millimhos/meter. With ten divisions across the track, there would be five divisions on the right half from 20 ohm-meters (50 millimhos/meter) at the middle of the track to zero millimhos/meter at the right side. Each division reading from right to left (in units of millimhos/meter) is then 0, 10, 20, 30, 40. Dividing each of these into 1,000 and recording the calculated resistivities starting at the first division to the right of center should give you the answer. Note that, although hybrid scales allow a full range of resistivity to be recorded without a backup scale, you lose completely any resolution in the higher resistivity ranges. To a lesser extent, this occurs with logarithmic scales, too.

CHAPTER 7

7–1. From the chart in Figure 7–9, the correct porosity is 23%. Enter the horizontal scale at the bottom with the limestone porosity of 20% and move vertically upward to the SNP sand line (line closest to, and above the limestone line), then move left to the vertical scale (*true porosity*) and read 23% porosity. Even though calibrated and working properly, the tool may be in error by a percent or two because of the presence of some small quantities of good thermal neutron absorbers such as boron or gadolinium, or perhaps small amounts of clay minerals will contribute to the tool response because of the hydrogen in the chemically bound water of the clay minerals.

7–2. ϕ = 9.5%. In Figure 7–9, enter the left, vertical scale with 15% and move right to the sandstone line closest to, and above the limestone line, then move down to the dolomite line (closest line below the limestone line) and back horizontally to the left, vertical scale and read 9.5%. This could be in error since all dolomites may not exhibit the same exact departure if trace elements are responsible for the dolomite lithology effect.

7–3.   16.8%. From Figure 7–4, ρ_b = 2.44 gm/cc in the interval of interest. A straightforward calculation from Eq. 7–7, using a limestone grain density of 2.71 gm/cc and a fluid density of 1.10 gm/cc, gives 16.8% porosity. Note that with a bed this thin (4 ft bed is approximately twice the tool spacing) the well-defined peak is probably the best value for the *whole* interval. For 50% dolomite and 50% limestone the grain density will be the average of 2.87 gm/cc and 2.71 gm/cc, or 2.79 gm/cc. Using this grain density in Eq. 7–7 with the 1.10 gm/cc fluid density will give 21.7% porosity. Thus, a −4.9% porosity error would be made. The error could be smaller if you were dealing with one of the dolomites with a grain density less than 2.87 gm/cc (dolomite actually has a variable mixture of calcite and magnesite, and hence different grain densities are possible).

The correction curve from 13,278 ft to 13,282 ft reads just over one scale division (approximately − .06 gm/cc), which is not too bad. The caliper (assuming a 6 in. to 16 in. scale in track 1) shows a slight enlargement relative to the hole size above and below the zone. Considering the correction curve, the density log likely maintained reasonably good contact within the slightly enlarged zone and the density data are probably acceptable.

If the pores contained only zero density gas, use Eq. 7–8 to obtain a 10% porosity. A more realistic estimate for both gas and oil or water in the pores would be perhaps 85% times ϕ_D(= 16.8%) or 14.3%. The actual porosity will vary according to S_{xo}, the water saturation, and the gas saturation in the flushed zone near the borehole.

7–4.   Neutron tools actually measure hydrogen concentration. Density logs measure electron density. Gas has no effect on density log *measurements* but may affect calculations if it is not accounted for as part of the pore fluid. Gas results in a lower hydrogen concentration in the pores, which results in a lower apparent neutron porosity measurement.

7–5.   Epithermal (sidewall neutron porosity) neutron tools can be run in air- or gas-drilled holes. The old style single-detector thermal neutron tools and modern compensated neutron logs can be run behind casing.

7–6.   Some dolomites are radioactive and may exhibit a gamma ray signature similar to shales. They will also have a neutron porosity reading higher than the density porosity reading, again similar to some shales.

7–7. You can run a casing collar–gamma ray correlation log behind casing and correlate the cased-hole gamma ray log to the open-hole gamma ray log run before casing was set in the hole.

7–8. Despite established calibration standards, there is always a possibility that one company's tool may not respond identically to that of another company's in a given formation over the entire possible range of observable porosities. This is due to design differences unique to each company's tool. Only if they use the same design will the tools necessarily read the same. Even then there may be some differences, depending on the calibration procedures and standards and whether the tools have sources with the same radioactive strength at the time of comparison.

You could try comparing the neutron tool readings in some zone of constant properties over a large area and shift the tool readings to read the same value (say the average for all tools in the common, standard zone), but there are pitfalls to watch for:

1. The common rock property varies over the area studied.
2. The logging speed is too high in the zone used for normalization since it may be above (or below) the zones of interest (it is common practice to pull the tool out of the hole a little faster as soon as the zones of interest are logged).
3. The tools may not be linearly calibrated, that is, a correction determined from the values in the normalization zone may not be the correct amount for a different property value in the zones of interest.

7–9. Calculate a range of porosities using the limits for grain density. Since the distribution of grain densities is uniform, all values of grain density are equally likely (unless you are able to relate them to some other measurable property).

7–10. Operate the radioactive logging tool at slower speeds in thin-bed sequences to improve both bed resolution and accuracy. The best tool in low porosity thin beds would be a neutron log operated at a slower-than-normal speed.

7–11. The sidewall neutron device will be a better choice in shaly sands and areas where trace elements with high neutron capture cross section are present in the sands if the borewall is relatively smooth and mudcake thickness is not excessive. On the other hand, you can use the compensated neutron tool with its large lithology effect to advantage in complex carbonate–evaporite depositional sequences.

7–12. The photoelectric effect curve reads slightly over three units and increases slightly at the bottom of the interval. Therefore, the zone is likely dolomite with some other mineral in the bottom of the zone (e.g., calcite or anhydrite, but not quartz, which has a lower PE value than dolomite).

7–13(a). Considering the logs to be recorded on limestone matrices, the lithology is likely dolomite. The neutron log reads higher than the density log and there is a low gamma ray count (therefore probably not a shale). Consulting Figure 7–11, say, the upper part where ϕ_D is about 3% and ϕ_N is about 15%, the lithology is either dolomite with some limestone (say, 20–25%) with ϕ about 9.5% or a pure dolomite with some gas in the pores and a porosity of about 10%. This uncertainty might be resolved with the addition of a PE measurement using the techniques described by Gardner and Dumanoir.

7–13(b). Tabulate your data and calculations as:

Depth (ft)	R_{LLD} (ohm-meters)	R_{LLS} (ohm-meters)	ϕ_N (%)	ϕ_D (%)	ϕ_{xplt} (%)	S_{w_1} (%)	S_{w_2} (%)
13,162–13,165	4	3	15	3	9.5	64	74
13,165–13,168	11	11	11	−.5	6	62	62
13,168–13,171	7.2	5	12	2.5	8	57	68

If you compare these answers to those of the corresponding problem using the sonic log porosities in Chapter 6, you will note that the sonic porosities were higher, which yields lower water saturations than using the neutron–density cross-plot porosity (ϕ_{xplt}). There is a possibility of a gas effect (see discussion in **[a]** above). If the density porosities were used without gas correction (by calculating the porosities from the bulk density reading using a 2.87 gm/cc grain density and 1.1 gm/cc fluid density), the porosity at 13,162–13,165 ft would be nearly 12% for a pure dolomite (much closer to the sonic porosity of 13.3%). It is also possible that the sonic log porosities are inaccurate, that the rock type was more like a chalky dolomite (larger lack-of-compaction correction or slope parameter required), and the sonic porosities used in Chapter 6 were too optimistic. Without other information, you may have to report a range of water saturations—an optimistic S_w from the sonic porosity and a pessimistic S_w from the neutron–density cross-plot porosity. Note that you will still get reasonably good movable hydrocarbon saturation calculations (20% to 30%) with the neutron–density cross-plot porosity. This zone may still be a good bet for completion, although somewhat marginal.

Index